Tobacco Mosaic Virus

One Hundred Years of Contributions to Virology

From the laboratory journal of M. W. Beijerinck

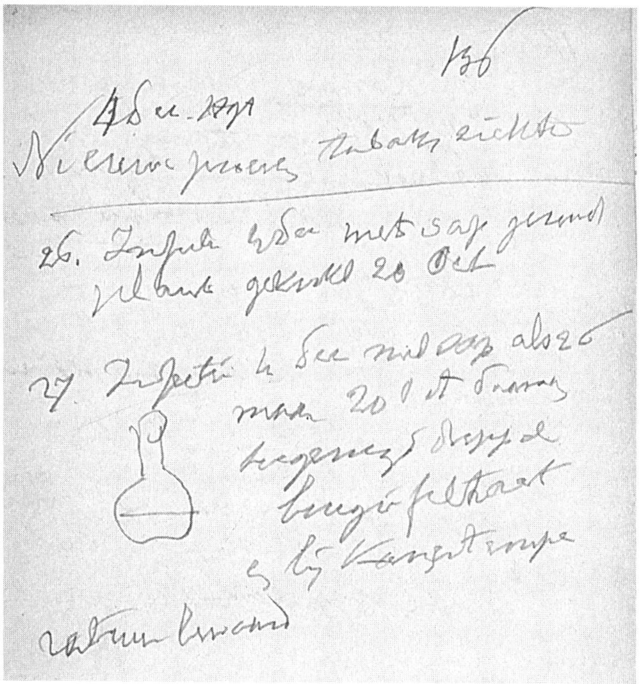

4 Dec 1898
<u>Nieuwe proeven tabaks ziekte</u>
26. *Infectie* 4 Dec met sap gezonde
    planten gekweekt 20 Oct
27. *Infectie* 4 Dec met sap als 26
    maar 20 Oct daaraan
    togevoegd druppel
    bougie filtraat
    en bij kamer tempe
    ratuur bewaard.

4 Dec 1898
<u>New experiments on tobacco sickness</u>
26. Inspection of plants from 4 Dec infected with
    sap from healthy plants grown 20 October
27. Inspection of plants from 4 Dec treated with
    sap as 26, but with a drop of bougie filtrate
    added 20 Oct, and kept at room temperature.

Tobacco Mosaic Virus

One Hundred Years of Contributions to Virology

Karen-Beth G. Scholthof
Texas A & M University, College Station

John G. Shaw
University of Kentucky, Lexington

Milton Zaitlin
Cornell University, Ithaca, New York

APS PRESS
The American Phytopathological Society
St. Paul, Minnesota

This book has been reproduced directly from copy submitted in final form to APS Press. No editing or proofreading has been done by the Press.

Library of Congress Catalog Card Number: 99-62928
International Standard Book Number: 0-89054-236-8

Printed in the United States of America on acid-free paper

The American Phytopathological Society
3340 Pilot Knob Road
St. Paul, Minnesota 55121-2097, USA

Contents

Introduction

The science of virology was founded one hundred years ago with the publication by Professor Martinus Beijerinck of a paper describing some experiments with diseased tobacco plants. The virus that was the subject of Beijerinck's report was tobacco mosaic virus (TMV) and, over the following decades, an astonishing number of "firsts" in fundamental research in virology were to be achieved in investigations with this virus. In marking the centennial anniversary of the birth of virology, we have selected many of the classic papers of the past century dealing with TMV and they are reproduced in this anthology. Accompanying the papers are commentaries written, in some cases, by the very persons who conducted the experiments described in the papers and, in others, by individuals who are familiar with the published works and their significance.

It is the hope of the editors that this collection will provide readers with an appreciation of the importance of TMV in the establishment of the virus concept and in the development of related fields of study such as biochemistry, biophysics, molecular biology, structural biology, genetics and plant pathology. After a century in such a pioneering role, the virus continues to be a model for studies of intra- and intercellular trafficking, pathogenesis, disease resistance, gene delivery and genetic modification and other areas of fundamental plant science. Our intention is not to provide an up-to-the-minute review of the status of virology or TMV studies but rather to celebrate the original contributions themselves and their impact on various branches of science at the time they were reported. Hopefully, this will also serve to reveal the extent to which the classic experiments with TMV led to our current understanding of viruses of all types and host specificities.

We wish to extend our deepest appreciation to the contributors of the commentaries. These individuals participated in the project with great enthusiasm and, in particular in those cases where they were discussing their own work, the emotions and feelings of wonderment that they brought to their efforts are quite apparent. Those contributors who provided comments concerning the work of others also provided personal insights and experiences that serve to enrich the collection. The editors made no attempt to influence the treatment, content or style of the commentaries for the simple reason that this would only have served to dilute the flavor and personal touch of the contributions. It is with much regret that we record the omission from our list of selections of a few papers that we would have wished to include but for which we were unable to solicit suitable commentaries.

Special thanks are extended to those who have provided original copies of the papers for reproduction. We are also grateful for the generosity of Lesley A. Robertson, curator of Beijerinck's papers at Delft University, and the Kluyver Institute for allowing us to reproduce the notebook page from 4 December 1898 with the experimental evidence that TMV is a *contagium vivum fluidum*.

The Editors
23 December 1998

On Beijerinck's Concept of a *Contagium Vivum Fluidum*

Milton Zaitlin
Department of Plant Pathology, Cornell University, Ithaca, NY

This 1898 paper is one of four plant virus "classics" translated into English in 1942 by James Johnson, a distinguished plant virologist in his own right. These translations are published in one volume as a Phytopathological Classic (Johnson, 1942). The first translation in the volume is a 1886 contribution from Adolf Mayer, who worked in Wageningen, Holland, as a contemporary of Beijerinck, who was there until 1887. Mayer described the tobacco mosaic disease in detail and was considered to be the first to transmit the disease experimentally; he named it the "mosaic disease of tobacco".

Beijerinck moved to Delft, Holland, in 1887; he spent the next thirty-four years there. His most important finding about the tobacco mosaic disease and the elucidation of its causal agent was based on experiments in which sap extracted from infected leaves could be passed through an unglazed porcelain filter and remain infectious. Such filters had been used by eminent scientists such as Pasteur, Lister and Koch to demonstrate the germ theory of disease. In their studies the infectious agent was retained in the filter, while the filtrate was pathogen free. The experiments described in the paper reprinted here document Beijerinck's virus studies and detail his work in which he concluded that he had discovered a truly unique infectious agent; i.e., a *contagium vivum fluidum*, which he named a virus.

Beijerinck was an accomplished scientist, having published about 140 papers in his active career between 1877 and 1927 (he died in 1931). His work with tobacco mosaic was in actuality only a small component of his accomplishments, but it is among the most lasting; the other very notable finding was his isolation of the symbiotic bacterium from nitrogen-fixing nodules of legumes. A partial list of his other accomplishments includes the determination of the etiologic agent of a number of bacterial and fungal plant diseases, the elucidation of the nutritional requirements for the culture of numerous bacteria, and the hereditary variations among bacteria. He also investigated luminous bacteria, free-living nitrogen-fixing bacteria, and studied the chemistry of fermentation and the properties of a number of enzymes.

Although we cite Beijerinck for his role in the concept of the virus, he was not the first to suggest that the causal agent of tobacco mosaic virus could be something unique. Beijerinck was, however, the first to conclude that he had a new type of infectious agent. Another paper translated in the Johnson volume is that of Dmitri Iwanowski, a Russian scientist who studied the disease also utilizing porcelain filters to try to ascertain the nature of the infectious agent. Iwanowski's publication predated that of Beijerinck by six years. Without going into details, some, such as Stanley (1944) and Lustig and Levine (1992) have given attribution to Iwanowski for the concept of the new class of pathogens. On the other hand, others (Bos, 1995; Zaitlin, 1998) maintain that because Iwanowski did not conclude from his own observations that he had something unique, he should not be given credit for establishing the concept of the virus.

In addition to the discovery that the tobacco mosaic disease was caused by a new type of infectious agent, Beijerinck's paper is particularly noteworthy because of the clever experimentation he used to discover the properties of the virus, its mode of transmission, and the effect of the disease on its host plant. His ability to transmit the disease with ease enabled him to accomplish these studies. His paper is worth reading for its insights, clever observation and experimentation. Innovative experiments include the determination of the diffusion rate of the agent through agar (important to his conclusion that the agent was not a bacterium), determination that the agent moved in the phloem (and possibly the xylem), and that the disease had an association with the chloroplast. He also studied the effect of temperature and formaldehyde on the infectivity of the agent, ascertained its storage properties, and investigated whether there were other diseases which could be caused by a similar "*contagium vivum fluidum*". He was a marvelous investigator!

References

Bos, L. (1995). The embryonic beginning of virology: unbiased thinking and dogmatic stagnation. *Arch. Virol.* **140**:613-619.

Johnson, J. (1942). Phytopathological Classics Number 7. American Phytopathological Society Press, St. Paul, Minnesota.

Lustig, A. and A. J. Levine. (1992). One hundred years of virology. *J. Virol.* **66**:4269-4631.

Stanley, W. M. (1944). Soviet studies on viruses. *Science* **99**:136-138.

Zaitlin, M. (1998). The discovery of the causal agent of the tobacco mosaic disease. *In*: Discoveries in Plant Biology, Volume 1. Shain-Dow Kung and Shang-Fa Yang, eds. World

Scientific Publishing Company, Singapore. pp. 105-110.

Biographical Sketch

Milton Zaitlin graduated from the University of California, Berkeley with a degree in plant pathology in 1949. He worked as a technician at the California Institute of Technology for one year where he met Sam Wildman, who was moving to a position in the Botany Department at UCLA. He entered graduate school in 1950, and received his Ph.D. under Wildman in 1954 in botany, with a minor in biochemistry. Irving Rappaport and Albert Siegel, postdoctoral fellows in the Wildman laboratory, were also significant in his training. He spent about 4 years working at the Commonwealth Scientific and Industrial Research Organisation in Canberra, Australia, with Keith Boardman, Don Spencer and Paul Whitfeld. He returned to the United States to a position in the Department of Horticulture at the University of Missouri with the late Dan Millikan. In 1960 he worked as a postdoctoral fellow with Albert Siegel at the University of Arizona, Tucson, and rose in the ranks to professor. In 1973 he accepted a position as Professor in the Department of Plant Pathology at Cornell University. He also served as both Associate Director and Director of the Cornell Biotechnology Program. He received Emeritus status in 1997.

Amsterdam.

Royal Academy of Sciences, November 26.—Prof. Van de Sande Bakhuyzen in the chair.—Prof. Beyerinck, on a contagium vivum fluidum, causing the spot-disease of tobacco leaves. This disease, also known as the mosaic disease of tobacco leaves, may be inoculated into healthy plants by injecting into the stem, near a bud, sap pressed from infected plants. The active virus passes completely through the pores of very dense porcelain, and can even penetrate into agar by diffusion; therefore it cannot be a contagium fixum in the usual sense, but it must be fluid. Out of the tobacco plant it cannot be made to multiply; but in the dividing tissues of the leaf-rudiments and the meristems of the buds it multiplies freely and over a great extent. A very small drop of the porcelain filtrate can render all the leaves of the infected plant entirely covered with spots, and the sap of these leaves would be sufficient for the contagion of an unlimited number of healthy plants. The virus is destroyed by boiling at so low a degree as 90° C.—Prof. Bakhuis Roozeboom, on the phenomena to be observed on the solidification of liquids, consisting of two tautomeric forms. In the case of equilibrium being established between these forms at the temperature of solidification, these phenomena have been treated by Bancroft. A new deduction was given for those cases in which solidification takes place at temperatures at which no equilibrium can be established any more in the liquid, and specially when supposing that one passes from the region of equilibrium through two regions of one-sided equilibrium to the region of non-equilibrium. All the various consequences of slow and quick heating and cooling may be graphically represented.—Prof. Van der Waals deduced from the phase equation for a mixture, given by himself, the laws for Δ_v (the volume contraction on mixing under constant pressure) and Δ_p (the pressure contraction on mixing in given volume), and compared the results, obtained by himself, with the observations of Kuenen and others in the case of mixtures of carbonic acid and methyl chloride. According to Amagat, Δ_v would be $=0$, and according to Dalton's law, $\Delta_p=0$. The results, arrived at by the author, may briefly be summed up as follows: Δ_v is small all along the course of the isotherm, and the amount may be considered a magnitude of the same order. On the other hand, Δ_p follows a course equal to the deviation from Boyle's law, and when the volume is small it approximates infinity.—Prof. Van Bemmelen presented for publication in the *Proceedings* a communication by Mr. F. A. H. Schreinemakers, entitled, "Equilibriums in systems of three components, variation of the temperature of solution of binary mixtures by the addition of a third component."—Prof. Van der Waals, on the errors that may be committed in the determination of the molecular weight from the vapour density in consequence of the deviations from Boyle's and Guy-Lussac's laws.

CONCERNING A CONTAGIUM VIVUM FLUIDUM AS CAUSE OF THE SPOT DISEASE OF TOBACCO LEAVES[1]

M. W. Beijerinck

In 1885 Mr. Adolf Mayer[2] showed that the mosaic or leaf-spot[a] disease of the tobacco plant is contagious. He pressed the sap from diseased plants, introduced it into capillary tubes, and pierced these into the leaves and stems of healthy plants growing out in the open. After a few weeks the latter were then attacked by the spot disease. He himself could not find any bacteria or other parasites in the diseased leaves through microscopic inspection. I was at that time Mr. Mayer's colleague at the Agricultural School at Wageningen, he showed me his experiments, and I, no more than he, could prove the presence of microbes in the diseased plants to which the disease could be ascribed. At that time, however, my bacteriological knowledge was so incomplete that I could not take my own direct observations as a conclusive proof.

Since that time I have been continually occupied with bacteriological experiments, and when I discovered the bacteria of the Papilionae nodules in 1887, I also took up the tobacco disease again. However, the result was also negative at that time. Since, however, in all my experiments at that time, the microscopic picture, on the one hand, had to be decisive, and on the other hand only cultural experiments pertaining to aerobes were carried out, the possibility still remained that anaerobes were vegetating in small numbers in the plant tissue, which defied direct observa-

[a] The term Fleckenkrankheit (spot disease) as contrasted to Mosaikkrankheit (mosaic disease) and Pockenkrankheit (pock disease) as used in German by many writers, is often confusing. Flecken (spots) may apparently be either chlorotic or necrotic spots. As chlorotic spots, "fleckenkrankheit" would be synonymous with either mottling or variegation.

[1] Beijerinck, M. W. Ueber ein contagium vivum fluidum als Ursache der Fleckenkrankheit der Tabaksblätter. Verhandelingen der Koninklyke akademie van Wettenschappen te Amsterdam. **65**: (2) 3-21, 1898. (With 2 plates).

[2] Landwirtschaftliche Versuchsstationen **32**: 450, 1886.

tion, but were affecting the surrounding plant tissue with poisons, like the tetanus bacteria do with a poison, which is soluble, non-living, i. e., unable to reproduce itself.

It is well known that reduced pigments, which become colored when exposed to the air, often appear[3] inside of cells of the organs of higher plants so that the possibility of the presence of anaerobes in the tobacco plant was from the very beginning not to be excluded. It is true that the presence of such microbes inside of the green organs of aerial plants, is highly improbable; yet, the discovery of "mikroaerophilie" in anaerobes[4] demands the greatest attention when facts of such far-reaching importance as those presented here are concerned, and these give special incentive to new experiments concerning the microbes appearing in the roots below the surface of the ground.

But after I had taken great pains to find anaerobes that could be causally connected with the disease, within or proximate to the diseased leaves and roots of the affected plants, but always with negative results, and I finally knew positively that these, too, were not present, the conclusion was no longer to be denied, that the spot disease is an infectious one that is not caused by microbes.

Then, in 1897, the resources of the newly erected bacteriological laboratory of the Polytechnical Institute at Delft were put at my disposal. This included a greenhouse with heating facilities, which I started using that same year for further experiments on the spot disease. I was, therefore, able to carry out a series of incontestable infection experiments the results of which I shall now briefly describe.

My experimental plants belonged mainly to the local variety from Amerongen and partly they came from seeds from Erfurt.[5]

1. *The infection is not caused by microbes, but by a contagium vivum fluidum.*

It soon became evident that the sap of diseased plants remains infectious when filtered through porcelain, through

[3] I remind you for instance of the presence of *Indigo white* in the labellum of Cattleya.

[4] On the relation of the obligatory anaerobes to free oxygen. Proceedings Royal Academy of Sciences. Amsterdam, May 28, 1898.

[5] Diseased plants were sent to me from various sources, for which I here wish to express my thanks.

which process all aerobes are held back. However, I was not only concerned with the search for aerobes alone, but I also carried out careful experiments to determine the presence of anaerobes in the filtered juice, but with negative results, so that the sap used appeared entirely sterile.

The quantity of candle filtrate necessary for infection is extremely small. A small drop put into the right place in the plant with a Pravaz syringe can infect numerous leaves and branches. If these diseased parts are extracted, an infinite number of healthy plants may be inoculated and infected from this sap, from which we draw the conclusion that the contagium, although fluid, reproduces itself in the living plant.

Since, however, experiments using the candle filtrate are still open to criticism, especially when the possibility of the presence of anaerobes is not excluded, and, because of this, the corpuscular nature of the contagium has not been entirely disproved, I have carried out the following diffusion experiments, which, it seems to me, have produced entirely incontestable results from both points of view.

Drops of the extracted juice of diseased leaves, as well as ground-up diseased leaves, were put on the surface of thick (poured) agar plates, and left to diffuse with water for several days. I hoped in this way to separate the virus from the raw leaf substance, as well as from all bacteria, through diffusion, since the virus, if at all capable of diffusion, could penetrate into the agar downwards and sideways, thereby leaving as a residue all discrete parts, aerobic and anerobic bacteria and their spores. The experiment was, therefore, decisive in determining whether the contagium was actually capable of diffusion and, accordingly, had to be considered as soluble in water, or if not capable of diffusion, therefore, as extremely minutely distributed, yet as corpuscular, that is, as *contagium fixum*. It appeared that the substance causing the infection may penetrate into the agar plate to no small depth, as may be seen from the following circumstances.

When I thought a sufficient time had elapsed for the virus to have penetrated the agar plate to a considerable depth, if diffusion takes place at all, the plate was first cleansed with water, then washed off with a sublimate solution, and finally a layer of agar about half a millimeter thick was removed by means of a sharp platinum spatula from the outer surface of the spot, where the raw material of the leaves or the extracted juice had lain. The mass lying immediately below was then removed in two successive layers and both parts used for the infection of healthy plants. The results left no room for doubt, in both cases the characteristic symptoms of infection were brought about, very intensively by the upper and more weakly by the lower layer of agar.[6] After ten days the distance covered by the virus may have been at least two millimeters, perhaps considerably more. Although a diffusion distance of only a few millimeters was involved here, it seems nevertheless proved that the virus must really be regarded as liquid or soluble and not as corpuscular. This result might be of special interest in so far as it points to the fact that a similar forward movement of specific vital bodies inside of the plant meristem must be considered possible.[7]

The candle filtrate has a somewhat weaker effect on the plant than the extracted juice that has not yet been filtered. This may be seen from the following circumstance. Fresh extracted juice not only produced the peculiar spots of the leaves, which later become necrotic and are characteristic of the disease, but, if considerable quantities are used, it also causes an actual malformation of the leaves, which also remain small since the midrib does not reach its full size, and are less deeply lobate through disturbances in the growth of the edges and often show palmate-veining through which they become quite different from the normal tobacco leaves. Such malformations also may be produced by means of candle filtrates, if one wishes to do so, only much more material must be used for this. From this we must conclude that the virus is held back in the filter pores, at least at the beginning of the filtration process. How incorrect it would be to conclude from this that the virus is of corpuscular nature may be shown by the following experiment.

As is well known, a malt diastase consists mainly of a mixture of two enzymes, granulase and maltase, which may

[6] Egg albumin and cooked potato starch also penetrate slowly into agar plates, which may be followed in the latter through the iodine reaction. Drops of soluble starch laid on gelatin plates diffuse much more easily than ordinary starch and also sideways to a relatively long distance.

[7] I came to an identical conclusion earlier concerning the cecidiogene bodies causing gall formation: these bodies, too, must be soluble in water and capable of diffusion inside of the meristematic tissue.

be separated by diffusion.[8] If, for example, a drop of malt extract is placed on a gelatin plate containing starch, the maltase soon precedes the granulase in diffusion. Maltase produces erythrodextrin and maltase from starch, while granulase produces only dextrin, aside from maltase, from starch as well as from erythrodextrin, which are not colored by iodine, so the action of iodine on the diffusion field of the diastase mixture will show the relative amount of maltase as compared with granulase through a red ring of erythrodextrin on a blue background, which ring surrounds the colorless field of granulase. If the same malt-extract is passed through a porcelain candle, a considerable widening of the maltase rings is found in carrying out the diffusion experiment with the first parts of the filtrate, from which may be concluded that the filter pores more easily retain granulase, which diffuses slowly, than maltase, which diffuses more quickly. Later, when the wall of the filter is saturated with granulase, the original width of the maltase ring returns.

Therefore it was to be expected that a substance like the virus, which does not diffuse easily would flow through in a diluted form at the beginning of the filtration process yet without being composed of corpusculor parts because of this behavior.[9]

Although I had known for a long time that bacteria were not directly concerned with infection, I performed many inoculations on my experimental plants with the forms that happened to occur on the diseased tobacco leaves, as well as with those that developed in the extracted sap of diseased leaves, in order to make this fact absolutely sure. I always had negative results when the experiments were carried out correctly; never did a bacterial culture, free from the virus, produce symptoms of infection. In section 9, however, I shall show that, under these circumstances, it is not easy to entirely separate from the virus the bacteria isolated from the sap of diseased leaves, for these same bacteria, even after

---

[8] The third enzyme of malt extract, glucase, is to be found only in small quantity in it.

[9] I, therefore, cannot agree with the conclusion of Mr. Loeffler as regards the corpuscular nature of the virus of the foot and mouth disease (Centralblatt. für Bacteriologie. Part I. Vol. 24, p. 570, 1898). It would be interesting to know if the watery solutions of gold and platinum, produced by Mr. Bredig by means of the electrical luminous arc between metal electrodes in water, would pass the pores of the Bougie and could diffuse into gelatin or agar jelly.

reinoculations, can contain enough virus to result in most marked chlorsis.

A proper experiment for the purpose of determining the fact that any microbe isolated from a diseased plant is not capable of causing the disease, therefore presupposes a carefully carried out colony culture, consisting of the isolated single germs that have been rinsed with much water and, under certain circumstances, even after repeated reinoculations, which are continued until the last traces of the virus, which has been absorbed or is clinging to the bacteria, have disappeared.

I believe that these remarks are not without importance. For I see in them an analogy to the experiences of the pathologists, according to which the organisms causing certain infectious diseases lose their virulence through culture outside of the organism and can increase it by repeated passage through susceptible animals. Although the analogy is not a very close one, it is certain that there is an analogy.

2. Only those organs of the plant that are growing and whose cells are dividing are capable of being infected; here only does the virus reproduce itself.

Only those tissues and organs of the tobacco plant are attacked by the virus that are not only in a state of active growth but in which the division of cells is still in full progress; all tissue that has reached its full growth is immune from it, but may under certain circumstances transport the virus. Leaves that are growing, but are beyond the expansion stage, can no longer be infected, although they are still suitable for the transmission of the virus to the stem.

If the stem is inoculated, only the young leaf-buds and the leaves that are newly developing from these growing points are infected. If young leaves are inoculated, the same thing takes place; the virus returns to the stem from the leaf and infects either the axillary buds or rises to infect the terminal bud. If fully matured organs are used for infection, be it stems or leaves, and very little virus is used, failure is certain;—obviously the virus then remains in the matured cells without having any effect. A larger quantity of the virus may, however, move out of the matured parts into the surrounding new tissues and affect them.

In any case, it is reasonably certain that the virus in the

plant is capable of reproduction and infection only when it occurs in cell tissues that are dividing, while not only the matured, but also the expanded tissues are unsuitable for this. Without being able to grow independently, it is drawn into the growth of the dividing cells and here increased to a great degree without losing in any way its own individuality in the process. In conformity with this, no ability of reproduction outside of the plant could be proved. It is true that a Bougie extract that was filtered clear and was entirely free from bacteria could be kept for over three months[10] without losing its virulence or even seeming to decrease it. But an increase of infectivity was not to be observed, not even in the first period of the experiment, even though the extract had been prepared in such a way that not only diseased parts but also healthy buds and leaves were extracted, so that if nourishment in the usual sense had been able to bring about reproduction, this should have taken place. Also, in transferring the virus to appropriate gelatin-media, the color and index of refraction of the latter apparently remains unchanged throughout.

Yet, judging the original quantity of the virus, which is reproducing independently in the plant used for infection, is after all, difficult, and, since the question naturally is of special importance, new experiments in this direction are to be expected. For the time being I must, as I say, take for granted that propagation results only when the virus is connected with the living and growing protoplasm of the host-plant.

The behavior of the virus in connection with the growing tissues reminds one of similar relationships in gall formation, for the cecidiogenen substance also can affect only growing parts. As far as the movement is concerned, the latter bodies behave differently from the virus; they must, in order to be effective, be brought into meristematic tissues and they move only through these.

The method of reproduction of the virus reminds one in certain ways of that of the amyloplasts and chromoplasts, which also grow only with the growing cell protoplasm, but can also exist and function independently.

[10] How long the extract may be kept I am not prepared to say as yet; in any case, longer than three months.

3. *The flow of the virus inside the plant. Various ways of infecting: Local and general infection.*

In artificial infection the virus may move with the flow of water through the xylem bundles. But I do not believe that the xylem is the normal path of the flow in a closed plant. For if young leaves are infected at the time they are in the seedling, or later, the mature stage, these leaves themselves remain entirely normal, as we have seen, but the virus from here returns to the stem in order to infect new tissues that are higher up. Now it seems to me, that such a movement can occur only when the so-called descending flow of sap is followed; but this is directed through the phloem. The existence of a flow of sap, which under normal conditions might be in a direction from the leaves toward the stem along the xylem or the parenchyma, is, however, improbable and, during strong transpiration, impossible, at least in the xylem.

If the virus really moves in the phloem bundles the direction of the movement should be both ascending and descending, and it should be subject to the laws followed by the usual food substances in the nutrition of new tissues or the depositing of reserve material. The flow necessary for this must, however, according to the circumstances, be directed either to the base or to the top of the organs. The symptoms of infection are in good agreement with this.

The slow flow of the virus along the phloem bundles is, in my opinion, shown in a strange manner in the order of the diseased leaves in one-sided local infection of the stem. Often (perhaps always) the leaf that first becomes diseased is situated directly above the wound left by the infecting needle. If the place of infection was closely circumscribed, for example to a single shallow puncture of the needle with the Pravaz syringe, the second diseased leaf, in a $3/8$ leaf position, may be exactly the ninth above the first one to become diseased. Then, or shortly before, an apparently fan-like spreading of the virus occurs through which at first the surrounding leaf series and finally everything around the stem appears to be infected. It is strange that the buds take up the virus less easily than its new tissues do, or perhaps, to speak more exactly, it is again able to rid itself of the virus, for it is a fact that under certain circumstances healthy organs may be again produced from it later.

[40]

I conclude that the virus may also move in the xylem and from there is able to infect the meristematic leaf-buds from the fact that many of my experimental plants produced leaves that were extremely malformed at their tips, even from the moment that they were first visible in the bud, while the general symptoms of the disease were then not yet noticeable on the lower part of the same leaflets. It seems as though in such cases a large quantity of virus is poured into the tip of the new tissues at once, which is probably only possible through the transpiration stream along the xylem bundles. Such leaves, however, later show the normal symptoms of the disease, also, in such a manner that they point in the usual way to the flow of the plastic nutritional substances as the carrier of the virus.

It was proved that the transportation of the virus for great distances through healthy and mature stems and roots areas is possible by attempts at infection with soil taken from the roots of potted tobacco plants. In order to approximate natural conditions, the experiments were arranged in a very simple manner, described in sections 6 and 7. It is sufficient to mention here that plants that were already two or more decimeters high and whose lower leaves had long since died off could be easily infected from the root by means of soil in which the dry virus was present and had therefore become infested. Just as in wound infection, all mature parts and even the leaves still in the process of expansion remain healthy while only the leaves newly formed from the terminal and axillary buds become diseased.

Under these circumstances, the movement takes place slowly and the symptoms may be expected only after a time of at least three weeks to a month. This period of time, however, depends on the phase of development of the plant, so that seedlings that are injured in the root show up the symptoms of the disease sooner after infection. My experiments with uninjured seedlings are not finished, so that I am yet unable to give the point of entrance of the virus into completely normal plants. For the tobacco grower, this question seems not to be an insignificant one, since the tobacco plants are transplanted after sowing whereby many open wounds on unavoidably severed fibrous roots at least make possible the entrance of the virus.

If one wishes to convince oneself in the shortest possible

time of the virulence of the contagium it is best to deeply wound with a knife the youngest part of the stem below the terminal bud, which still may be easily treated without injury, and to place into the wound a piece of fresh, diseased tissue. The newly formed leaves will then plainly show the first traces of the disease after ten to twelve days; after three weeks the disease symptom is clearly distinguishable, even to the layman.[11]

The difference between the plants infected from the soil and those infected through wounds in the stem is notable. While the former show a general infection from the moment that the disease makes its appearance, which means that the diseased leaves are all around the stem, in the latter the infection, as has been discussed before, is at first confined to a stem segment as a local infection, which later develops into a general infection.

4. The virus may be dried without loss in strength of infection.

Pieces of dried diseased leaves put into healthy plants showed themselves capable of inducing infection even after they had been kept in my herbarium for two years. The same was true of the pieces of filter paper that had been moistened with the expressed sap of diseased leaves and carefully dried at 40°C. Extremely small pieces of leaves may be used for infection, which shows infinitely small quantities of the contagium are sufficient for infection. The virulence of the dried materials is, however, always less than that of the fresh materials, and for the present, I shall ascribe this difference to the partial destruction of the virus in drying and not to a change into a modified, weaker form, for, with a small amount of fresh virus, I obtained symptoms apparently identical to those resulting from use of a much greater amount of the dried virus. It, therefore, still seems doubtful whether the word virulence is really applicable here. I placed the dried virus into parts of the stem, as well as into the central veins of young leaves, and obtained the usual results.

I wish to mention at this time that the alcohol precipitate of virulent extracted sap retains its virulence after drying

[11] Recently, I was able to reduce to three days the time interval necessary to appearance of the disease by treating parts of buds that were yet much younger than those used before.

at 40° C. However, strong alcohol is also not injurious to many bacterial spores.

5. The virus may winter in its dry state outside of the plant in the soil.

In the fall of 1897 I let a diseased plant in a large flower pot in the shed die through lack of water. The plant was pulled up, the soil shaken from the root system into the pot, and the latter was kept in a dry place. In the following spring I divided the soil between four pots, partly filled with fresh soil. One pot was larger and was set with three plants; the three smaller pots received a plant apiece, all of which were bearing several leaves of which the lower ones had already dried off. All these plants were without a doubt entirely healthy. After about four weeks the conditions were as follows: Of the three plants in the large pot, one had become diseased; the two others were healthy and remained so until the end of the experiments. The plants in the three small pots all became diseased. One of them developed poorly from the very beginning, became strongly diseased, and soon showed the peculiar malformed leaves so characteristic of the more pronounced, artificial-wound infection. The plant also produced several chlorotic leaves. Since the other typical symptoms of the disease were also very pronounced in this plant, it is certain that the virus can retain its full virulence after wintering in air-dried soil. The other plants showed the normal development of the disease. Since I had subsequently stirred the soil of some of these plants in the pots with a piece of wood, I presume that the malformed plant had received large root injuries in the process, through which many entrances may have been opened to the virus.

6. Other attempts at infection through the roots.

On the 6th of June, 1898, a series of healthy plants, growing in pots and several decimeters high, were infected in the following manner. A severely diseased plant was taken out of the ground with a lump of soil, and the soil shaken out of the root system and strewn in small quantities on the soil close to the main stem of potted plants. The plants were then sprinkled with tap water and the soil partly dug under, avoiding injury to the roots. After almost four weeks all the experimental plants showed general infection in the newly formed leaves. Then, to my astonishment, came a period of recovery, so that at the end of August I considered the plants to be healthy. Later, however, their leaves again became diseased, although not in very virulent form.

I feel that I must conclude from these experiments that normal roots are capable of taking up the virus from the soil through their closed outer epidermis. I admit, however, that this conclusion may not be reliable, for animals living underground may have made possible or facilitated the entrance of the virus by means of root injuries. Only experiments with plants grown in nutrient solutions would, in my opinion, yield absolute certainty on this question.

7. *The virus becomes ineffective in boiling temperatures. The effect of Formalin.*

For a long time I considered it possible that some anaerobe might have something to do with the infection. Especially when I learned of the characteristics of a group of these organisms occurring in manure and faeces, to which I will give the name of "skatolbacteria," did I think I had reason for this conception. The "microaerophilie" is of such a nature in these forms that one is forced to infer a relatively large consumption of oxygen. More than that, some varieties have such extremely small, either spherical or oblong spores that the possibility of their passing through the Bougie pores is not entirely unthinkable. For this reason, I carried out several experiments with freshly extracted heated juices, and juices after filtration through the candles.

These experiments produced a result that was definitely negative so far as concerned the presence of bacterial spores. Boiling completely destroys the virus. It cannot even stand 90° C.; the shortest period of heating is sufficient to destroy the virus. I have not as yet ascertained the minimum temperature, but do not doubt that it is only a matter of a pasteurization temperature. Through these experiments the phantom of the anaerobes and their spores was forever banished.

This is perhaps the place to say a word about the sterilization of the utensils used in the experiments, especially of the Pravaz syringe. The modification by Koch, it is true, is better suited for sterilization, but its rubber ball does not allow the injection of the fluid with as great a pressure as does the original form. I therefore tried to sterilize the latter with Formalin, since it does not stand the heat very well. The success of this, however, is only partial and only with greater quantities of Formalin. Weak solutions of Formalin mixed with the virus do not destroy it, but they do so lengthen the period of incubation that the disease cannot become apparent until six or more weeks after infection. In any case, one must be sure that the last traces of Formalin have completely evaporated from the syringe before using it again, for it has become apparent that Formalin is very poisonous for the tissues of the tobacco plant, much more so than to the virus itself. In section 9 I shall come back to this matter. Here I wish to emphasize that Formalin, once it has entered the vessels, is carried with noticeable rapidity through the leaf veins, whose living cells soon die off. Through careful injection of Formalin into the midribs, in such a way that the xylem bundles are not mechanically injured, it is possible to destroy all living tissues, without disturbing the flow of water in the least, so that the leaf remains fresh and can continue growing.

8. Different symptoms of the disease. Development of malformed leaf tissues through large quantities of virus.

In my opinion the spot disease (mosaic) of the tobacco leaves in its milder form is a disease of the chlorophyll particles and, in its more intensive forms, a general disease of the living protoplasm.[12] The course of the milder form is as follows: With artificial infection of the virus in stem wounds below the terminal bud, those leaves that unfold within ten days remain healthy. The leaves that develop later present a yellow-spotted appearance, which in itself is not particularly characteristic and often occurs in healthy leaves. After two to three weeks a characteristic symptom of the disease appears (Table II, Fig. 1). In the neighborhood of the lateral veins of the 2nd or 3d division, the color becomes very dark-green in places, namely in rectangular areas that are cut in half by the veins; in the rest of the leaf the coloring process is somewhat slower than under normal conditions and sometimes even recedes to complete albi-

[12] For the time being I must skip the anatomical relationships because they have not as yet become entirely clear to me.

[45]

nism. In any case, dark-green spots develop on a light-green background. The border between the two colors is either sharply defined and sudden or it is diffused as if diluted. Since the darker parts grow more rapidly than the lighter ones, they soon stand out over the even surface of the leaf more or less convexly, whereby, in the acute cases, pronounced bullate blisters develop on the upper surface of the leaf. Later (Plate II, fig. 2) one can observe a necrosis of the leaf cells developing at the edge or even in the middle of the dark spots, which soon leads to the small, tan, dead, and dried spots feared so much by tobacco growers because the leaf thereby becomes unsuitable for use as a wrapper for cigars. Although most of the dead tissue spots develop in the manner described near or in the dark-green fields near the veins, the origin of some of them remains uncertain; apparently, they also may develop in the yellow spots. The symptoms in the tobacco fields are usually not of as great an intensity as in artificial infection, especially the blistery outgrowth of the dark-green parts on the leaf blade is entirely lacking. In contrast to this, the necrosis and drying of leaf spots were not observed in some of the greenhouse plants.

With artificial injection of fresh extracted sap, or with inoculation with diseased tissue the disease may reach a higher stage of intensity than I have as yet observed under natural conditions.[13] I mean the abnormal tissues of the newly formed leaves (Plate I b, c, d, Plate II, fig. 4 and 5). This is no doubt connected with the quantity of infectious material used for the experiment. Therefore, it is much easier to produce leaf monstrostities with fresh extracted juice than with the Bougie filtrate, since, as has been remarked earlier, more of the latter must be injected in order to obtain the same effect, which certainly is remarkable for a contagium that increases through growth.

The first noticeable symptoms of the striking appearance of leaves becoming malformed is the retardation of the growth in the direction of the midrib and of the principal lateral veins. Oval or circular leaf surfaces develop because of this. Later the intensively green spots are to be observed, which rise blister-like and are peculiarly contrasted to the

[13] Probably because severely diseased plants are soon noticed and taken out.

rest of the leaf-surface, which remains a much lighter color and has a tendency towards albinism, particularly at the veins. Once, I got a small beautifully formed "ascidium" instead of an abnormality of the kind described. Such entirely unrecognizable leaves always remain much smaller than the leaves that develop later and are, by the way, healthy and fresh, as, on the whole, the symptoms of the disease never become particularly detrimental to the plant. Even specimens that have been severely attacked produce stalks of normal height and thickness and, finally, at the end of the vegetation period often entirely healthy leaves, bloom and fructify normally, and, as far as is known, their seeds are always healthy. I do not yet know whether it will be possible to artificially infect the blossoms and seeds, because I began the experiments in connection with this too late.

The symptoms of the disease in plants growing in the open are so variable in intensity that one involuntarily thinks of individual predisposition. If this impression is correct, and if it is a matter of something other than the uneven quantity of the virus originally introduced, it probably will be easy to produce an immune race in which, of course, artificial infection would have to be used as a criterion. The facility of such infection experiments seems to place success within reach.

9. Albinism or "bunt" as an incidental result of artificial infection.

In many of my experimental plants, spots appeared on the leaves in which chlorophyll was entirely lacking. In a few cases the spots were spread over the surface by the hundreds and in such elegant order that really decorative colored plant-leaves resulted (Plate II, fig. 3). Up to now I do not have the thing in hand; it is questionable whether it will be possible to make any experiment out of it with constant results. I wish to mention a few cases here, where it appears that there is a causal connection.

Albinism in a Mixture Infection of a Bacterium with the Virus

Extract of diseased leaves after having stood for a day at room temperature was plated out on a culture medium for

[47]

the isolation of the bacteria that had developed in it. The following mixture was used: Decoction of 20 grams of clover leaves in 100 grams of water and 2 grams of cane sugar solidified after filtration and boiling with 10 per cent of gelatin.

Mainly two types of bacteria developed of which the one, a non or weakly-liquefying and non-fermenting bacterium, which I have called *B. anglomerans*,[14] and which occurs very commonly on plants in general, occurred in millions per cc. In the first plating, the extracted juice was just poured over the gelatine plate, so that it had to be taken for granted that each bacterial colony was infected with the virus. Without further isolation, parts of these colonies were transferred into test tubes on the culture medium mentioned above, so that a trace of the virus might, however, have been transmitted, even if it had not reproduced itself in the bacterial colonies. Since the bacteria grow quickly, much material was soon formed with which a suspension was prepared in tap water and a plant was amply infected on the 30th of September. At first I thought that the plant would remain entirely healthy; but, on the 15th of October, I noticed a beginning of the disease, which, however, did not continue to develop in the regular manner but produced a beautiful albino plant.

The second variety of bacteria treated exactly as the first remained without effect in the infection experiment.

Since I have saved the bacterial cultures, I shall be able to repeat the experiment in the future. At present I am mainly interested in the question of the virus occurring only as a contamination in the colonies or in its increase either between the bacteria or in the bacteria proper. In the latter case a variation in the characteristics of the virus is not impossible.

Albinism through Infection with Virus Combined with Formalin

The observation to be discussed here was an entirely incidental one. When I cleaned my Pravaz syringe with Formalin before using it, a trace of Formalin remained in the hollow needle in one case and entered the experimental

[14] Bot. Zeit. 1888 p. 749.

plant along with the virus, which I immediately noticed because of the necrosis of cells adjacent to the wound. Later the plant showed the injury only indistinctly, but, afterwards, several leaves became mottled.[15]

Albinism through Infection from the Soil

Some of the plants, which had become diseased very late in the season in the greenhouse through infection from the pot-infested soil are to be described more as variegated than as spot-diseased. The dark-green spots near the veins had become barely visible, while the discoloration in the rest of the leaf parenchyma had appeared particularly early and intensively. However, only part of the areas of the colored leaves became white, the majority of them remained yellowish. In one of these plants the lower leaves remained very small and became malformed in the manner described before.

Of the three cases of bunt (mosaic) that I have mentioned in this section the first two may agree in that the virus entered the plant in a greatly diluted state; but I do not believe that dilution is here an essential factor, because the third case gave the impression that especially much virus had been active. However, I consider it, if not proved, at least highly probable that there is some connection between the virus of the spot disease and the bunt (mosaic), and the old question whether the bunt (mosaic) is always of the same origin has come back into discussion with these experiments.

10. Other infectious plant diseases caused by a contagium fluidum and not by parasites.

Even if the symptoms of the spot disease coincide so closely with certain forms of albinism or bunt (mosaic) that both may unhesitatingly be put under the category of infectious diseases of the chlorophyll bodies, there yet remains, according to known observations, a difference in the principle of the mode of transmission of the contagium, a difference that leads one to consider both of them as separate kinds of diseases, each with its particular virus. The form of albinism (variegation) suitable for transmission is

[15] I repeated this experiment, but found only the usual symptoms of the disease, although very late. If the virus remains long in contact with even strongly diluted Formalin, it is completely destroyed.

namely transferred only when the living albino tissues grow together with the living-tissues of the green plant by means of grafting or budding, while simple inoculation of green plants with the tissues or the extracted juice of variegated varieties of the same kind, remains entirely without results[16] according to my several-times-repeated experiments with *Ulmus campestris, Acer negundo, Pelargonium zonale,* and *Utrica diocia.* It appears, therefore, that the contagium of variegation is transmissible but that it stands in a much closer relationship to the protoplasm of the plant than the contagium of the spot disease, in that it cannot exist, like the latter, outside of the plant and dies when the plant cells that carry or continue it themselves die. My preceding observations, however, sufficiently show that the last word has not yet been spoken on this subject. Since the question of the contagiousness of variegation is important, for the evolution theory as well as for the theory of variability, further experiments on the subject would be very desirable.

Another disease, which surely belongs here, is that known in America as "peach yellows."[17] The symptoms of this disease consist mainly in immaturity of the fruits, growth of the latent buds at unusual places into thin brooms, which are often colorless, and yellow discoloration of the leaves, which is followed in a few years by the death of the tree. According to Mr. Smith, bacteria and parasites are definitely not the cause. Nevertheless, it was easy to transmit the disease to healthy trees simply through grafting or budding with a bud of a diseased tree. This experiment showed that it is necessary for the bud to unite if the disease was to be transmitted, for the virus is not capable of infecting healthy trees without the connection of the living tissues, according

[16] It is true that some researchers have doubts as to the transmissibility of variegation as such and have expressed the opinion that those green plants that become variegated themselves through grafting with variegated ones would have become so without any grafting whatsoever, i.e., through spontaneous bud variation. They remark that the stocks used (Abutilon, Jasminum, Pelargonium) are garden plants whose green specimens have a strong tendency towards variegation anyway. Such objections, however, are not sufficiently grounded (see Lindemuth, Vegetative Bastardenerzeugung durch Impfung, Landwirtschaftliche Jahrbücher 1878, no. 6 and Vochting, Transplantation, p. 13, 22, 92, and 112, Tübingen 1892.)

[17] Erwin F. Smith, Peach Yellows and Peach Rosette, U. S. Department of Agriculture. Farmers' Bulletin No. 17, Washington, 1894. I know this short but interesting discussion only through the separate which the author kindly sent. Much to my astonishment, I was not able to find a word about it in the scientific literature available to me.

to Mr. Smith. He neglects to point to the agreement of this observation with the mode of transmission of variegation in Abutilon and Jasminum.

"Peach rosette" is, according to Mr. Smith, another non-parasitic disease, closely related to "peach yellows," which is easily transmissible through budding and root grafting. The disease manifests itself in that all buds, dormant as well as active ones, grow into small rosettes that consist of single large leaves and several hundred small leaves. The color of the leaves is yellow. The fruit does not ripen but dries and falls to the ground prematurely. Here, too, we find the peculiarity I have described in the spot disease, that the virus moves laterally with difficulty but upwards with ease, so that a tree may become diseased on the side on which the rosette bud was grafted, while the opposite side remains healthy for years.[18]

Smith says that the epidemic character of yellows, as well as rosette, leads one to the conclusion that there must exist another mode of transmission than that of tissue intergrowth, but he does not believe that the virus can come from the soil; however, he notes that, particularly in rosette, a whole tree may become diseased in almost all its parts at the same time, which, as we saw earlier, is not compatible with local infection but rather points to a general infection similar to the spot disease when the tobacco plant is infected from the soil.

Since Smith did not carry out any experiments with artificially transferred sap, the possibility remains, even the probability, that these, too, could give a positive result. If this should really be the case, the virus would probably also be capable of existence outside of the plant, and infection from the soil through the roots would be possible, and yellows and rosette would then approximate much more closely the spot disease than is apparent from the descriptions given.

I consider it highly probable that many other non-parasitic diseases of unknown cause may be ascribed to a *contagium fluidum*. It seems useful to me in further research on this matter to distinguish sharply between the two forms

---

[18] The latter observation seems to exclude entirely the possibility that in peach rosette we are concerned with a "phytoptus" invasion, although the other symptoms of the disease seem to point to it.

in which, according to the available knowledge, such a *contagium* may appear, namely, firstly as an independent *contagium*, which is capable of existence outside of the plant, even if only for a time, as in the leaf-spot disease of the tobacco plant, and secondly as a *contagium* that exists only in living tissues as in the form of variegation, which is transmissible through graftage only.

Explanation of the Illustrations

Plate I

A young tobacco plant that has become diseased through artificial infection with much virus. The virus was introduced through a wound at *a*, which penetrated through the whole stem. The diseased leaves, *b*, *c*, *d*, which first developed, are malformed; those following, *e*, *f*. are diseased, though not malformed.

Plate II

Fig. I. A young tobacco leaf, in the first stage of the disease, with a moderate amount of virus. The dark-green spots are visible next to the vein, the local changes, by the way, in the color of the chlorophyl did not produce any distinct contrasts on the photographic plate.

Fig. 2. A mildly diseased tobacco leaf in the second stage of the disease with a few brown spots that were produced through the premature necrosis of the tissues. The most important stage of the disease in which the dead, brown spots are increased by hundreds or thousands is not pictured.

Fig. 3. A vari-colored tobacco leaf of a plant that had become vari-colored through the mixed infection of the virus with *Bacillus anglomerans*.

Fig. 4 and 5. Small malformed tobacco leaves, produced by the introduction of large amounts of virus into the stem.

[These plates, in color in the original, are here presented in black and white because of the high cost for reproduction in color.]

Plate I

Plate II

H. H. McKinney and Cross Protection

Robert W. Fulton
Department of Plant Pathology, University of Wisconsin, Madison

The article by McKinney deals mainly with the occurrence of mosaic symptoms on various plant species he encountered in West Africa and the Canary Islands. Its most notable contribution, however, is contained in a single paragraph (pages 566-567) which now seems almost casual.

Some consideration must be given to the scientific climate of the time (1929) in regard to plant pathology. Little was known about plant virus diseases. The word "mosaic" was almost synonymous with the word "virus." This word was applied partly because no pathogen could be visualized and whatever it was that caused tobacco mosaic could pass a bacteria-proof filter. A favorite topic for discussion was "the mosaic problem": speculation on the nature of viruses. Such discussions often centered on the question of whether or not viruses were living, and usually ended with someone demanding a definition of "life". There was an idea that all mosaic symptoms were due to the same agent, although some evidence to the contrary had been published, aphid transmissibility, for example.

McKinney recognized differences in severity of mosaic symptoms in some of his collections and knew that differences persisted in successive transfers. From small yellow areas that appeared in mosaic leaves he isolated virus which produced infected plants that were entirely yellow mosaic. In the significant paragraph he noted that yellow mosaic did not develop when it was inoculated to tobacco showing mild *light* green mosaic although yellow symptoms did appear when mild *dark* green mosaic-infected tobacco was inoculated. This is what became known as cross protection and has been the subject of continuing research ever since, by others (see Fraser, 1998) as well as McKinney.

The time may have been ripe for a new idea. Control of tobacco mosaic virus (TMV) had usually been unsuccessful. Many years ago I once heard TMV referred to from a scientific podium as a "laboratory curiosity." It may have been a curiosity in that laboratory; in tobacco and tomato fields it was a scourge. Cultural practices, largely emphasizing sanitation, did not give satisfactory control. A Columbian tobacco variety (Ambalema) that was tolerant of the virus did not impart this quality to acceptable commercial types in conventional breeding trials. Thus the use of cross protection as a control was considered.

Two years after McKinney published, such trials were described by T. H. Thung (1931) who cited McKinney and suggested that the immunizing characteristic might be useful in virus differentiation. McKinney pointed out later that this was not a reliable test. Suggestions for using cross protection for control were made by various plant pathologists for eleven other virus diseases (see Fraser, 1998). Among commercial successes were citrus tristeza, papaya ringspot, and, despite difficulties with an annual seed-propagated crop, tomato mosaic. Control of this last now mainly involves resistant varieties, illustrating the requirement that plant disease controls be cost effective.

McKinney had a wide base of knowledge of biological subjects and did not take criticism lightly from those he believed had not carefully considered the experimental evidence. A case in point came up when, early in his career, he was assigned by the U. S. Department of Agriculture to work at Wisconsin on a new disease of wheat in Illinois. It was thought the disease was "take all" introduced from Australia where it had been first described. McKinney pointed out that descriptions of the Illinois disease did not match Australian information. When virus inclusion bodies associated with the disease were interpreted as amoebae and the probable cause of the disease, McKinney disagreed vigorously, pointing out that the internal structure of these bodies was not that of an amoeba.

Dr. L. R. Jones, head of the plant pathology department at Wisconsin, was disturbed that McKinney disagreed with his administrative superiors in Washington on the nature of the disease. McKinney made microscopic observations and worked with Dr. Sophia Eckerson at Berkeley, California to show that the inclusion bodies differed cytologically from amoebae. A theory that "mosaic" diseases were caused by protozoa enjoyed a brief popularity at the time. A published report linking bean and potato mosaics and amoebae went out of favor when McKinney, Eckerson, and Webb showed that the presumed amoebae were not cytologically similar to protozoa.

Observations by McKinney on the absence of specimens showing only yellow mosaic in the areas surveyed led him to the idea that the small yellow areas in green mosaic leaves were the result of mutation of the virus. McKinney's use of the term "mutant" was criticized because it appeared to imply that the virus was living. This irritated him because some of the criticism came from administrators who did not have "hands on" knowledge or sufficient information to make judgments. McKinney believed firmly that scientists should have freedom to investigate fundamental aspects of their problems without being bound by administrative restrictions. The U. S. Department of Agriculture (USDA) eventually adopted this policy and allowed considerable freedom for personnel to do research outside of areas defined strictly by particular crops.

Such a policy was to McKinney's liking and he spent most of the remainder of his life with the USDA (McKinney, 1972). He worked with, among others, Dr. E. E. Clayton (Division of Tobacco and Special Crops) who was involved with tobacco mosaic virus control.

McKinney was continually involved with basic investigations of TMV and other viruses. He devised ways of estimating virus concentration by measuring infectivity of highly diluted sap of infected plants. He was interested for many years in virus characterization, serology, and purification. His suggestion that centrifugation might be used to concentrate and purify plant viruses was met with a published objection, with data, that seems amusing now. It pointed out that centrifuges did not spin fast enough to sediment viruses. Among other contributions, McKinney devised a rapid drying method for preserving cultures of unstable plant viruses. This technique was widely used and of much value, as it made possible comparisons between new isolates and cultures that had been described years before.

Cross protection has come a long way (Miller and Hemenway, 1998) since the first description, but I don't think data implicating the involvement of coat protein (Sherwood and Fulton, 1982) and other virus components would have surprised McKinney very much.

References

Fraser, R. S. S. (1998). Introduction to classical cross-protection. *In*: Methods in Molecular Biology, vol. 81, Plant Virology Protocols (G. D. Foster and S. C. Taylor, eds.). Chapter 2, pp. 13-24, Humana Press, Totawa, N. J.

McKinney, H. H. (1972). Over fifty two years as a government servant in plant disease and virus research. 88 p. Privately published.

Miller, E. D. and C. Hemenway. (1998). History of coat protein-mediated protection. *In*: Methods in Molecular Biology, vol. 81, Plant Virology Protocols (G. D. Foster and S. C. Taylor, eds.). Chapter 3, pp. 25-38, Humana Press, Totawa, N. J.

Sherwood, J. L. and R. W. Fulton. (1982). The specific involvement of coat protein in tobacco mosaic virus cross protection. *Virology* **119**:150-158.

Thung, T. H. (1931). Smetstof en plantencel bij enkele virusziekten van de tabaksplant. Z. Ned. Indisch. Naturwetensch. Cong. Bandoeng, Java. pp. 450-463.

Biographical Sketch

Robert W. Fulton graduated from Wabash College in 1935 and was elected to Phi Beta Kappa. He taught at Wabash for two years then began graduate work with tobacco viruses under James Johnson at the University of Wisconsin. He obtained his Ph.D. in plant pathology with a minor in botany in 1940. After being employed in a postdoctoral position working on tobacco diseases until 1942, he entered the army. He was an assistant professor at the University of Wisconsin 1947-55; associate professor 1955-60; professor from 1960-84; visiting professor, University of California, Berkeley in 1964, and in 1970 was elected a fellow of the American Phytopathological Society.

MOSAIC DISEASES IN THE CANARY ISLANDS, WEST AFRICA, AND GIBRALTAR [1]

By H. H. McKinney [2]

Senior Pathologist, Office of Cereal Crops and Diseases, Bureau of Plant Industry, United States Department of Agriculture

INTRODUCTION

The Allison V. Armour expedition was conducted primarily for the purpose of collecting living plants and seeds and introducing them into the United States. However, as the yacht used on the expedition was fitted for doing laboratory work and for growing potted plants, it was possible to conduct certain types of experiments coincident with the work of collection; and other experiments have been conducted since the return of the writer to the United States. On the expedition the writer's time was devoted to the search for plant viruses suitable for experimental studies and to the accumulation of information concerning the virus diseases of plants in the regions visited.

This report deals with the observations made in the Canary Islands, west Africa, and Gibraltar, and with some of the experimental results relating both directly and indirectly to the mosaic viruses from *Nicotiana glauca* R. Grah. collected on the Canary Islands and at Gibraltar. Limited comparative studies have been made with some of the viruses collected in the United States.

MOSAICS IN CANARY ISLANDS

The expedition reached Santa Cruz, Teneriffe, the capital of the Canary group, on December 22, and on December 28 it went to the island of Grand Canary. Several days were spent making observations and collections on each of these islands. During December the islands are rather cool and are rich in growing plants, including many kinds of garden vegetables.

Tomatoes and potatoes were growing in many localities, and various species of cucurbits were found in gardens and along the roadsides. Many of the potato plantings were located on narrow terraces, sometimes high above the highways and very inaccessible. In one of these terrace plantings which overlooked the sea near the port of Orotava, Teneriffe, green mosaic was found on most of the potatoes and on a garliclike plant [3] (fig. 1) which was growing as a

[1] Received for publication Mar. 2, 1929; issued October, 1929. The work described in this paper was done by the Allison V. Armour expedition of 1926–27. A full account of the two expeditions sponsored by Armour has been published by Fairchild (*1*).[4]

[2] The writer wishes to express his appreciation to Mr. Armour and to Doctor Fairchild for the opportunity of collecting and studying the mosaic diseases reported in this paper and for personal assistance rendered in various ways during the voyage. Credit is due to Dr. J. M. Dalziel, of Kew Botanical Gardens, Kew, England, who accompanied the expedition, and also to Oliver M. Freeman, of the United States Bureau of Plant Industry, for identifying some of the plants collected.

[3] When this plant was collected it was thought to be a wild garlic, and the writer referred to it as such in an abstract (*6*). After being studied further in a quarantine greenhouse, this plant appears to be one of the false garlics (*Nothoscordum fragrans* Kunth).

[4] Reference is made by number (italic) to "Literature cited," p. 577.

Journal of Agricultural Research, Washington, D. C.

Vol. 39, No. 8
Oct. 15, 1929
Key No. G–675

61922—29——1

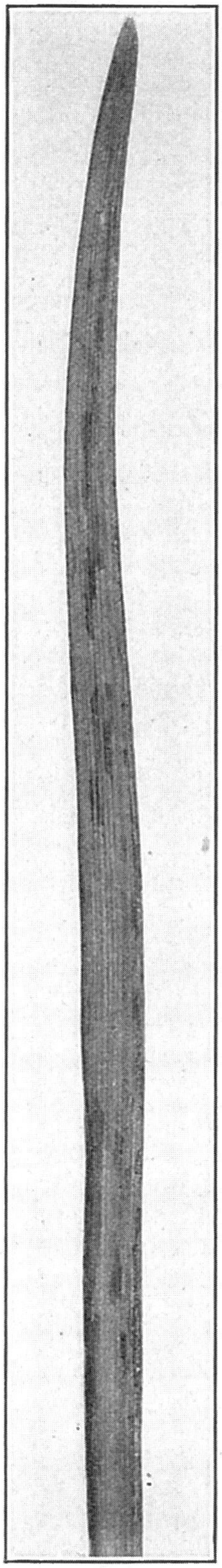

FIGURE 1.—Green mosaic on false garlic (*Nathoscordum fragrans* Kunth). Collected on the island of Teneriffe, Canary Islands. × 1

FIGURE 2.—Mosaic on *Nicotiana glauca:* A, Mosaic-free leaf; B, mild mosaic; C, light-green mosaic; D, yellow mosaic. Collected on the island of Grand Canary, Canary Islands. × 1

weed among the potatoes. In one garden at Orotava, fuchias (*Fuchsia magellanica gracilis*) and cultivated peppers (*Capsicum frutescens* L. var. *grossum*) had typical green mosaic. Many tomato plants were examined on the islands, but no mosaic was observed. Also no mosaic was observed on the cucurbit species which were found.

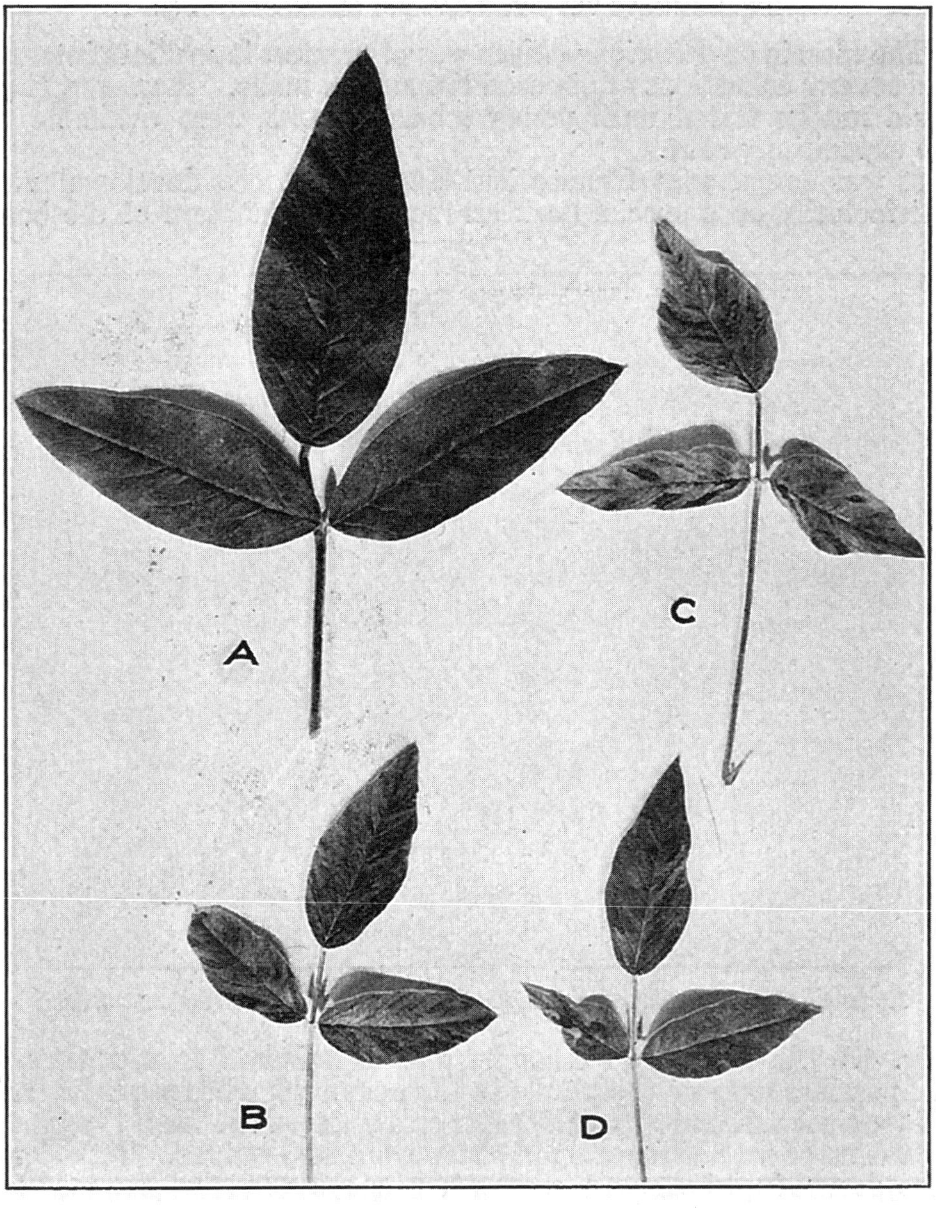

FIGURE 3.—Green mosaic on *Psoralea bituminosa:* A, Healthy leaf; B–D, diseased leaves. Collected on the island of Teneriffe, Canary Islands. × $\frac{9}{10}$

Bananas are grown in all suitable areas on both of the islands. Many plantations were visited and hundreds of plants were examined for bunchy top, a destructive virus disease (7) which occurs in Australia, Fiji, Egypt, Ceylon, and the Philippine Islands. None of the plants examined gave evidence of this disease, and no mosaic was observed.

Psoralea bituminosa L., a legume which is a native of Arabia and the Mediterranean districts, and *Nicotiana glauca*,[5] a native of South America, were found growing wild in all the waste places and barrancos or dry creek channels. Both species had green mosaic. (Figs. 2 and 3.) In fact, it was exceptional to find mosaic-free plants of either of the species on Teneriffe or Grand Canary.

EXPERIMENTS WITH THE CANARY ISLANDS MOSAICS

The mosaic on *Nicotiana glauca* was of greatest immediate interest, and several collections of diseased tissue were made. However, these could not be tested until young tobacco plants were available on the expedition yacht.

It was found that Connecticut-Havana tobacco developed very satisfactorily on a special deck arranged near the stern of the boat.

FIGURE 4.—Method of growing experimental tobacco plants on the deck of the yacht *Utowana*

(Fig. 4.) As no serious difficulty was experienced from storms, it was possible to grow the plants in the open. The intense light from the tropical sun and any salt-water spray were kept from the plants by means of properly arranged cheesecloth and canvas. The plants were grown in 4-inch pots arranged in a large meshed net made of heavy cord and wire. This prevented the pots from rolling about.

An inoculation test was made with an extract from dried leaf tissue obtained from a mosaic-affected plant of *Nicotiana glauca* collected in Teneriffe. This mosaic was easily transmitted to tobacco. From the beginning, the symptoms were of the light-green type. After the plants had developed 12 to 18 leaves, yellow spots and

[5] This species grows wild in many districts in the southwestern part of the United States, and mosaic-diseased specimens have been sent to the writer by David Fairchild from the Boyce Thompson Southwestern Arboretum at Superior, Ariz.

patches appeared on some of the foliage, as indicated in Figure 5, A. These results were of especial interest as they are similar to those that had been obtained with all the mosaic viruses that had been tested on tobacco previously (*4*, *5*).

Another test was carried out with virus from *Nicotiana glauca* collected in Grand Canary. The green-mosaic symptoms were slightly different from those obtained in the previous experiments, but yellow spots occurred in the same manner.

When the ship returned to the Canary Islands for fuel in March, one day was spent collecting and observing mosaic-affected *Nicotiana glauca* growing in the regions of the upper and lower highways between Las Palmas and Telde on the island of Grand Canary. Several hundred diseased plants were examined, and of these one had what

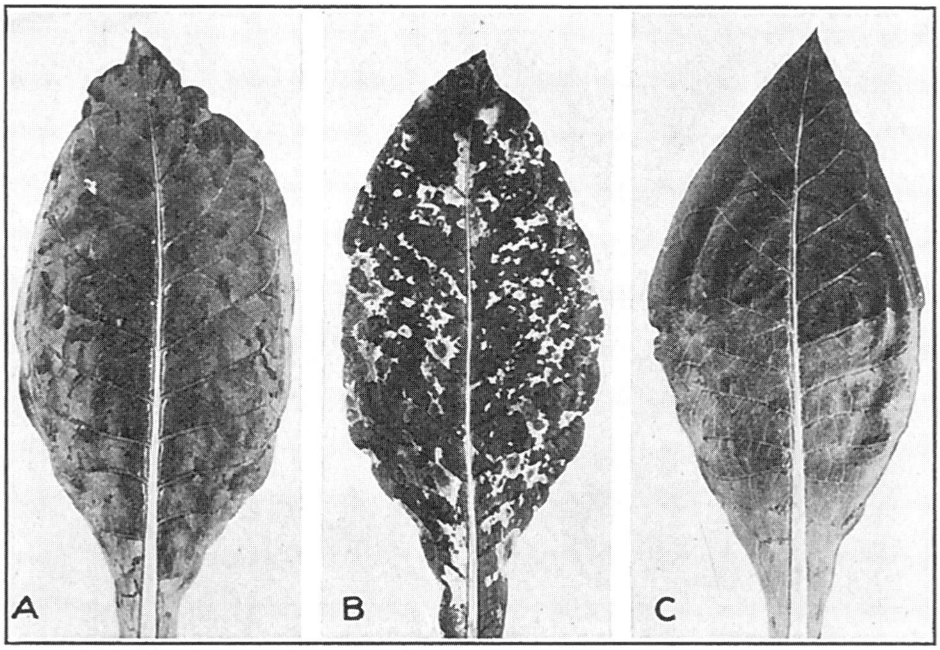

FIGURE 5.—Mosaic on Connecticut-Havana tobacco: A, Light-green mosaic showing a yellow spot; B, yellow mosaic produced by a concentrated virus obtained from spots similar to that shown in A; C, mild dark-green mosaic obtained from *Nicotiana glauca*. × ⅓

appeared to be pure yellow mosaic (fig. 2, D) and two had a yellow and green combination with yellow predominating. The remainder had either a light-green (fig. 2, C) or a mild dark-green mosaic (fig. 2, B). In some cases leaf abnormality was associated with green mosaic.

Mosaic leaf tissue was collected from eight specimens which had somewhat different symptoms ranging from the mild dark-green mosaic to the yellow type. These tissues were handled with great care to prevent cross contamination after collection. The fresh leaves were cut into fine pieces and dried in a warm closet, after which they were carefully stored in envelopes separated from each other by means of heavy oiled paper.

All these dry tissues have been tested on a strain of Connecticut-Havana tobacco in a quarantine greenhouse, and each produced mosaic one year after collection. Ten or more plants were inoculated with each sample. Plants inoculated with virus from different sources were protected from one another by means of frames having galvanized wire screen (12 meshes to the inch each way) tacked on each side, as shown in Figure 6. A frame of single screen was placed in front of the plants. Extensions were arranged on top of the partition screens if the plants became very tall. All openings in the greenhouse were covered with copper screen having 30 wires to the inch each way. The legs of the greenhouse benches were coated with "tanglefoot" paste. The benches did not touch the walls of the house at any point.

Thus far it seems that the mottling symptoms produced by these viruses on tobacco fall into three general classes: (1) Yellow mosaic, a type which has little or no chlorophyll in the lightest-colored areas; (2) light-green mosaic, a type characterized by light-green areas which occupy the greatest portion of the leaf surface, normal dark-

FIGURE 6.—Method used in conducting greenhouse experiments with different viruses. One front screen was removed for the purpose of making the photograph

green areas, and occasional yellow spots; and (3) mild dark-green mosaic, a type in which the dark or normal green is usually in excess of the light-green portion of the leaf, and yellow spots occur rarely. In young plants mild dark-green mosaic is very inconspicuous.

The yellow mosaic, thought to be pure when collected, proved not to be free from the green type. In fact, it appears that the three viruses were present in some of the diseased plants collected. All the green mosaics collected have been studied on maturing tobacco plants, and all have produced yellow spots. Some of the light-green types developed yellow spots in all the inoculated plants. One of the mild dark-green forms has produced yellow spots on 2 of the 52 plants which have been inoculated.

When yellow spots were removed from mosaic-affected leaves and inoculated into young plants, a yellow type of mosaic developed soon after the inoculation. (Fig. 5, B.) By making several successive subinoculations from the increasing number of yellow spots, strains of concentrated virus of yellow mosaic have been developed. From three to seven isolations and subinoculations have been required to accomplish this. Figure 7 shows a plant inoculated with such a concentrated virus. The yellow mosaics isolated thus far from *Nicotiana glauca* are similar in appearance to the yellow mosaic isolated by the writer (*4*) from cultivated tobacco having light-green mosaic.

Yellow mosaics are very attractive and ornamental, but they are more destructive than the green types. When they are inoculated into tomato plants, a high percentage of killing frequently results. In some cases young tobacco plants have been killed, and in all cases tobacco plants are markedly stunted, and the lower leaves become very yellow and frequently are killed completely.

One of the outstanding characteristics of the viruses of yellow mosaic is their ability to produce a general chlorosis on the old foliage.

FIGURE 7.—Yellow mosaic on Connecticut-Havana tobacco. The virus used was concentrated by isolating yellow spots from tobacco leaves (fig. 5, A) infected with light-green mosaic induced from virus collected on *Nicotiana glauca*. × ¼

Tobacco plants 30 inches tall at the time of inoculation have developed chlorosis of the type shown in Figure 8 on all the leaves which were over 3 inches in length at the time of inoculation.

The yellow mosaic pattern on the older leaves is quite distinct from that which occurs on the young leaves (fig. 7), and it is very distinct from the yellowing associated with senility. This chlorotic pattern occurs in a milder form on detached leaves which are inoculated with the virus of yellow mosaic and cultured in sand according to the methods of Purdy (*9*). Mottling is very pronounced on the stems and fruit of tobacco plants having yellow mosaic. (Figs. 9 and 10.)

Many combinations of yellow and green mosaic symptoms have been produced by inoculating tobacco and tomato plants with extracts consisting of different proportions of viruses from yellow mosaic and light-green mosaic plants. Several of these synthetic combinations resembled yellow and green mosaics which the writer observed on *Nicotiana glauca* in the Canary Islands, and on tomato plants in the southern United States.

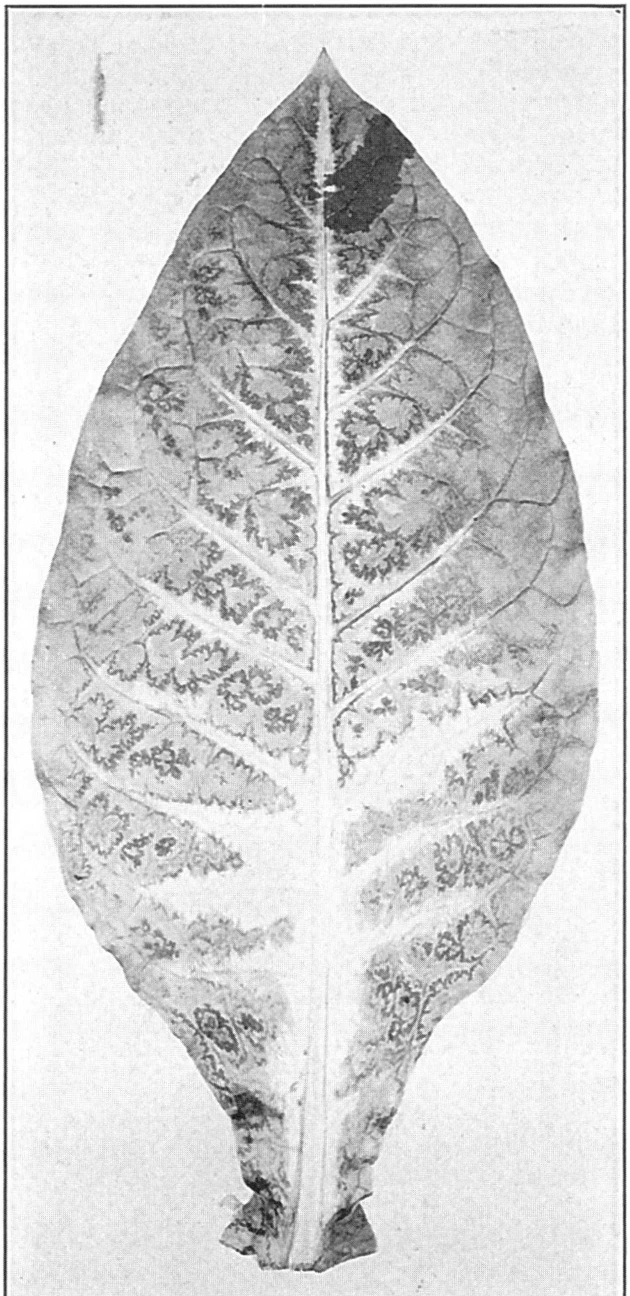

FIGURE 8.—Mature leaf of Connecticut-Havana tobacco. This leaf was about 3 inches long and the plant was 36 inches tall at the time of inoculation with a virus of yellow mosaic. The inoculation was made at the base of the stem. All leaves below the illustrated leaf gradually developed this type of chlorosis and died prematurely. Leaves above this one developed a mottling more nearly like that shown in Figures 5, B, and 7. The dark area near the apex was normal green as compared with leaves of the same stage of development on healthy plants. × ⅔

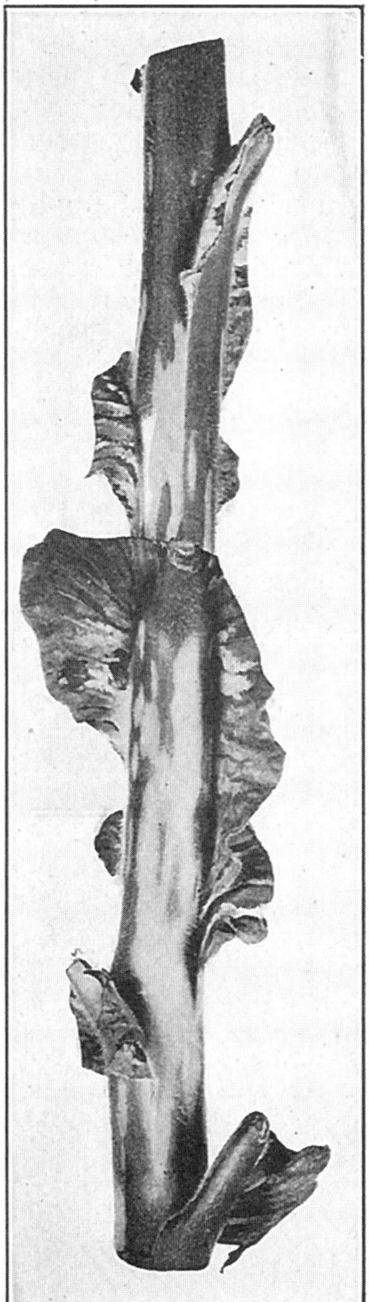

FIGURE 9.—Yellow mottling on a stem of Connecticut-Havana tobacco, induced by a concentrated virus of yellow mosaic. × 1

The light-green type of mosaic from *Nicotiana glauca* is very similar in appearance, when transferred to tobacco, to the light-green mosaic which occurs on tobacco in the United States. It is especially like the light-green mosaic which the writer has studied on tobacco (*5*). Differences in the properties of the viruses may exist, however.

The mild dark-green mosaic collected on *Nicotiana glauca* produces very little mottling on tobacco in the early stages of the disease. The first indication of disease consists of a bleaching of the veins. The leaves may become slightly stunted and more or less deformed. Sometimes a very faint green mottling may be noted at this period, but in most cases no mottling is evident. The later leaves show less

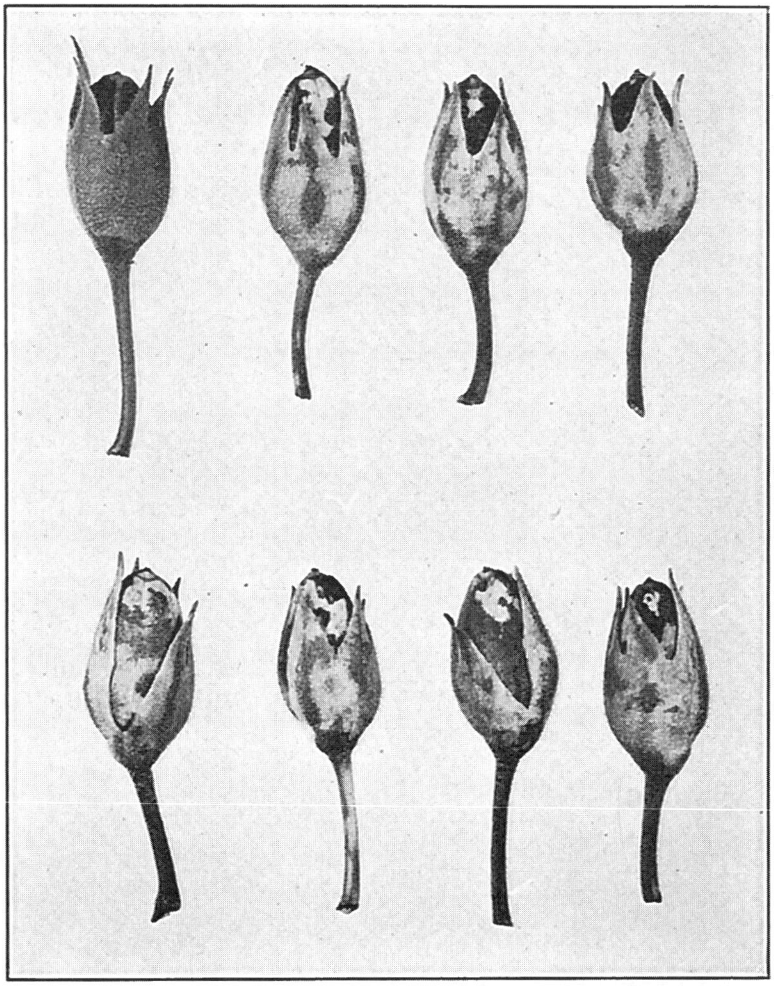

FIGURE 10.—Yellow mottling on the calyx and fruit of Connecticut-Havana tobacco, induced by a concentrated virus of yellow mosaic. Mosaic-free specimen at upper left. × 1

deformity. They are normal green in color, and no mottling can be detected. From three to five such leaves may develop. These are followed by three to five leaves which show very small, faint light-green mottlings as shown in Figure 11. This condition becomes more intense on subsequent leaves until a very pronounced mottling occurs. This mottling is characterized by very large light and dark green areas (fig. 5, C), and sometimes more than half of the leaf is solid dark green and the remainder solid light green. The margins of the leaves frequently turn downward, causing an inverted spoon effect as shown in Figure 12. This stage may be followed by one in

61922—29——2

which a few leaves develop very small, faint, light and dark green mottlings as in the case of the earlier stage. These observations are based on more than 50 plants which grew to maturity.

FIGURE 11.—Mild or faint mottling on Connecticut-Havana tobacco leaf, induced by a virus of mild dark-green mosaic collected on *Nicotiana glauca* in the Canary Islands. This mottling occurs in conjunction with the type illustrated in Figure 5, C. × ⅔

Thus far the mild dark-green mosaics have shown much less evidence of the yellow type than have the light-green forms, and at present it appears that the mild green types offer the best opportunity for obtaining a green mosaic free from yellow mosaic.

The writer has isolated a mild dark-green form from a slightly mixed infection occurring in *Nicotiana glauca* collected on Grand Canary, and this virus seems to be free from viruses causing the yellow and the light-green types of mosaic.

Inoculations have failed to transmit the mild dark-green mosaic of *Nicotiana glauca* to tomato plants grown from Canary Island seed as well as from American seed. The light-green and the yellow mosaics from *N. glauca* have been transmitted to tomatoes. Some of the tobacco mosaics of American origin have failed to produce symptoms in *N. glauca* collected on Teneriffe, but in some cases *N. glauca* has proved to be a carrier of these viruses after being inoculated. These same mosaics were transmitted by Johnson (*2*) to *N. glauca* from Italy, and symptoms were produced. From correspondence with Roy E. Clausen it is evident that all collections of *N. glauca* are not alike genetically. It appears, therefore, that the successful use of wild and domesticated hosts for differentiating viruses depends on the use of strains known to be homozygous for resistance or for susceptibility to a given mosaic. Obviously, workers in this field can not accurately check one another's results until a stock of pure-line seeds is available.

When tobacco plants with mild dark-green mosaic were reinoculated with the virus of one of the writer's yellow mosaics, the former type became obscured and the plants showed the typical symptoms of yellow mosaic. When plants with one of the writer's light-green mosaics were reinoculated with the same virus of yellow mosaic no change in the symptoms occurred. The plants continued to produce typical light-green mosaic after three reinoculations with the virus of yellow mosaic.

Leaf deformity is very frequently associated with mosaic on tobacco and *Nicotiana glauca*. Some of these deformities (fig. 12) occur with

FIGURE 12.—Extreme leaf deformations produced on Connecticut-Havana tobacco inoculated with a virus obtained from mosaic-diseased *Nicotiana glauca* collected at Gibraltar. One midrib had no lamina and another possessed only a small portion of a lamina. × ⅓

considerable regularity regardless of special environmental conditions, whereas others (fig. 13, A) occur only under certain limited conditions and do not continue throughout the life of the plant.

Some of the long narrow leaves shown in Figure 13, A, have distorted vascular elements. The veins are fused in many cases. Also the lamina is often distorted on the underside of the leaf forming liplike structures as shown in Figure 13, B. These leaf deformities are seemingly accentuated by reduced light, and they occur most frequently in tobacco plants which are inoculated when very small. Woods (*13*) found that mosaic-affected tobacco produced deformed leaves after the plants had been cut back. The writer has obtained the same results with mosaic-affected tobacco plants and also with

FIGURE 13.—Mosaic on Connecticut-Havana tobacco: A, Deformed leaves of tobacco produced by a light-green mosaic virus obtained from domestic tobacco. This plant was grown in shade during its early development after inoculation. This deformity does not continue as the plants develop, especially if they are given full sunlight. × ½. B, A portion of the underside of a deformed leaf showing the liplike structure in the lamina and the fusion of laterals with the midrib. × 2

healthy plants. However, in both instances the leaf deformities did not continue throughout the life of the plants.

Other species of mosaic-free Solanaceae behave in the same manner. This phenomenon was observed very commonly in slashings of solanaceous perennials in the bush and the mountain jungles in Africa.

MOSAICS IN WEST AFRICA

The expedition reached Bathurst, capital of Gambia, early in January. Almost immediately a trip was made up the Gambia River to McCarthy's Island, which is located in the savanna type of vegetation.

In a garden located in the native village of Sankule, near Georgetown, mosaic was found on tabasco pepper and on eggplants. It was found again on cultivated eggplants in Monrovia, Liberia.

Colonial agricultural workers informed the writer that the rosette disease is very destructive on peanuts (*Arachis hypogaea* L.) in Gambia and in Sierra Leone. In some seasons this disease is the limiting factor in peanut production. The peanut crop had been harvested, and therefore it was not possible to observe rosette. This virus disease causes considerable damage in several regions in East, West, and South Africa (*12*). It has not yet been reported in the United States.

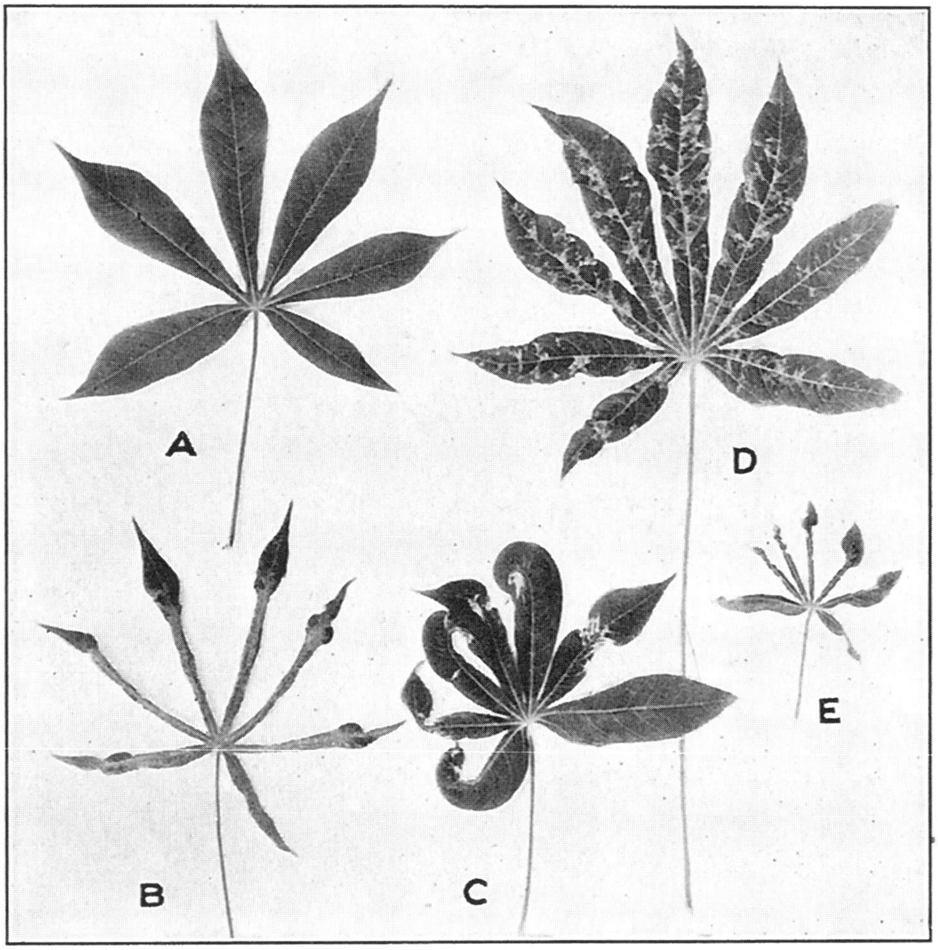

FIGURE 14.—Light-green mosaic on leaves of cassava (*Manihot* sp.): A, Healthy leaf ;B–E, diseased leaves. Collected in Sierra Leone. × 1/4

In all the colonies visited, from Liberia to French Cameroon, mosaic was prevalent on Cayenne, chili, and tabasco types of pepper (*Capsicum frutescens* L.) and on several types of cassava (*Manihot* sp.).

The mosaic on pepper is of the green type. It is identical in appearance with the mosaic commonly found on peppers in the United States, and it has been transmitted successfully by the writer to healthy chili pepper plants. Doubtless this pepper was taken to the West Coast in the days of the slave trade with the West Indies and South America.

The mosaic occurring on cassava is a yellowish green type and produces a considerable number of deformed leaves, as shown in Figure 14. Although this disease has not been studied, it has the appearance of an infectious mosaic, and it is considered as such by agricultural workers who are familiar with it. In many localities the diseased plants were very much stunted, and the crop undoubtedly was reduced as a result.

The tuberous roots of the cassava supply much of the starchy food of the inhabitants of the Tropics of both hemispheres, and the tapioca of commerce is made from the starch derived from this plant. It seems apparent, therefore, that this disease should be studied intensively. Several slightly different kinds of cassava were observed in west Africa, and it is possible that some of these may offer an opportunity for making resistant selections.

FIGURE 15.—Yellow mosaic on *Asystasia coromandeliana:* A, Healthy leaf; B, diseased leaf. Collected near Buea, British Cameroon. × 1

The disease was noticeably less abundant on the lower elevations of Cameroon Mountain in the vicinity of Buea. A larger number of mosaic-free pepper plants also were found there than elsewhere. The district is somewhat cooler than the others visited, and it is possible that this has some bearing on the reduced amount of mosaic.

In Duala, French Cameroon, a yellow mosaic was found on *Asystasia coromandeliana* Nees (fig. 15), a weed which is rather abundant in that district. *Fleurya podocarpa* Wedd., a nettle, found on a jungle roadside near Buea, British Cameroon, had yellow mosaic.

The first evidence of mosaic on a cucurbit was found below Buea on Cameroon Mountain. A typical light-green type was found on *Physedra barteri* Cogn., a native wild species. (Fig. 16.) This mosaic resembled the one which occurs on the commercial cucumber in the United States. The leaves on the mosaic-free plants were much larger than the diseased leaves.

Yellow mosaic (fig. 17) was found on a native cucurbit (*Momordica charantia* Linn.) at the Aburi Agricultural Experiment Station, Gold Coast, and on *Kedrostis foetidissima* Cogn. (fig. 18), a native cucurbit growing in a "bush island" (fig. 19) on the Winneba Plain in the Gold Coast. Plants of *Hibiscus calycinus* (fig. 20) growing 20 miles north of Accra, Gold Coast, also were found to have a yellow mosaic.

Many plants of the calabash gourd (*Lagenaria leucantha* Rusby), which were cultivated by the natives, were examined in all the colonies visited, but no suggestion of mosaic were found on this group of varieties.

FIGURE 16.—Green mosaic on *Physedra barteri*. Collected near Buea, British Cameroon. × 2/3

The pepper and the eggplant were the only solanaceous plants that were found affected with mosaic in west Africa. Tomatoes cultivated near Buea, British Cameroon, were free from mosaic. In Gambia a few tomatoes of the cherry type, found in a native garden, had what appeared to be a fernleaf type of disorder. Attempts to transmit this condition to tobacco failed.

Many wild species of Solanaceae were found in the Futa Jallon highlands, French Guinea, and on the lower elevations of Cameroon Mountain, but none showed mosaic.

Bananas and plantains were growing in many of the colonies visited. When possible, the plants were examined for indications of mosaic and bunchy top, but there was no evidence of these diseases.

Wild grasses were growing abundantly in the British and French Cameroons, in Fernando Po, and in parts of the Gold Coast. However, many observations failed to reveal any suggestion of mosaic on a grass.

Several types of sugar cane were found in all the colonies from the Gambia to the Cameroons. For the most part these plants were in native gardens. Some patches were found on the agricultural stations and farms, and one large planting was found near Cape Coast in the Gold Coast. Cane plants were carefully examined in all the localities, but no mosaic could be found. Infection experiments show that cane plants, from stalks collected in west Africa, are susceptible to the cane mosaic prevalent in Louisiana.

FIGURE 17.—Yellow mosaic on *Momordica charantia:* A, Healthy leaf; B, diseased leaf. Collected at Aburi, Gold Coast. × 1

FIGURE 18.—Yellow mosaic on *Kedrostis foetidissima.* Collected in a "bush island" on the Winneba Plain, Gold Coast. × 1

While in the Gold Coast the writer was informed by the Colonial mycologist that the streak disease occurs on maize in that colony. The disease could not be observed, as the crop had been harvested some time previously. This is a destructive virus disease which occurs on the East Coast of Africa, in Egypt, and also in South Africa. It has not yet been reported from the Western Hemisphere.

From the literature and verbal descriptions obtained from workers who are familiar with streak, it appears to produce a stripe or streak which is similar in many particulars to the yellow mosaic associated with the green mosaic of wheat and other small grains. This yellow mosaic was mentioned and one type was illustrated by the writer several years ago (*3*). From the illustrations published by Storey (*11*), maize streak also resembles the Cuban maize stripe described by Stahl (*10*).

It is possible that Stahl considers streak and stripe distinct, because he could not transmit stripe to sugar cane, and because Storey failed

FIGURE 19.—Typical "bush islands" in the Winneba Plain, Gold Coast

to transmit streak with *Peregrinus maidis* Ashm., the vector for stripe. Illustrations of the two diseases are so similar that one is tempted to raise the question as to their possible identity in spite of vector differences. On the basis of published information it seems equally plausible that streak and stripe are identical and that *Peregrinus maidis* has biologic forms or races differing in their host preferences or in their ability to transmit a given virus. Certainly there is ample opportunity for such strain differences to manifest themselves in regions so widely separated as Africa and Cuba.

There is no information on the problem of biologic forms within species of insects serving as plant-disease vectors, and it is believed that this problem must be thoroughly explored in order that there may be a better understanding concerning vector relationships and possible differences among certain viruses.

EXPERIMENTS WITH A MOSAICLIKE DISEASE OF PEANUTS

When the writer was in Sierra Leone, E. Hargreaves, the colonial entomologist, called attention to a so-called mosaic on a few peanut plants he was carrying through the dry period in his compound at the agricultural experiment station at Njala, located more than 100 miles inland from Freetown. The writer took parts of diseased and healthy plants back to the yacht for inoculum and for obtaining photographic records. (Fig. 21.)

FIGURE 20—Yellow mosaic on leaf of *Hibiscus calycinus*. Collected near Accra, Gold Coast. × 1

The tissue was placed in the ship's cool room until young peanut plants were available for inoculation. Plants were ready for inoculation in a few days, and at this time it was found that many aphids had been increasing on the diseased tissue. These were carefully removed to several of the young healthy plants. In from four to five days after this transfer was made the new leaves on the aphid-infested plants showed mottling identical with that on the original material. The control plants, which had no aphids, were free from mottling. As only one lot of aphids was available it was not possible to determine the effects from insects which had fed on healthy plants. However, it was considered that in case the mottling were an infectious mosaic it would persist after the removal of the aphids. If it were the result of aphid attack, the plants would recover after the removal of the insects.

The aphids were removed from part of the mottled plants by means of a soapsuds spray. On these plants the new foliage was free from mottling, whereas the mottling continued on the new leaves of the plants which still harbored aphids. The mottlings developed again on the new foliage of recovered plants when aphids were allowed to feed on them again. The control plants which were kept free from aphids remained healthy.

The mottling differed somewhat from that of the ordinary types of green mosaic, and, from the behavior of the disease, it seems likely that it is not a mosaic of the virus type. It seems to behave more nearly like some of the diseases of forage legumes which presumably are produced by the toxic products of certain species of leaf hoppers (*8*).

Unfortunately, the aphids were lost before they were identified, and further effort is necessary before identification is possible.

MOSAIC IN GIBRALTAR

On returning to Gibraltar late in March, the writer spent several days collecting. On one of the collecting trips a plant of *Nicotiana glauca* was found growing from a crevice in the side of the rock facing the Mediterranean Sea near Europa Point. This plant was just within reach, and on examination it was found to have a mild form of green mosaic. The virus from the dried leaves of this plant produced mosaic in tobacco plants one year after the collection was made. This mosaic has been studied only in a preliminary way. It appears to be very similar to the mild dark-green types collected

FIGURE 21.—Leaves of peanut (*Arachis hypogaea*): A, Healthy; B, mosaiclike mottling caused by aphids. Collected in Sierra Leone. × 7/10

in the Canary Islands. It frequently produces extreme leaf deformity, as shown in Figure 12.

DISCUSSION

The writer recognizes the importance of making inoculation tests before concluding definitely that a chlorotic condition is an infectious mosaic. It has not been possible to make inoculation tests with all of the collections, and it is possible that some of the mosaics described are not caused by viruses. However, in all the cases described, the symptoms were very typical for green or for yellow types of mosaic.

Mosaics of *Nicotiana glauca* obtained on Grand Canary, Teneriffe, and Gibraltar are of considerable experimental interest.

The collections from the Canary Islands have been studied more thoroughly than the others, and it has been found that each mosaic-diseased specimen was affected by more than one type of mosaic. In all cases a yellow mosaic is associated to a greater or less extent with the green type, and in some cases it appears that two green mosaics and the yellow type are associated in the same plant.

Evidence obtained thus far indicates that the yellow mosaic is associated more generally with the light-green mosaics, and at the present time the mild dark-green types seem to offer an opportunity for obtaining a single virus of green mosaic.

It is of especial interest that each of the 17 viruses of light-green mosaic which have been tested on tobacco, and in some cases on tomato, contains virus of yellow mosaic. These viruses were obtained from seven different States in the United States and from Hawaii, England, and the Canary Islands.

If these associations of green and yellow mosaic, which appear to be rather general, are due to simple admixtures of two independent types or strains of virus, it would seem that there would be more evidence of a relatively pure yellow mosaic in fields of tobacco and tomatoes and in wild *Nicotiana glauca*. The writer has observed yellow mosaic most frequently in southern-grown tomatoes, but even in this case the number of plants has been exceedingly small in comparison with the large number in which green mosaic predominated. The fact that a yellow mosaic was not reported on field tobacco until recently by Johnson (*2*) is good evidence that this form is much less general than the green types.

A satisfactory interpretation of the yellow and green mosaic associations can not be given at this time because so few data are available. It is possible that these are a simple admixture of independent viruses comparable to a mixed infection of two or more species of fungi. However, it seems entirely possible that viruses may become altered locally in the plant, thus producing mutations, to use this term in its broadest meaning.

No attempt has been made to classify the viruses causing the mosaics described in this paper. This can not be done until studies now under way have been advanced. Any method of classification depends largely on the purity of the viruses under study. Johnson's work shows that differential hosts, heat, and chemical treatments assist greatly in separating types of viruses. However, there are instances in which these methods have not yet proved effective, as in the case of the association of the yellow and green mosaics described previously (*4*).

The writer's work shows that the studies on virus purification and classification should not be confined to young plants. In many cases plants must be carried to maturity in order to determine mixtures of viruses and to obtain a full knowledge of the symptom produced by a given pure virus.

It appears also that classification work should be based on host material which is homozygous for resistance or susceptibility to the mosaics being tested. Also, attention should be directed to the influence of toxic or other plant products on the behavior of a given virus after it has been passed through several species of hosts. The importance of homozygous host plants has been strikingly evident in the work on wheat mosaic. In the case of tobacco, the writer obtains seed from a single plant of known history and propagates it vegetatively in the greenhouse under conditions which prevent its receiving pollen from other plants.

It is not known how general the association of yellow and green types may be among the mosaics of other groups of plants. This association was observed in winter wheat (*3*) in 1920 near Granite

City, Ill., and it has been evident in experimental plot tests conducted since that time. Experiments show that yellow mosaic occurs more frequently in certain species of small grains and in certain varieties within a given species. Further study is required to determine whether the yellow and green mosaic association in the small grains is in any way analogous to the one occurring in the Solanaceae. Results obtained thus far seem to indicate that the symptoms of mosaic types in the small grains are influenced in part by genetic factors in the hosts.

Yellow mosaics are more devastating than the green types that have been observed, and it is fortunate that they are rare.

SUMMARY

In the Canary Islands green mosaics were found on *Solanum tuberosum*, *Nothoscordum fragrans*, *Capsicum frutescens* var. *grossum*, *Fuchsia magellanica gracilis*, *Psoralea bituminosa*, and *Nicotiana glauca*. Yellow mosaic was found on three plants of *N. glauca*.

Experimental studies have been carried out with Connecticut-Havana tobacco inoculated with viruses from the 10 mosaics collected on *Nicotiana glauca* in the Canary Islands. The results of these tests show that the green forms of mosaic are of two types, a light green and a mild dark green. The green forms that have been studied thoroughly were found to contain, to a greater or less degree, traces of a yellow mosaic virus which can be concentrated by successive spot isolations and subinoculations. The light-green mosaics develop more yellow mosaic spots than do the mild dark-green forms, and it appears that a strain of the latter has been isolated which is free from viruses of the yellow and the light-green mosaics.

In west Africa green mosaics were found on eggplant, pepper, *Manihot* sp., and *Physedra barteri*. Yellow mosaics were observed on *Fleurya podocarpa*, *Asystasia coromandeliana*, *Momordica charantia*, *Kedrostis foetidissima*, and *Hibiscus calycinus*.

The rosette disease occurs on peanuts in Gambia and Sierra Leone, and maize streak occurs in the Gold Coast.

In Sierra Leone a mosaiclike mottling occurs on peanuts. This is produced by aphids and seems not to be caused by a virus.

In Gibraltar mild dark-green mosaic was found on *Nicotiana glauca*.

LITERATURE CITED

(1) FAIRCHILD, D.
1928. TWO EXPEDITIONS AFTER LIVING PLANTS. Sci. Mo. 26: 97–127, illus.

(2) JOHNSON, J.
1927. THE CLASSIFICATION OF PLANT VIRUSES. Wis. Agr. Expt. Sta. Research Bul. 76, 16 p., illus.

(3) MCKINNEY, H. H.
1925. A MOSAIC DISEASE OF WINTER WHEAT AND WINTER RYE. U. S. Dept. Agr. Bul. 1361, 11 p., illus.

(4) ———
1926. VIRUS MIXTURES THAT MAY NOT BE DETECTED IN YOUNG TOBACCO PLANTS. Phytopathology 16: 893.

(5) ———
1927. QUANTITATIVE AND PURIFICATION METHODS IN VIRUS STUDIES. Jour. Agr. Research 35: 13–38, illus.

(6) ———
1928. VIRUS DISEASES OBSERVED BY THE ALLISON V. ARMOUR EXPEDITION. (Abstract) Phytopathology 18: 155.

(7) MAGEE, C. J. P.
1927. INVESTIGATION ON THE BUNCHY TOP DISEASE OF THE BANANA. Aust. Council Sci. and Indus. Research Bul. 30, 64 p., illus.

(8) MONTEITH, J., JR.
1928. LEAFHOPPER INJURY OF LEGUMES. (Abstract) Phytopathology 18: 137–138.

(9) PURDY, H. A.
1928. MULTIPLICATION OF THE VIRUS OF TOBACCO MOSAIC IN DETACHED LEAVES. Amer. Jour. Bot. 15: 94–99, illus.

(10) STAHL, C. F.
1927. CORN STRIPE DISEASE IN CUBA NOT IDENTICAL WITH SUGAR-CANE MOSAIC. Tropical Plant Research Foundation Bul. 7, 12 p., illus.

(11) STOREY, H. H.
1928. TRANSMISSION STUDIES OF MAIZE STREAK DISEASE. Ann. Appl. Biol. 15: 1–25.

(12) STOREY, H. H., and BOTTOMLEY, A. M.
1928. THE ROSETTE DISEASE OF PEANUTS (ARACHIS HYPOGAEA L.). Ann. Appl. Biol. 15: 26–45, illus.

(13) WOODS, A. F.
1902. OBSERVATIONS ON THE MOSAIC DISEASE OF TOBACCO. U. S. Dept. Agr., Bur. Plant Indus. Bul. 18, 24 p., illus.

Francis O. Holmes and the Local Lesion Assay

John G. Shaw
Department of Plant Pathology, University of Kentucky, Lexington, KY

In developing a method for measuring the relative content of virus in extracts of plants infected with tobacco mosaic virus (TMV), Dr. Francis O. Holmes provided one of the most important contributions to the work that led to an understanding of the physical and chemical nature of viruses. Efforts to reach this goal were based mainly on the effects of various treatments on the infectivity (and, therefore it was presumed, the structural integrity) of a particular sample of virus; the hope was that such treatments would eventually reveal the chemical composition and physical structure of the agents of diseases such as tobacco mosaic. However, during the first thirty years of virus research, the only method for determining the effects of the various treatments on plant viruses was manual inoculation of very large numbers of plants that were susceptible to systemic infection with various dilutions of each test sample. Some days later, the plants would be checked in the hopes that an infectivity end-point could be identified. With the publication of his classic paper of 1929, Holmes put an end to such requirements for those working with TMV and, eventually, many other viruses. In describing the use of the leaves of plants that would react to TMV infection by producing local lesions instead of systemic mosaic symptoms, he converted plant virus assays from a laborious, time-consuming and unreliable process to a simple and much more accurate procedure. In a sense, each spot on one of his inoculated leaves was equivalent to a systemically infected plant in the earlier types of assay. His achievement stands as one of the most important contributions to plant virology and his technique is practiced today in much the same way as he developed it 70 years ago.

Francis Holmes began his professional career in 1923 as a protozoologist at the Boyce Thompson Institute for Plant Research, but his association there with Dr. L. O. Kunkel soon resulted in his attention being directed to diseases thought to be caused by viruses. Some idea of the infancy of plant virology at that time may be gained from the fact that it took him only one week to read every existing publication dealing with plant viruses and virus diseases.[1] In describing his discovery of the local lesion method, Holmes made the interesting observation that, while there had been previous reports of necrotic lesions in the leaves of *Nicotiana* plants inoculated with extracts of TMV-infected plants (e.g., Allard, 1914), these lesions had not been recognized as symptoms of disease. Part of his 1929 paper was devoted to proving that the virus is itself responsible for the production of the local lesions and the rest to the more widely known work in which Holmes demonstrated "the usefulness of these lesions in measuring the concentration of mosaic virus samples."

To achieve these goals, Holmes surveyed seventeen species of *Nicotiana* and found five to produce conspicuous local lesions in response to inoculation with TMV. His first tests involved pricking of the surfaces of leaves with insect pins that had been wetted with sap from an infected tobacco plant or that had pricked a leaf of an infected plant. The pin-pricking method had been in use for several years but was regarded by some investigators as being too laborious and of limited use. Nevertheless, Holmes, who must have been a person of remarkable dexterity, was able to make 500 inoculations per hour with this technique and to do so without touching the assay plants (Holmes, 1928). The wounds produced by this method were regarded by Holmes as being so slight that they would not themselves be the cause of localized necrosis. He further demonstrated that the local lesions were caused by the virus and not by physical damage by inoculating the leaves of *N. rustica* plants, which react to TMV by producing necrotic spots followed by severe systemic symptoms, with single pin-pricks and observing that only those plants that developed a single local lesion subsequently displayed the systemic disease.

In experiments with *N. rustica* plants that were designed to test the relationship between local lesions and virus concentration, Holmes described two sets of experiments in which, with a device that held five pins, he made a total of some 38,000 pin pricks (250-500 per leaf) in the leaves of this species. The more significant experiments were conducted with *N. glutinosa* which was a more suitable species because the lesions were produced more rapidly, and of a distinctive appearance and because of the absence of systemic spread of the virus. In tests with this species, he gently rubbed the leaf surfaces with a piece of cloth that had been wetted with inoculum and this procedure, much less laborious and exacting than the pin-pricking method, resulted in the production of 60-120 local lesions per leaf with undiluted sap. This technique, which was probably more efficient than a previously described rubbing procedure in which a soft brush was used (Clinton, 1915), and

[1]This and some other facts concerning the life and work of Francis O. Holmes were kindly provided by Holmes' son, Dr. Francis W. Holmes, to whom the Author expresses his deep gratitude.

which was later improved by the addition of an abrasive substance such as Carborundum or Celite to the inoculum or to the surface of the leaf to be inoculated, became the standard method of local lesion assay that is still in use today.

Holmes claimed that, with just a few *N. glutinosa* plants, he could detect two-fold differences in virus concentration "with satisfactory accuracy". In earlier work, looking for a dilution end-point, he had calculated that 1350 plants would be required to demonstrate a two-fold difference in TMV concentration (Holmes, 1928). Over a period of some four decades after his 1929 paper appeared, there were numerous investigations of mathematical features of the local lesion assay and of the many factors that can affect the results obtained with it. Some of these resulted in useful improvements in sensitivity of the procedure, while others were directed to attempts to understand the nature of virus particles and the infection process.

Holmes also raised the possibility that the local lesion method might be used to obtain pure isolates of plant viruses. He reckoned that, at dilutions where samples would give rise to only one lesion per plant, the infection might have been the result of a single virus particle and thus represent a single form of the virus. Though this is sometimes not the case, the single-lesion technique has been widely used by plant virologists since the time of Holmes' paper.

The final remarks in the paper include a comparison of Holmes' system with Koch's plating method for counting bacterial cells. A more interesting comparison might have been with the plaque-counting procedure that had been developed for the assay of bacteriophage concentration prior to Holmes' report of the local lesion assay (d'Herelle, 1926). The parallels between the two types of assay, both of which involved the counting of areas in which cells had been killed by a virus, are interesting, and one wonders if Holmes was aware of d'Herelle's work at the time he was preparing his paper.

Holmes went on to exploit the local lesion response as a means of developing virus-resistant plants. He introduced the N gene from *N. glutinosa* that is responsible for the local necrotic reaction to TMV into *N. tabacum* (Holmes, 1938) and others were later able to utilize the N gene in plant breeding programs to produce commercially acceptable tobacco cultivars.

Francis O. Holmes made a number of other notable contributions to plant virology, but his 1929 achievement remains his most important and celebrated effort. In this paper, as with his others, the elegance of his writing style coupled with the extent of his vision bespeak a person of intellect and scholarship. He retired from his position at the Rockefeller University in 1965 and moved to a small town in New Hampshire where he devoted much effort to community activities and to his interests in beekeeping and the taxonomy of the genus *Salix*. Tragically, in 1990, at the age of nearly 91 years, he was struck down by a vehicle while walking near his home.

References

Allard, H. A. (1914). The mosaic disease of tobacco. *U. S. Dept. Agr. Bull.* 40.

Clinton, C. D. (1915). *In:* Chlorosis of plants with special reference to calico of tobacco. *Conn. Agr. Exp. Sta. Rpt. 1914*, pp. 357-424.

d'Herelle, F. (1926), The Bacteriophage and its Behavior. Williams and Wilkins Co., Baltimore, 629 pp.

Holmes, F. O. (1928). Accuracy in quantitative work with tobacco mosaic virus. *Bot. Gaz.* **86**:67-81.

Holmes, F. O. (1938). Inheritance of resistance to tobacco mosaic disease in tobacco. *Phytopathology* **28**:553-561.

Biographical Sketch

John Shaw received a bachelor's degree in botany and chemistry from the University of British Columbia in 1956 and went to the University of Wisconsin for graduate studies in plant pathology. His Ph.D. research involved an investigation of the coat protein of potato virus X and was supervised by Russell Larson. In 1959, he took a position with the Canada Department of Agriculture in Ottawa in a research group headed by John Slykhuis and, in 1968, he moved to his present position in the Department of Plant Pathology at the University of Kentucky.

Reprinted for private circulation from
THE BOTANICAL GAZETTE, Vol. LXXXVII, No. 1, February, 1929
PRINTED IN THE U.S.A.

LOCAL LESIONS IN TOBACCO MOSAIC[1]

FRANCIS O. HOLMES

(WITH ELEVEN FIGURES)

Introduction

The literature concerned with the virus diseases of plants repeatedly emphasizes the systemic nature of these infections. MAYER (6), in his original description of tobacco mosaic in 1886, stated that symptoms do not develop on the leaf inoculated, but appear on all of the young developing leaves. Essentially the same account of the course of the disease was given by BEIJERINCK (2) in 1898, by IWANOWSKI (5) in 1903, by ALLARD (1) in 1914, and by others in more recent years.

No detailed descriptions of local lesions developing at the points where tobacco mosaic virus has been introduced have been published. This has probably been due in part to the fact that the local lesions are not conspicuous in commercial tobacco, *Nicotiana tabacum*, which has been used extensively. Another factor which has made the recognition of the local development of the disease more difficult has been the practice of inoculation by scratching and severe wounding. This tends to obscure the primary lesions by producing dead areas mechanically.

A few references in the literature indicate that local lesions have been observed, although their real nature has not been understood. ALLARD (1) referred to *N. langsdorffii* as follows:

> Plants of this species when inoculated through the stalk and petioles seem particularly susceptible to a very destructive and progressive rot, which begins at the point of inoculation and finally kills the plant by slowly involving the surrounding tissues. *Nicotiana viscosum* (*N. glutinosa*) is sometimes killed in exactly the same manner.

FERNOW (3) said of *N. rustica:* "The leaves generally turn yellow and then brown near each point of inoculation." The briefness of

[1] Contributions from the Boyce Thompson Institute for Plant Research, Inc., Yonkers, N.Y., published at the expense of the Institute out of the order determined by the date of receipt of the manuscript.

these references to the local lesions developing at the site of inoculation with tobacco mosaic virus indicates that these lesions were not recognized as symptoms of the disease.

There is in the literature one account of a plant virus which produces lesions at the point of inoculation. This is the virus of ring-spot of tobacco, which PRIODE (7) describes as forming typical rings of necrotic tissue around needle punctures used to introduce it into tobacco and petunia. These local lesions are exactly like the lesions later produced when the disease becomes systemic; but tobacco mosaic has never been shown to act in this way. Most writers have specifically stated or implied that the inoculated leaf if fully developed never shows symptoms of the infection.

In the present paper the local lesions caused by the introduction of tobacco mosaic virus into the tissues of a number of *Nicotiana* species will be described, and the usefulness of these lesions in measuring the concentration of mosaic virus samples will be shown.

In a former paper (4) the writer has described a method of inoculating plants by means of very small needle punctures. This method is well suited to the demonstration of the occurrence of local lesions. The absence of extensive dead tissue around such needle punctures when transmission does not take place makes the slightest deviation from the normal condition of the leaf tissues conspicuous. The present study was undertaken because evidence was obtained that changes take place at the site of inoculation of tobacco mosaic virus into *N. tabacum.* In this plant it was noticed that pale yellow areas sometimes develop around one or more of the five pin pricks made in each plant. Since in some cases a very dilute virus was being measured, not many plants were expected to take the disease. This fact made the pale yellow areas particularly noticeable, and led to a later examination of the plants which showed them. It was found that every plant which showed such local changes around one or more of the pin pricks developed mottling within a few days. Most of the plants which did not show pale yellow areas near the inoculation punctures remained healthy. This suggested that these yellow areas might represent localized symptoms of a primary infection, since they were consistently followed by the familiar systemic symptoms. Unfortunately the pale yellow areas were very inconspicuous and could not be detected in all cases in which systemic symptoms developed.

It seemed possible that some species of *Nicotiana* might show local symptoms more conspicuously and more consistently than *N. tabacum*. A survey of a number of species[2] was therefore made. These were *N. rustica*, *N. trigonophylla*, *N. plumbaginifolia*, *N. longiflora*, *N. tomentosa*, *N. suaveolens*, *N. quadrivalvis*, *N. paniculata*, *N. sylvestris*, *N. sanderae*, *N. glutinosa*, *N. nudicaulis*, *N. langsdorffii*, *N. clevelandii*, *N. acuminata*, *N. glauca*, and *N. multivalvis*. Of these species, five showed pronounced necrotic local lesions instead of pale yellow areas. These species were *N. rustica*, *N. langsdorffii*, *N. acuminata*, *N. sanderae*, and *N. glutinosa*. A description of the local lesions in each of these five species follows.

In *N. rustica*, as in all of the species studied, pin pricks which fail to transmit the virus heal perfectly with no macroscopic trace of necrosis; successful transfers, on the other hand, are marked by necrotic spots. These are easily counted eight to ten days after the time of inoculation, and may be distinguished even when many are on the same leaf (fig. 1). These necrotic spots on *N. rustica* are brown in color, circular in some cases, but frequently irregular in outline. They increase slowly in size during the first week after they appear. Later they spread rapidly, especially in young leaves. They may involve a considerable area of the leaf surface before the leaf concerned becomes old and drops off.

The lesions on *N. langsdorffii* (fig. 2) and *N. sanderae* (fig. 3) are more conspicuous than those on *N. rustica*, because they are blackish with concentric rings of dead tissue. They grow larger day by day after their first appearance, and may become very extensive. Sometimes veins and even the stem of the plant are involved in the slowly spreading necrosis of the tissues.

In *N. acuminata* (fig. 4) the lesions are irregular in outline and brown in color. They appear one or two weeks after the introduction of the virus, and increase in size rather more slowly than is the case in the other species.

[2] Thanks are due to Dr. S. A. Wingard of the Virginia Agricultural Experiment Station, Dr. R. E. Clausen of the University of California, and Dr. E. M. East of the Bussey Institution, Forest Hills, Massachusetts, for their kindness in supplying seeds of these species.

N. glutinosa exhibits a very different response (fig. 5). The lesions begin to be noticeable on the second or third day and are all completely formed by the fifth day, when in other species the

FIGS. 1–5.—Local necrotic lesions in: (1) *N. rustica*, (2) *N. langsdorffii*, (3) *N. sanderae*, (4) *N. acuminata*, (5) *N. glutinosa*. Lesions in *N. rustica* and *N. acuminata* are brown, in *N. langsdorffii* and *N. sanderae* almost black, and in *N. glutinosa* pale brown surrounded by rings of darker brown.

first indications of breakdown appear. These rapidly developing necrotic lesions appear first as tiny glistening dark spots. The centers soon dry down and develop a light brown color. Around them darker brown rings form concentrically. This gives the lesions an easily recognized appearance not to be confused with other dead spots in the leaves.

In order to discover whether these necrotic local lesions consistently appear whenever a successful transfer of the mosaic virus is accomplished, a test with small *N. rustica* plants was arranged. In a number of experiments each plant was inoculated with a single pin prick in a young leaf. The inoculation was carried out with a no. 00 black enamel insect pin previously wet with juice from a mosaic leaf of commercial tobacco. Altogether 582 plants were thus inoculated. Of these, 136 took the disease, as shown by the development of the severe systemic symptoms characteristic for this species. Every one of these plants showed a local necrotic lesion at the site of inoculation. These lesions were similar to those shown in fig. 1, except that there was but one on each plant. The fact that every plant which took the disease showed a local lesion is evidence that the local lesion is a definite symptom of the mosaic. Six plants in the series showed local lesions without a later development of the systemic disease. Apparently the virus was not able to spread in these individual cases. Their number is so small, however, that the diagnosis of transfer of the disease is not much changed if local lesions are taken as evidence of the transfer without waiting to examine the later systemic symptoms. Later experiments, in which more than one pin prick was used for each plant, have substantiated these conclusions.

Further evidence that the successful transfer of virus can be detected by examination of the local necrotic lesions is afforded by the fact that concentrated virus samples cause the development of large numbers of these local lesions when they are used as inoculum, whereas known dilutions of the virus samples in water cause the development of smaller numbers of the lesions. Graphs showing the accuracy with which the numbers of lesions are dependent upon the known concentrations of virus samples will be presented later in this paper, when the use of the local lesions in measuring virus concentrations is discussed.

No bacteria or other visible microorganisms have been found upon microscopical examination of the local lesions. Fluids from plants not affected with mosaic do not produce the lesions when used as inoculum. Yet it appeared that formal experiments ought to be carried out with virus suspensions free of bacteria to show that the production of the lesions was not due to the introduction of foreign organisms. Three samples of virus were therefore prepared by adding large numbers of a small bacterium, *Aplanobacter michiganensis*, to virus previously diluted with seven volumes of water. These three samples were then passed through Berkefeld W filter candles. The bacteria were entirely removed from the mixture, as was shown by plating from the filtrates. The virus passed through the filter with no observable decrease in its concentration. The lesions developing on *N. glutinosa* plants inoculated with these sterile filtrates were identical in appearance with those produced by the use of unfiltered samples of virus.

In further support of the view that the virus of mosaic itself is responsible for the production of these local lesions, it may be stated that exactly similar lesions have been produced by the use of a virus sample which had been frozen solid for three years, and by the use of several field samples of virus collected at widely separated points. It is improbable that any second virus or any other organism would be so closely associated with tobacco mosaic virus as to be present in each of these cases.

In order to prove beyond doubt that the lesions are not caused by any foreign organism, they should be produced in a sterile plant by the use of virus freed from bacteria. But the fact that the test plants are grown in sterilized soil from seeds seems sufficient protection against the presence of any organisms on their surface, especially as the individual plants are invariably capable of showing the lesions.

Use of local lesions in measuring virus concentrations

It has long been the custom to use the appearance of the systemic disease in *N. tabacum* plants as an indicator of the successful transfer of virus from mosaic plants or extracts from them. Several methods of this kind are in use for the more or less accurate measurement of the relative concentrations of mosaic virus in different samples. In such work a whole plant is necessary for each successful transfer. As the accuracy of the measurement depends largely upon the number of successful transfers, the numbers of plants required usually prove a limiting factor in the number of measurements which can be made. Obviously it would be a great advantage if many successful transfers could be distinguished on a single plant or even on a single leaf. This advantage is given by the use of the local necrotic lesions here described.

It has been found that the number of successful transmissions of virus can be learned by counting the local necrotic spots at the points of inoculation in such species as *N. rustica* and *N. glutinosa* as well as by using whole plants as indicators. Moreover, since the lesions appear locally before systemic symptoms appear, economy of time is secured.

Since *N. rustica* offers the advantage of very large leaves, a method of using it for measuring virus concentrations will first be described. Dilution curves and a discussion of the accuracy attainable will be given. *N. glutinosa* has proved of great value, because on it necrotic lesions, which can readily be counted, appear very soon after inoculation. The disease may be transmitted by wiping the leaf surface gently with a cloth saturated in virus extract.

In *N. rustica* lesions are most readily counted when produced by pin prick punctures. A set of five insect pins held in a temporary handle has usually been used by the writer for introducing the virus. The pins are alternately dipped in a sample of virus and used to puncture the test plant leaf. In this way large numbers of punctures are rapidly made in a series of leaves. Usually each leaf will accommodate 250 or 500 punctures. The number of necrotic lesions developing is small in comparison with the number of punctures, but is sufficient to allow a fairly accurate reading of virus strength to be made with a few plants. In fig. 1 a leaf inoculated as described is shown. It is necessary to make inoculations of samples on opposite sides of the midvein of the same leaf or to use a large number of leaves for a single test. This is because leaves of different ages have been found to differ somewhat in susceptibility. In general the younger leaves are the more susceptible.

A curve showing the effect of dilution over a considerable range of virus concentrations is presented in fig. 6. The graph is based upon two series of experiments, one with a mosaic virus of usual

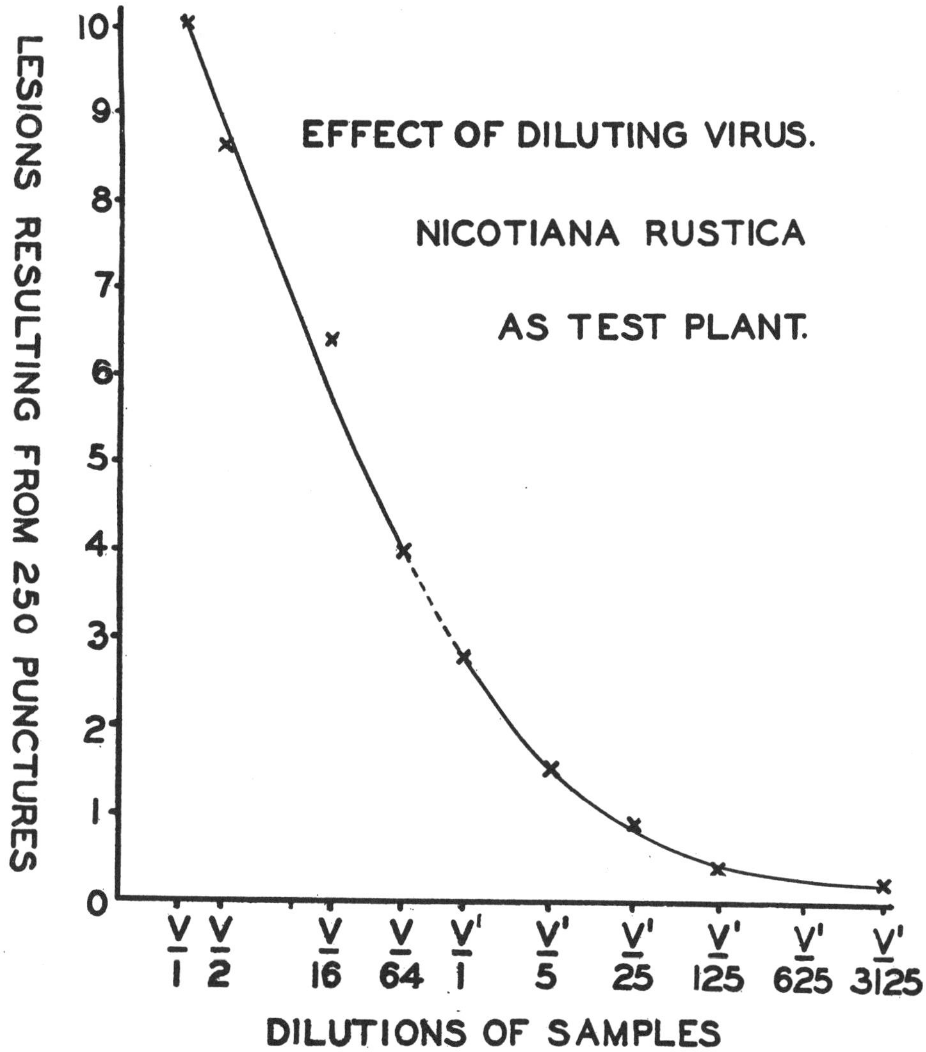

FIG. 6.—Effect of diluting virus when *N. rustica* is used as test plant. Two samples of virus, V and V′, used in obtaining the data represented. Virus V was more concentrated than virus V′, which was exceptionally weak. The lesions resulting from 38,000 pin prick punctures are represented by the nine points determining the curve.

strength and one with a weak sample. Thirty-eight thousand punctures were made to obtain the transfers represented by the figure. It will be observed that in the region in which four to ten lesions were obtained in each set of 250 punctures, the line appears nearly straight as drawn to a semilogarithmic scale. This part of the curve is similar to that known from earlier work with commercial tobacco. It is interesting to note the direction of the line when more dilute virus is used. With *N. rustica* it has been possible to study the lower range, because greater numbers of measurements can be made by the use of local lesions than by the use of the systemic disease as an indicator of successful transmission. The upper range is the portion most frequently dealt with, and the range in which the greatest accuracy can be obtained.

Nicotiana glutinosa as test plant

The characteristics of *N. glutinosa* make it a very useful test plant for measuring virus concentrations. It has a low virus content when diseased, which gives it an advantage over *N. tabacum*, in that contaminations do not readily occur when the plants are handled. The rapidity of development of the local lesions makes it possible to have preliminary results of measurements on the second or third day after inoculation, according to the season. On the fourth or fifth day final results may be noted and the plants discarded. Large numbers of lesions can be distinguished on individual leaves; thus a high degree of accuracy may be obtained in comparing virus concentrations. The use of pin punctures is unnecessarily slow in the use of this species as a test plant. In its stead a much more rapid method of inoculation may be used, allowing tobacco mosaic virus to be measured as readily and as rapidly as bacteria are counted by plating methods.

The procedure is as follows. *N. glutinosa* plants are grown in 4-inch clay pots until flower buds begin to appear. At this stage at least five leaves on each plant are of good size. These five leaves are used for the inoculation; and for convenience in manipulation all the remaining leaves of the plant, as well as the growing point, are pinched off. This leaves a sturdy stem supporting five large leaves. Virus from any source to be tested is now taken up on a small piece of white cheesecloth and rubbed once firmly but gently over the entire upper surface of the five leaves. A full stream of tap water is used to wash away excess virus at once after this inoculation. After a little practice the operations become quite uniform. The results justify the belief that approximate uniformity of inoculation can thus be obtained. Undiluted tobacco mosaic extracts result in the production of about 300–600 lesions on each test plant when applied in this way. If they are diluted with water, the decrease in the number of lesions is at first very rapid, later more gradual. A wide range of concentrations can be studied accurately.

As an illustration of the effectiveness of this method of inoculation, a series of leaves showing the decrease in the number of lesions appearing with decreasing concentrations of virus in the inoculum is shown in fig. 7. A dilution curve showing the effect more accurately is shown in fig. 8. Fig. 9 shows the lower range of this curve more clearly. A curve showing the probable errors of counts of lesions on single test plants is shown in fig. 10. By reference to these curves it will be observed that a very slight dilution, as by a single volume of water, can be detected with satisfactory accuracy by the use of a small number of test plants. In most preliminary measurements a single test plant is sufficient to give an excellent idea of the strength of the virus sample in hand. Even the most important measurements are usually made accurately enough if sixteen test plants are used. Fig. 11 shows two test plants, one five days after inoculation, the other one day after inoculation.

FIG. 7.—Five leaves from plants of *N. glutinosa* used to measure the effect of dilutions (1:1, 1:3.16, 1:10, 1:100, 1:1000). Numbers of lesions do not correspond to the more exact averages shown in dilution graph in fig. 8, but decrease in lesions with serial dilution is plainly shown.

The value of this method of measuring the concentration of tobacco mosaic extracts can best be appreciated when it is compared with the old method for estimating the strength of a sample by

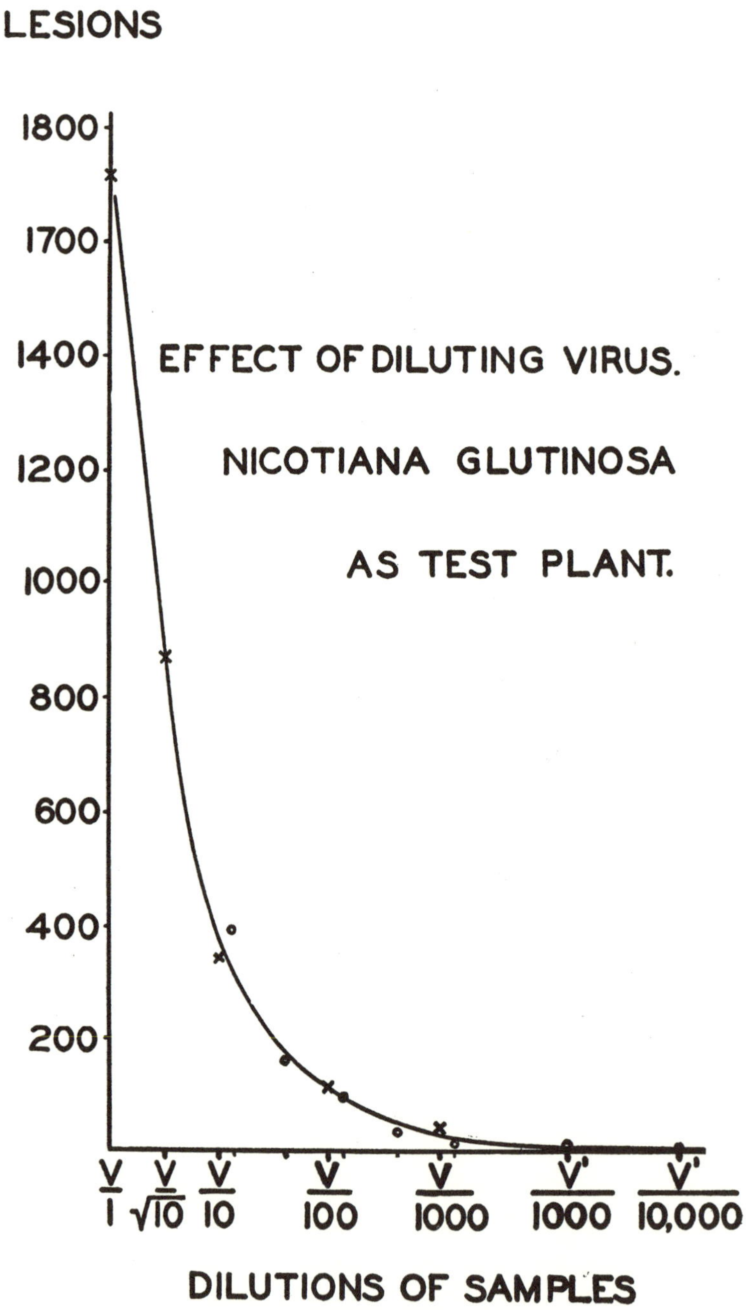

FIG. 8.—Effect of diluting virus sample, *N. glutinosa* as test plant. Counts of lesions in every case represent average number of lesions appearing for each test plant, when five leaves of each plant had been wiped with the virus sample in question. Accuracy attained may be noted by closeness of determined points to the smooth curve. Nine plants used for each determination; probable error of each point as determined is therefore one-third the error shown in fig. 10 for single observations of similar numbers of lesions.

inoculating potted plants of *N. tabacum.* Ten plants of this species were frequently used in a single test. On the average five successful inoculations and five failures would be observed on this number of test plants. On an equal number of *N. glutinosa* test plants in a

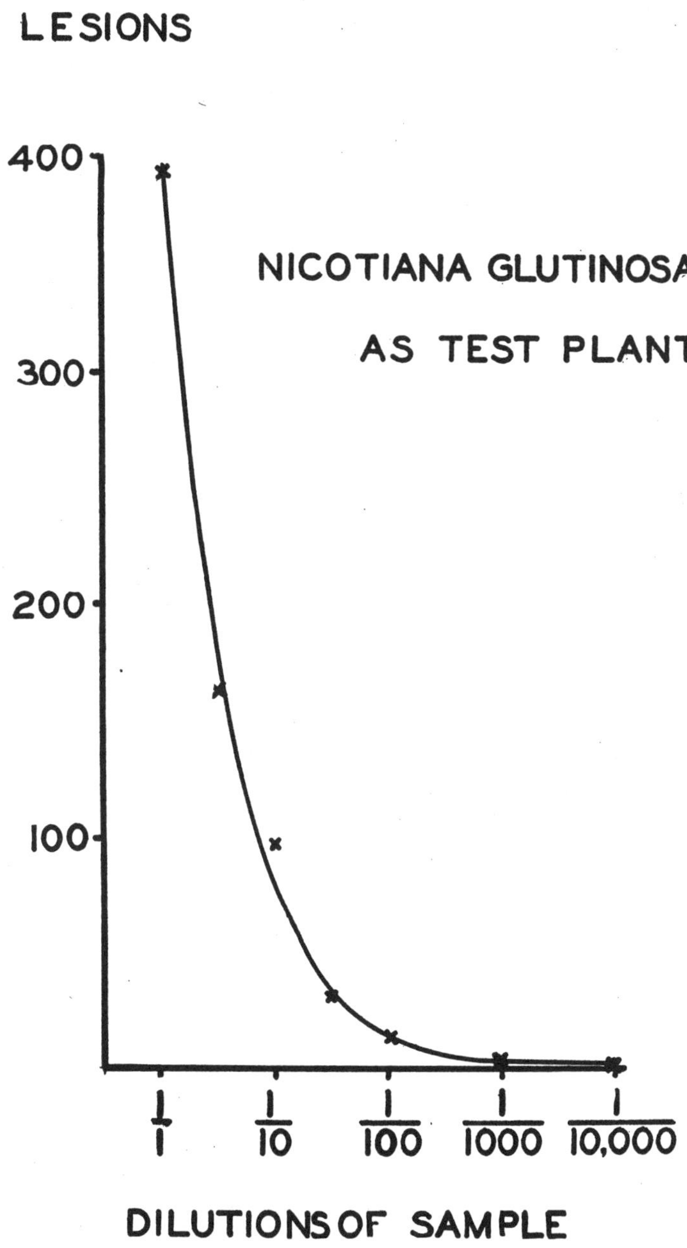

FIG. 9.—More dilute range of curve shown in fig. 8, drawn to such a scale that readings may be made accurately (see fig. 10 for probable errors of single observations in this range of measurements).

typical case 5000 successful inoculations would be observed because of the localized nature of the lesions counted. The accuracy of both

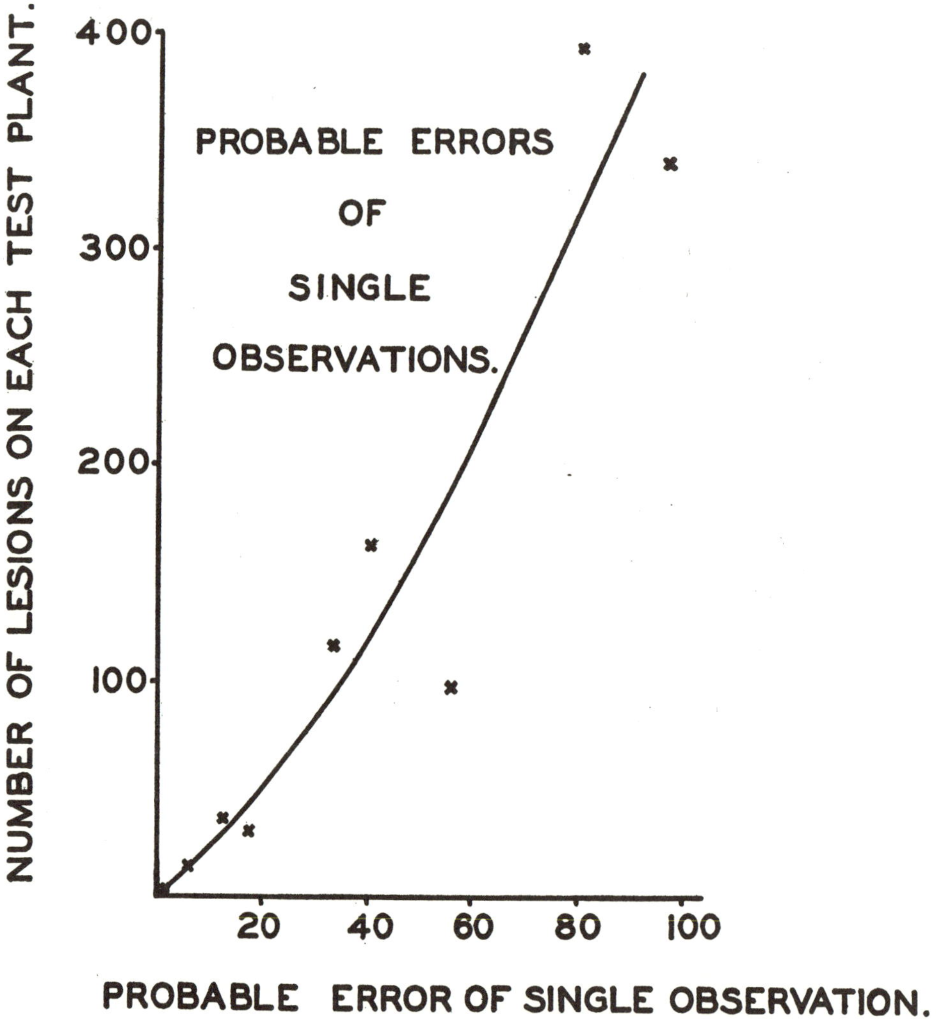

FIG. 10.—Probable errors of single observations for test plants having numbers of lesions between 0 and 400. The error in determining these probable errors is naturally large, and information was insufficient to extend curve beyond 400. At 1714, however, the probable error of a single observation has been found to be 414.

methods depends largely on the number of successful inoculations counted. It is estimated that in the average case for the same accuracy one test plant of *N. glutinosa* used as described serves the purpose for which at least several hundred *N. tabacum* plants are required with the customary methods of using that species.

N. glutinosa differs from the other species of *Nicotiana* in two ways. It produces very little virus when successfully inoculated, and symptoms appear on the upper parts of the plant to a very limited extent. Measurements of the amounts of virus produced by this plant have shown that the concentration present does not exceed the strength of an ordinary sample of commercial tobacco mosaic virus diluted with water to one-five hundredths of its original strength. The systemic course of the disease is not like that of most

FIG. 11.—Test plant of *N. glutinosa* (at left) as used to determine virus concentrations, inoculated five days before it was photographed. Similar plant (at right) inoculated one day before it was photographed. Lesions had not yet appeared on this recently inoculated plant.

species. Generally the growing tip soon shows the same type of symptoms which have occurred where the virus was introduced. The writer has never seen a case in which the developing leaves of *N. glutinosa* were affected with necrotic spots. The local lesions enlarge day by day, laying down ring after ring of dead tissue. After a week or so new secondary spots are sometimes formed on the peripheries of the extended primary lesions. Later lesions may appear on leaves younger than those inoculated, but they do not affect the developing leaves. Sometimes veins and portions of the stem are killed by the extending lesions. The green portions of the leaves between the primary lesions are not suitable sources of virus. No successful transfers have been secured when juice from them has been used as inoculum. The virus seems to be confined to the visible lesions.

N. glutinosa, like *N. rustica*, shows a gradient of susceptibility when successively older leaves are considered. The young leaves tend to produce more local lesions when inoculated with the same source of virus as neighboring older leaves. This condition is partially remedied by removing the growing point of the plant, as is done in using the plant for measurements. In two series of plants, one with the growing tops attached, the other with the tops removed, this condition of affairs was demonstrated. In the first series, in which the growing points remained attached, the totals from top leaf to lowest leaf were 6066, 5117, 4292, 3222, and 3198. The gradient is very marked. In the second series, in which the growing points were removed, the totals were 4795, 4778, 4227, 4320, and 4343. The gradient is much less. The practice of removing the tops insures less error in comparing leaves of different age, since the chance choice of an older or a younger leaf in any plant will introduce less variation when the difference in the susceptibility is decreased.

Discussion

The recognition of symptoms which appear at the site of inoculation opens a new field of investigation in connection with tobacco mosaic. The first development of the disease seems to be very strictly localized. It is only the later course of the infection which is marked by changes in appearance of the developing leaves. The species of *Nicotiana* differ markedly in the conspicuousness of their local symptoms. In some cases necrotic spots are formed, in others a pale yellow area marks the point of entrance and early development of the virus.

Most of what is known regarding virus diseases of plants has been learned from inoculation studies. The behavior of the virus under experimental conditions, such as filtration, purification by chemical processes, and contact with disinfectants has always been judged by its reinoculation into living plants. Such tests have been difficult to carry on with a high degree of accuracy, because of the large numbers of plants required for each determination. The use of local necrotic lesions makes it possible to recognize very large numbers of successful transmissions on single plants. This reduces the amount of labor in many experiments. A degree of accuracy never before possible can be attained in this way with the use of moderate numbers of plants. *N. glutinosa* lends itself particularly to use in measuring virus concentrations, since the necrotic lesions in this plant develop in from two to five days. These lesions are easily counted and their relative numbers show variations in the concentrations of samples with great clearness.

When very small numbers of lesions develop on *N. glutinosa* because of the use of highly diluted samples of virus, there occur many cases in which a plant shows a single local lesion. Under such circumstances it seems possible that the infection has resulted from the entrance of a single virus particle. Transfers from this type of plant may be assumed to develop virus produced from this single particle. The process of isolation may be repeated if desired. Possibly a pure line of the causal agent is thus secured. Such single particle strains have not yet been carefully studied; their isolation and use may be expected to furnish material for future work. The use of *Nicotiana* species which show local necrotic lesions is as helpful in the study of tobacco mosaic virus as KOCH's plate method is in the study of bacterial cultures.

Summary

1. Five *Nicotiana* species were found to develop necrotic lesions wherever virus of the common field type of tobacco mosaic successfully entered leaf tissues. These species were *N. rustica, N. langsdorffii, N. sanderae, N. acuminata*, and *N. glutinosa*.

2. The local lesions developing in *N. rustica* can be used to measure accurately the potency of tobacco mosaic virus.

3. The local lesions in *N. glutinosa* are exceptional in the rapidity of their development. They sometimes begin to appear thirty hours after inoculation. In four or five days they are well developed. Large numbers of them may be distinguished on a single plant. This allows comparisons to be made between virus samples, since the number of lesions developing depends on the virus concentration of the inoculum.

4. A standardized method for using *N. glutinosa* as a test plant for measuring the concentration of mosaic virus gives as rapid and as

accurate results as the determination of bacterial numbers by plating methods.

Boyce Thompson Institute for Plant Research, Inc.
Yonkers, N.Y.

LITERATURE CITED

1. Allard, H. A., The mosaic disease of tobacco. U. S. Dept. Agric. Bull. 40. 1914.
2. Beijerinck, M. W., Über ein Contagium vivum fluidum als Ursache der Fleckenkrankheit der Tabaksblätter. Verhandelingen der Koninklijke Akademie van Wetenschappen te Amsterdam, Sect. 2. **6**:3–22. 1898.
3. Fernow, K. H., Interspecific transmission of mosaic diseases of plants. Cornell Agric. Exp. Sta. Memoir 96. 1925.
4. Holmes, F. O., Accuracy in quantitative work with tobacco mosaic virus. Bot. Gaz. **86**:67–81. 1928.
5. Iwanowski, D., Über die Mosaikkrankheit der Tabakspflanze. Zeits. Pflanzenkrankheiten **13**:1–41. 1903.
6. Mayer, A., Über die Mosaikkrankheit des Tabaks. Die Landwirthschaftlichen Versuchs-Stationen **32**:451–467. 1886.
7. Priode, C. N., Further studies in the ring-spot disease of tobacco. Amer. Jour. Bot. **15**:88–93. 1928.

The Chemical Nature of Tobacco Mosaic Virus Particles

Bryan D. Harrison
Scottish Crop Research Institute, Invergowrie, Dundee, U.K.

Wendell Stanley's paper in *Science* in 1935 that established the proteinaceous nature of particles of tobacco mosaic virus (TMV) was a landmark in virology. What was the background to this discovery? Since the turn of the century, when the agent causing tobacco mosaic disease had been found to pass through bacteria-retaining filters, progress in discovering its nature had been slow. In 1928 Helen Purdy Beale had detected a specific antigen in TMV-infected plants that was not found in healthy ones, in 1929 C. G. Vinson and A. W. Petre showed that infectious material in sap from infected plants could be precipitated by protein precipitants such as lead acetate, and in 1932 W. N. Takahashi and T. E. Rawlins reported that sap from TMV-infected plants, but not that from healthy ones, showed birefringence caused by anisotropy of flow, suggesting that it contained elongated particles. Moreover, Francis O. Holmes had revolutionized the quantification of infectivity in 1929, when he introduced the local lesion method for assaying TMV. In addition, around 1930 the science of biochemistry was evolving, and enzymes were being purified and found to be proteins. In the light of these pointers and developments, it must have seemed reasonable to take seriously the possibility that TMV was a protein.

One way to seek evidence on this question would be to ascertain the effects of proteolytic enzymes. Indeed, one of Stanley's first actions after joining the Rockefeller Institute for Medical Research at Princeton was to study the effect of trypsin on TMV infectivity. In what proved to be a pioneering study of a virus inhibitor, Stanley (1934a) found that trypsin inhibited infectivity reversibly, but it did not inactivate the virus. However, soon afterwards he reported (Stanley, 1934b) that TMV infectivity was inactivated by treatment for several days with pepsin at 37°C and pH 3. Others have claimed that TMV particles are resistant to pepsin, or are only digested when partially denatured, but the effect of pepsin in his tests may well have been the lead that encouraged Stanley to look further into the possibility that TMV was a protein. At Princeton, his laboratory was close to that of J. H. Northrop, whose group were in the full flood of their seminal studies on purification and crystallization of enzymes, all of which were proving to be proteins. Stanley applied some of their methods and, as reported in *Science* on June 28, 1935, obtained a purified product, consisting of needle-like 'crystals' (more correctly *paracrystals*), which he estimated to contain 20% nitrogen but no carbohydrate. The material was highly infective, but lost infectivity at pH 1 or 11.8, or on heating at 94°C. Loss of infectivity on heating roughly paralleled the loss of the native protein. In the full-length paper (Stanley, 1936), a more detailed chemical analysis is given but no phosphorus was detected and carbohydrate is not mentioned. However, the scale of the purification work is now evident. Plots of TMV-infected plants were grown in the field and the virus was obtained by processing their sap on a (literally) bucket scale to yield about 20 g of purified product. Stanley concluded that TMV was, he thought exclusively, an autocatalytic protein.

However, Stanley did not have the field completely to himself. In Australia, Rupert Best, also working with TMV, found that the infectious agent had an isoelectric point of pH 3.4, at which it precipitated, and he obtained partially purified preparations containing 14% nitrogen and reacting positively in tests for protein (Best, 1936). More importantly, in Britain, F. C. (Fred) Bawden and N. W. (Bill) Pirie had turned from purification studies on potato virus X, a tougher nut to crack, to TMV. Subsequent progress was rapid. As Pirie has remarked, somewhat magisterially, "in a few weeks using the methods that had been standard in protein chemistry for half a century, [we] got liquid-crystalline preparations of an infective nucleoprotein". In their 1936 paper in *Nature*, written jointly with the X-ray crystallographers J. D. Bernal and I. Fankuchen, they reported that preparations purified from TMV-infected plants were liquid crystalline, highly infectious, and contained 16.7% N, 0.5% P and 2.5% carbohydrate of the ribose type (i.e. 95% protein and 5% RNA). Moreover, X-ray diffraction analyses showed the presence of material with regular substructure (now known to represent the repeat at three turns of the particle helix) and size (now known to be the 15 nm side-to-side packing distance of particles in dried gels). Bawden and Pirie's (1937) subsequent full account of TMV purification is a masterly paper, which should be read by all aspiring plant virologists. In addition to the remarkably accurate elemental analyses, good estimates were made of parameters such as density and partial specific volume, the work was repeated with different TMV strains giving the same results, much information was reported on the properties of purified preparations in different conditions, well chosen control treatments were included, and the work was conducted and results interpreted in a careful, critical manner.

Stanley's initial failure to detect phosphorus or carbohydrate (and therefore RNA), and the failure of Bawden and Pirie to confirm some of his other observations led to continuing correspondence, mostly between Bawden and Stanley, in which Bawden's views were often expressed bluntly. Stanley's replies were always polite. He eventually conceded that his virus preparations did contain phosphorus but at first considered it to be a contaminant and not essential for activity (Stanley, 1937). However, in 1938 he admitted that the phosphorus was a virus constituent after all. Bawden and Pirie undoubtedly found this slowness to recognise the correctness of their findings irksome, especially as Stanley received much recognition for his initial demonstration that TMV particles contained protein, often expressed as 'for discovering the chemical nature of viruses'. In 1941, for example, Bawden wrote "what does it feel like (to receive a medal[1])...for discovering that tobacco mosaic virus was a liquid crystalline nucleoprotein after the stubborn fight you put up to maintain that it was a...crystalline globulin?"

Stanley, however, continued to prosper. He went on to establish the Virus Laboratory in the University of California at Berkeley, where he attracted many able colleagues. It became a powerhouse of basic virus research in the United States, with studies on TMV at the forefront. He also wrote or edited several authoritative books on viruses. In 1946 he was awarded the Nobel Prize for Chemistry for his work on TMV. One is left with the impression of a man with great drive, as befits a former All-State football player, who was in a hurry to get results but had more flair for organising than for bench work. He died in 1971 at the age of 66.

Bawden and Pirie continued their collaboration for a total of almost 40 years. In the late 1930's they went on from TMV to purify the particles of cucumber green mottle mosaic virus, potato viruses X and Y, tomato bushy stunt virus (TBSV) and tobacco necrosis virus (TNV). All were ribonucleoproteins and those of TBSV and TNV were shown to form true three-dimensional crystals. Stanley purified tobacco ringspot virus and found it too to be a ribonucleoprotein. However, all three failed to discover the infectivity of viral RNA. Bawden and Pirie certainly made TMV RNA preparations by a variety of still acceptable methods and tested their infectivity. However, as the two were by then in laboratories about 50 miles apart, Pirie's samples had to be transported by car or post for testing by Bawden, with the labile RNA presumably losing its infectivity in transit. By this circumstance, the coming of molecular biology was probably delayed by more than a decade.

Bawden built up a powerful group of plant virologists as Head of the Plant Pathology Department at Rothamsted Experimental Station, which became a center for both basic and applied research in plant virology. Subsequently, through his writings about viruses and as Director of Rothamsted, he was very influential in agricultural research in Britain and overseas, especially the Commonwealth, in which he was perhaps the most distinguished agricultural scientist of his day. He served as Treasurer and Vice-President of the Royal Society, was knighted in 1967 and died in 1972 at the age of 64.

Pirie was Head of the Biochemistry Department at Rothamsted from 1940 until his 'retirement' in 1973. In addition to his continuing virus research, he pioneered methods for the extraction from green leaves of protein to be used as food for humans and farm animals. Visitors to his cottage in the New Year's season were apt to be plied with tidbits made from leaf protein, and laced with flavourings, as publicity for its edibility. He was also deeply interested in, and wrote extensively about, the origin of life and on matters of scientific philosophy. In addition to being a meticulous experimenter, he was a penetrating thinker. Not a man to build up a large research group, he preferred to work at the bench, but without ever becoming a practising molecular biologist. In 1971 he received the Copley Medal, the highest award of the Royal Society. He died in 1997, aged 89.

These three scientists made seminal contributions to establishing the chemical nature of viruses, deemed to be 'at the threshold of life', and two of the three discovered in viruses the material that was later shown to endow them with genetic continuity. Their work stimulated a great surge of research on virus particle structure and function, and transformed ways of thinking about viruses.

[1] Presumably the 'Gold Medal' of the American Institute of the City of New York presented on 6 February 1941 (*Science* **93**:143-151).

References

Bawden, F. C. and N. W. Pirie. (1937). The isolation and some properties of liquid crystalline substances from solanaceous plants infected with three strains of tobacco mosaic virus. *Proc. Royal Soc.* **B123**:274-320.

Best, R. J. (1936). Precipitation of the tobacco mosaic virus complex at its isoelectric point. *Australian J. Exper. Biol. Med. Sci.* **14**:1-13.

Stanley, W. M. (1934a). Chemical studies on the virus of tobacco mosaic. I. Some effects of trypsin. *Phytopathology* **24**: 1055-1085.

Stanley, W. M. (1934b). Chemical studies on the virus of tobacco mosaic. II. The proteolytic action of pepsin. *Phytopathology* **24**:1269-1289.

Stanley, W. M. (1936). Chemical studies on the virus of tobacco mosaic. VI. The isolation from diseased Turkish tobacco plants of a crystalline protein possessing the properties of tobacco-mosaic virus. *Phytopathology* **26**:305-320.

Stanley, W. M. (1937). Chemical studies on the virus of tobacco mosaic. VII. The isolation of a crystalline protein possessing the properties of aucuba mosaic virus. *J. Biol. Chem.* **117**:325-340.

Biographical Sketch

Bryan Harrison's first degree (B. Sc. with Honors in Agricultural Botany, 1952) was from Reading University, where he was a Wantage Scholar and his major professor was W. B. Brierley, who previously had been the first Head of the Plant Pathology Department at Rothamsted Experimental Station, Harpenden. Fascinated by plant viruses, he obtained a Postgraduate Studentship to study them at Rothamsted under F. C. Bawden, and was awarded a Ph.D. in 1955 for a thesis on 'The multiplication of plant viruses in inoculated leaves'. From 1954-1957 he was a research virologist at the Scottish Horticultural Research Institute. Following a posting at Rothamsted as a research virologist (1958-1966), he was appointed the Head of the Virology Department at the SCRI (1966-1991), and was Professor of Plant Virology at Dundee University from 1991-1996. He was elected to Fellowship of the Royal Society of Edinburgh in 1979, and that of the Royal Society (London) in 1987. In 1998 he was elected as a Foreign Associate of the U. S. National Academy of Sciences. Presently he is a Professor Emeritus at Dundee University and Honorary Research Professor at SCRI.

ISOLATION OF A CRYSTALLINE PROTEIN POSSESSING THE PROPERTIES OF TOBACCO-MOSAIC VIRUS

A crystalline material, which has the properties of tobacco-mosaic virus, has been isolated from the juice of Turkish tobacco plants infected with this virus. The crystalline material contains 20 per cent. nitrogen and 1 per cent. ash, and a solution containing 1 milligram per cubic centimeter gives a positive test with Millon's biuret, xanthoproteic, glyoxylic acid and Folin's tyrosine reagents. The Molisch and Fehlings tests are negative, even with concentrated solutions. The material is precipitated by 0.4 saturated ammonium sulfate, by saturated magnesium sulfate, or by safranine, ethyl alcohol, acetone, trichloracetic acid, tannic acid, phosphotungstic acid and lead acetate. The crystalline protein is practically insoluble in water and is soluble in dilute acid, alkali or salt solutions. Solutions containing from 0.1 per cent. to 2 per cent. of the protein are opalescent. They are fairly clear between pH 6 and 11 and between pH 1 and 4, and take on a dense whitish appearance between pH 4 and 6.

The infectivity, chemical composition and optical rotation of the crystalline protein were unchanged after 10 successive crystallizations. In a fractional crystallization experiment the activity of the first small portion of crystals to come out of solution was the same as the activity of the mother liquor. When solutions are made more alkaline than about pH 11.8 the opalescence disappears and they become clear. Such solutions are devoid of activity and it was shown by solubility tests that the protein had been denatured. The material is also denatured and its activity lost when solutions are made more acid than about pH 1. It is completely coagulated and the activity lost on heating to 94° C. Preliminary experiments, in which the amorphous form of the protein was partially digested with pepsin, or partially coagulated by heat, indicate that the loss in activity is about proportional to the loss of native protein. The molecular weight of the protein, as determined by two preliminary experiments on osmotic pressure and diffusion, is of the order of a few millions. That the molecule is quite large is also indicated by the fact that the protein is held back by collodion filters through which proteins such as egg albumin readily pass. Collodion filters which fail to allow the protein to pass also fail to allow the active agent to pass. The material readily passes a Berkefeld "W" filter.

The crystals are over 100 times more active than the suspension made by grinding up diseased Turkish tobacco leaves, and about 1,000 times more active than the twice-frozen juice from diseased plants. One cubic centimeter of a 1 to 1,000,000,000 dilution of the crystals has usually proved infectious. The disease produced by this, as well as more concentrated solutions, has proved to be typical tobacco mosaic. Activity measurements were made by comparing the number of lesions produced on one half of the leaves of plants of Early Golden Cluster bean, *Nicotiana glutinosa* L., or *N. langsdorffii* Schrank after inoculation with dilutions of a solution of the crystals, with the number of lesions produced on the other halves of the same leaves after inoculation with dilutions of a virus preparation used for comparison.

The sera of animals injected with tobacco-mosaic virus give a precipitate when mixed with a solution of the crystals diluted as high as 1 part in 100,000. The sera of animals injected with juice from healthy tobacco plants give no precipitate when mixed with a solution of the crystals. Injection of solutions of the crystals into animals causes the production of a precipitin that is active for solutions of the crystals and juice of plants containing tobacco-mosaic virus but that is inactive for juice of normal plants.

The material herein described is quite different from the active crystalline material mentioned by Vinson and Petre[1] and by Barton-Wright and McBain,[2] which consisted, as Caldwell[3] has demonstrated, largely of inorganic matter having no connection with the activity. These preparations were less active than ordinary juice from diseased plants, and the activity they possessed diminished on further crystallizations.

The crystalline protein described in this paper was prepared from the juice of Turkish tobacco plants infected with tobacco-mosaic virus. The juice was brought to 0.4 saturation with ammonium sulfate and the precipitated globulin fraction thus obtained was removed by filtration. The dark brown globulin portion was repeatedly fractionated with ammonium sulfate and then most of the remaining color was removed by precipitation with a small amount of lead subacetate at pH 8.7. An inactive protein fraction was removed from the light yellow colored filtrate by adjusting to pH 4.5 and adding 2 per cent. by weight of standard celite. The celite was removed, suspended in

[1] C. G. Vinson and A. W. Petre, *Contrib. Boyce Thompson Inst.*, 3: 131, 1931.

[2] E. Barton-Wright and A. McBain, *Nature*, 132: 1003, 1933.

[3] J. Caldwell, *Nature*, 133: 177, 1934.

water at pH 8, and the suspension filtered. The active protein was found in the colorless filtrate. This procedure was repeated twice in order to remove completely the inactive protein. Crystallization was accomplished by adding slowly, with stirring, a solution containing 1 cubic centimeter of glacial acetic acid in 20 cubic centimeters of 0.5 saturated ammonium sulfate to a solution of the protein containing sufficient ammonium sulfate to cause a faint turbidity. Small needles about 0.03 millimeters long appeared immediately and crystallization was completed in an hour. Crystallization may also be caused by the addition of a little saturated ammonium or magnesium sulfate to a solution of the protein in 0.001 N acid. Several attempts to obtain crystals by dialyzing solutions of the protein gave only amorphous material. To date a little more than 10 grams of the active crystalline protein have been obtained.

Although it is difficult, if not impossible, to obtain conclusive positive proof of the purity of a protein, there is strong evidence that the crystalline protein herein described is either pure or is a solid solution of proteins. As yet no evidence for the existence of a mixture of active and inactive material in the crystals has been obtained. Tobacco-mosaic virus is regarded as an autocatalytic protein which, for the present, may be assumed to require the presence of living cells for multiplication.

W. M. STANLEY

THE ROCKEFELLER INSTITUTE FOR
MEDICAL RESEARCH,
PRINCETON, N. J.

Vol. 138, p. 1051, December 19, 1936.)

Liquid Crystalline Substances from Virus-infected Plants

STANLEY[1] has described the preparation of a crystalline protein possessing the properties of tobacco mosaic virus from the sap of infected tobacco and tomato plants. The crystals were small needles made

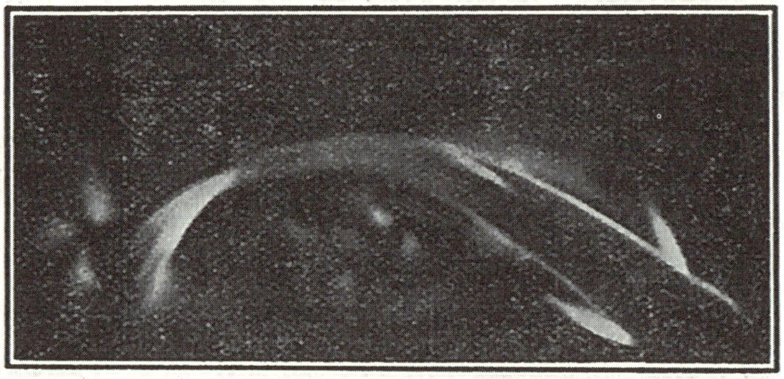

FIG. 1. Wake of a goldfish swimming in a dilute solution of protein from infected sap (observed between crossed nicols). *Photo by Ramsey and Muspratt.*

by precipitation with acid ammonium sulphate. We have confirmed these results, but have found that by further purification the protein in neutral aqueous solution can be obtained in liquid crystalline states.

The sap of tobacco and tomato plants infected with strains of tobacco mosaic virus, after clarification by centrifuging, contains five to ten times as much protein as sap from similarly treated uninfected plants. This extra protein can be precipitated from dilute salt solutions at around *p*H 3·4 and from neutral solutions with from 10 to 12 per cent ammonium sulphate. The protein in uninfected sap is not precipitated under these conditions. Using these properties, 1–2 gm. of protein can be isolated from a litre of sap; the yield varying with the age of the infected plant and the duration of infection. No enzyme preparation has yet been found which can attack the protein at an appreciable rate, and some impurities can be conveniently removed by incubation with trypsin.

We have worked with three strains of tobacco mosaic virus, those causing common tobacco mosaic, aucuba mosaic and enation mosaic. No gross chemical or physical differences have been found between the three proteins isolated, but each reproduces its characteristic disease when inoculated to susceptible plants. Differences in detail, which will be described in a later publication, have been noticed. Solutions of the protein are antigenic, and the antisera produced in rabbits give specific precipitates with 1 c.c. of solution containing 10^{-7} gm. The three proteins are serologically related. Plants inoculated with 1 c.c. containing 10^{-9} gm. usually become infected, and occasional infections have followed inoculations of 10^{-11} gm.

These proteins, when precipitated with acid and dried, have the usual analytical figures: C 51 per cent, H 7·1 per cent, N 16·7 per cent. The sulphur contents vary from 0·2 to 0·7 per cent and there is 0·5 per cent phosphorus and 2·5 per cent carbohydrate. The last two constituents can be isolated as nucleic acid of the ribose type from protein

FIG. 2. Highly purified protein from infected sap, when stronger than 2 per cent, separates into two layers on standing. If there is very little of the lower layer, it arranges itself in spindles. Photographed between crossed nicols, × 30.

denatured by heating. Neutral aqueous solutions of the proteins are almost colourless and faintly opalescent. The protein sediments in a centrifugal field of 23,000 times gravity at about 5 mm. per hour and is deposited as a very viscous layer containing up to 30 per cent of dry matter. Highly purified protein solutions, if stronger than 2 per cent, separate into two layers on standing. The lower layer, which may be water clear, is liquid crystalline. The birefringence depends on the concentration, but may rise to 0·002. The upper layer shows, on gentle agitation, the phenomenon of anisotropy of flow. This was noticed by Takahashi[2] in clarified sap. We have, however, been unable to confirm his later claim that the effect could be observed in healthy tobacco sap. The orientation will persist for several seconds in purified solutions, but if the solutions are impure or dilute the persistence is much shorter. This effect is illustrated in Fig. 1, where the wake of a goldfish swimming in a dilute solution is clearly shown ; the cross of isocline will be noticed in the eddies.

Solutions can readily be orientated by electric currents, but not by magnetic fields of 6,000 gauss. The lower layer orientates itself by flow like the top layer and also tends to orientate parallel to glass or air surfaces. On flowing through small capillary tubes it can be thrown into a state of reversed spiral vibration[3]. When small portions of lower layer are in equilibrium with the top liquid, they appear as spindle-shaped bodies (see Fig. 2) which are characterized by an approximately constant meridional curvature. Large spindles are consequently nearly spherical and small ones practically linear. On standing there appear holes, filled with top liquid of precisely the same spindle shape. All these phenomena indicate the presence of long fibres.

If any of these liquids be allowed to dry, a film forms on the surface consisting of three sharply distinguishable layers. The first formed is an extremely soft gel well orientated and with much higher birefringence, 0·007, than the liquid. The outer part of this wet gel shrinks by 50 per cent and forms a layer of higher refractive index, but lower birefringence, 0·003. Finally, on placing in air, extensive cracks appear in this dry gel and it becomes translucent and rather feebly doubly refractive. The dry gel is usually well-orientated but in certain conditions shrinks longitudinally to give a herring bone pattern, which indicates that little longitudinal shrinkage of the fibres themselves takes place.

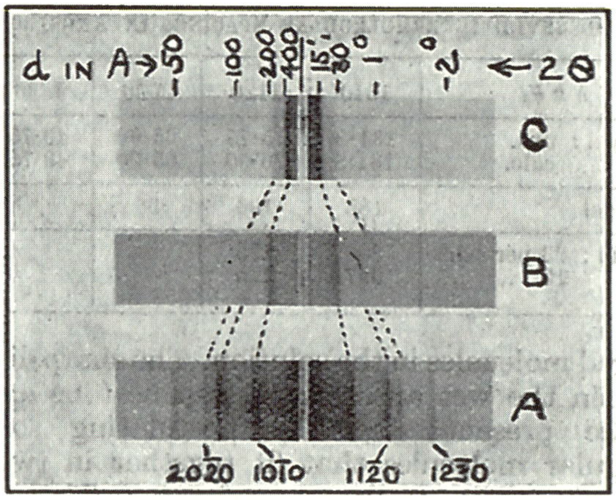

Fig. 3. X-ray patterns of dry gel (*A*), wet gel (*B*) and 13 per cent solution (*C*).

The X-ray patterns of these different forms show remarkable similarities and differences. For the large angle scattering there appears to be little difference, except in general intensity, between patterns given by all the different forms from the top liquid, orientated by flow in a Lindemann glass capillary, to the dry gel and the 'crystals', from ammonium sulphate solutions. This pattern is therefore entirely due to the protein molecules themselves and may be called the intramolecular pattern. It appears to have about the same order of complexity as that produced by feather keratin, with a repeat unit in the fibre direction of $3 \times 22 \cdot 2 \pm 0 \cdot 2$ A.

Near the centre of the photographs, spots were observed which varied from one preparation to another. A camera of 40 cm. length was constructed employing an X-ray beam of copper-$K\alpha$ radiation with an angular width of 3′ and capable of showing spacings up to 1,200 A. All the intermolecular pattern so far observed lies in the equatorial plane, that is, refers to the relative sideway positions of the rod-like molecules. The dry gel gives the most definite pattern (Fig. 3) showing five lines (see table) corresponding to the first five possible reflections of a hexagonal close-packing with intermolecular distance of $152 \cdot 0 \pm 0 \cdot 5$ A. The wet gel gives three distinct lines which seem to correspond to hexagonal close-packing with intermolecular distance of 210 A., but this varies slightly with the composition. The bottom layer liquid gives a pattern of three lines; two sharp lines the position of which depends on concentration, indicating a mean molecular distance ranging from 300 A. to 470 A., and a diffuse line at a spacing of 100 A. which is independent of the concentration and is thus plainly of intramolecular origin. The pattern of the liquid suggests that there exists a quasi-regular Debye-Hückel arrangement of parallel rod-like

OBSERVED INTERMOLECULAR SPACINGS IN ANGSTROMS.

$h\,k\,i\,l$	$10\bar{1}0$	$11\bar{2}0$	$20\bar{2}0$	$12\bar{3}0$	$30\bar{3}0$
Dry gel: obs. calc.	131·8 131·8	75·75 76·00	65·90 65·90	49·75 49·75	43·5 43·9
Wet gel	188	106	93		
Liquid: 23 per cent 13 „ „	300 397	175 225			

charged molecules in the solution. The sharp difference between the wet and dry gel can best be explained at the present stage by postulating somewhat triangular molecules that fit together in two ways, containing different amounts of water. The reflection at 100 A. supports this hypothesis. So far it has proved impossible to get any indication of intermolecular reflections in the other directions of the molecule, though any regularity of less than 1,200 A. would have been detected. This suggests strongly that the arrangement is regular only in cross-section and that no molecular sheets are formed.

From the results already gained, it is legitimate to make certain conclusions as to the nature of the protein molecules. First, the molecules seemed to be identical in cross-section. Secondly, each molecule has quasi-regular structure and thus may be considered to be built up of sub-units of approximately the same character. The physical properties of the substance can best be explained by postulating rod-shaped molecules. The minimum cross-section area of these is 20,100 sq. A. for the dry gel. The molecular length is more uncertain. The extreme character of the orientation phenomena and the X-ray data point to a minimum length of not less than ten times the width, or greater than 1,000 A. This gives a minimum molecular weight in reasonable agreement with Svedberg's[4] estimate of 17×10^6, though there is nothing to show that the lengths of the molecules are uniform.

These results have a certain intrinsic interest, but this would naturally be greatly enhanced could it be shown that these rods are in fact virus particles. This conclusion seems to us both reasonable and probable, but we feel that it is still not proved, nor is there any evidence that the particles we have observed exist as such in infected sap.

Note added Dec. 3.

Wyckoff and Corey have published[5] an X-ray study of the ammonium sulphate crystals of tobacco mosaic and aucuba protein. Their measurements of the intramolecular spacings obtained with unorientated material agree with ours, notably the lines they record at 11·0, 7·44, 5·44 and 3·7 A. correspond to our measurements of the planes (0006, 9, 12, 18) respectively.

F. C. Bawden.

Rothamsted Experimental Station,
Harpenden.

N. W. Pirie.

Biochemical Laboratory,
Cambridge.

J. D. Bernal.
I. Fankuchen.

Crystallographic Laboratory,
Cambridge.

Nov. 17.

[1] *Phytopathology*, **26**, 305 (1936).
[2] *Science*, **77**, 26 (1933).
[3] van Iterson, *Proc. Roy. Akad. Wetensch.*, **37**, 367 (1934).
[4] *J. Amer. Chem. Soc.*, **58**, 1863 (1936).
[5] *J. Biol. Chem.*, p. 51, Nov. 1936.

Printed in Great Britain by Fisher, Knight & Co., Ltd., St. Albans.

Commentary on Bernal and Fankuchen

Gerald Stubbs
Department of Molecular Biology, Vanderbilt University, Nashville, Tennessee

The diffraction patterns obtained by Bernal and Fankuchen in 1936 from tobacco mosaic virus gels, and published as Figures 24 and 25 of their historic *Journal of General Physiology* paper, appear to have puzzled the discoverers deeply. The patterns displayed a level of intricate detail that could only come from macromolecular components; this was not unexpected. The perfection of the ordering displayed was truly remarkable, far greater than anything previously seen with TMV (Wyckoff and Corey, 1936). But all previously understood diffraction patterns, certainly all patterns with anything approaching this degree of order, had come from crystals, with well-defined unit cells repeating in three dimensions. This was true even of previously observed fiber diffraction patterns; fiber patterns are cylindrically averaged because of the random orientation of the diffracting particles about the fiber axis, but many fibers are themselves crystalline, that is, the randomly oriented 'particles' are in fact tiny crystals. It is evident that Bernal and Fankuchen spent a long time trying to interpret the TMV patterns in terms of crystalline fibers. But crystalline diffraction patterns are confined to lattice points, so that all the observed reflections form a regular array. No conceivable lattice could fit the TMV diffraction pattern.

Even to reach the point of recording these patterns, Bernal and Fankuchen had to tackle engineering challenges that would deter most modern researchers. X-ray exposures of "nearly 400 hours" were required, using the strongest x-ray sources available and cameras built in the laboratory. Thirty to forty-five years later, when we recorded the diffraction patterns that were used for the high-resolution structure determination, the exposure times were down to a much more convenient 100 hours; today we would find a 10-hour exposure in the laboratory to be tedious, and at a synchrotron x-ray source a few minutes (or for some purposes, seconds) would suffice. To preserve the hydrated specimens and to induce orientation of the particles, Bernal and Fankuchen introduced the gels into thin-walled (0.02 mm) capillary tubes: again, these were made in the laboratory. By this means, they aligned the virions to an average disorientation of less than 1.5°, comparable to the best diffracting specimens made later. Their method is still used, but today we use commercially available capillary tubes and cameras. They made tiny specimen containers from brass, sheets of glass of "bubble thinness", and sealing wax. More elaborate still was the flow cell, in which a peristaltic pump circulated virus solution through a thin-walled capillary, allowing solutions that oriented only while actively flowing to be examined. They tried magnetic orientation, but with a field of only 0.5 Tesla did not observe any effect. Our later work with fields of 3.5 Tesla and more showed that TMV does indeed respond to magnetic fields, but the effect is, however, sharply dependent upon solution conditions.

The mystery of the non-crystalline diffraction patterns derives from the nature of the TMV gels. The particles in the gels are parallel to a remarkable degree, but their orientations about the particle axes are completely uncorrelated (as shown by Figure 23), and the distances between particles are regular only to very low resolution. As a result, the x-ray diffraction is not confined to sharp reflections like those produced by true three-dimensional crystals. The diffraction is sharply sampled in the direction of the particle axis because the TMV particles themselves contain many repeating units along the direction of the particle axis; Bernal and Fankuchen recognized this feature of the structure, and measured the repeat distance as 68 Å (later refined to 69 Å). The overall effect is to confine the diffraction to layer lines, but for diffraction to be continuous along those layer lines. Bernal and Fankuchen recognized that their specimens were not crystalline, but old ways of looking at diffraction patterns were deeply ingrained, and they continued to look for some sort of limited repeat in the direction at right angles to the particle axis so that the diffracted intensities could be described in terms of 'integral' and 'fractional' indices. The ideas related to fractional indices were remarkably creative, and might indeed have been significant had the virus structure turned out to be different, but advances in the theory of fiber diffraction eventually showed that for the TMV pattern, such indices simply have no meaning.

Bernal and Fankuchen realized that the hexagonal packing of the virus in dried gels yielded a minimum value (150 Å by their measurements) for its cross-sectional diameter, and they observed that in solution, the particles never approached each other more closely than 173 Å (185 Å at the isoelectric point), although it was left for later workers to connect these observations with the fact that the true diameter is 180 Å, and that the deep grooves and marked ridges on the viral surface fit together when a virus gel is dried, allowing the virions to pack closer together than their diameters would appear to allow.

Even without any theory of scattering from a helix, or indeed, without any realization that TMV was a helix (both of these insights would appear about a decade later, in the early 1950s), Bernal and Fankuchen were able to draw some conclusions about the internal structure of the molecules based on the locations of the strongest diffracted intensities. They speculated (and were very careful to emphasize that they were only speculating) upon the existence of subunits, having dimensions 44 Å x 44 Å x 22 Å. The first of these dimensions is an artifact caused by the lack of a proper treatment of helical diffraction, but the latter two are indeed the dimensions of the TMV coat protein subunit. They also suggested that these subunits were further divided into smaller subunits, having dimensions 11 Å x 11 Å x 11 Å. Again, the first dimension is an artifact, but again, the remaining two dimensions reflect a real feature (although not what we would now call a subunit) of the structure.

Perhaps the key observation, at least from the point of view of future researchers who would contemplate these same patterns, was that the TMV gels were not crystalline, and that a new way of analyzing diffraction data would be required before a detailed structure could be described. The theory of helical diffraction did eventually allow the analysis not only of TMV, but of DNA, RNA, and many other helically symmetric structures. Even without this theory, Bernal and Fankuchen recognized that subunits repeated along the viral axis, and measured this repeat as well as the radius of the virus. The production of the patterns themselves was, of course, a *tour de force*, the inspiration for all of the structural studies of TMV and other helical viruses that would follow.

Unlike most investigations, this one (or at least, the phase pursued by these investigators) had a well-defined beginning and ending. The first diffraction patterns were obtained in Cambridge shortly after Bawden and Pirie provided Bernal and Fankuchen with TMV (Bawden *et al.*, 1936), and the Second World War effectively brought an end to their work. In fact, with Bernal occupied by wartime matters and Fankuchen returned to America, it was left to Dorothy Crowfoot (later Hodgkin), as she recounted to me 35 years later, to put the manuscript of the *Journal of General Physiology* paper together and get it to publication, a task that included preparing many of the figures!

But of course, the story did not end there. Shortly before the war, Bernal had moved to Birkbeck, where in later years Rosalind Franklin took up the project. Franklin realized that the new protein crystallographic method of isomorphous replacement held the key to a high-resolution structure determination, and unlike most of her contemporaries and many of her successors, she believed that the structure would some day be determined at 3 Å resolution. She and her last student, Kenneth Holmes, used isomorphous replacement to determine the correct symmetry of the virus (Franklin and Holmes, 1958). She established that the RNA lay at a radius of 40 Å in the virion. The advance slowed a little after her untimely death in 1958, aged only 38, but Holmes continued the work. I joined him in 1972, and we published the structure at 6.7 Å in 1975. Bernal and Fankuchen's 11 Å "subunits" were at last revealed as α-helices! We were joined by Steve Warren for the 4 Å structure, published in 1977, and Steve and I, encouraged by Don Caspar, continued the work in America. Keiichi Namba joined me there in 1981, and in 1989, Bernal and Fankuchen's remarkable patterns finally yielded the fulfillment of Franklin's vision with the publication of the 2.9 Å resolution structure (Namba *et al.*, 1989).

References

Bawden, F. C., Pirie, N. W., Bernal, J. D., and I. Fankuchen. (1936). Liquid crystalline substances from virus-infected plants. *Nature* **138**:1051-1053.

Caspar, D. L. D. (1956). Radial density distribution in the tobacco mosaic virus particle. *Nature* **177**:928-928.

Franklin, R. E. (1956). Location of the ribonucleic acid in the tobacco mosaic virus particle. *Nature* **177**:928-930.

Franklin, R. E. and K. C. Holmes. (1958). Tobacco mosaic virus: application of the method of isomorphous replacement to the determination of the helical parameters and radial density distribution. *Acta Cryst.* **11**:213-220.

Namba, K., Pattanayek, R., and G. Stubbs. (1989). Visualization of protein-nucleic acid interactions in a virus. Refined structure of intact tobacco mosaic virus at 2.9 Å resolution by x-ray fiber diffraction. *J. Mol. Biol.* **208**:307-325.

Wyckoff, R. W. G. and R. B. Corey. (1936). X-ray diffraction patterns of crystalline tobacco mosaic proteins. *J. Biol. Chem.* **116**:51-55.

Biographical Sketch

Gerald Stubbs received his undergraduate degree from the Australian National University, and his doctorate from the University of Oxford, where his advisor was A. C. T. North. He did post-doctoral research with Kenneth C. Holmes at the Max-Planck-Insitut für medizinische Forschung in Heidelberg, where he began his structural studies of TMV. From 1976 to 1983, he was at the Rosenstiel Center at Brandeis University with D. L. D. Caspar after which he moved to his present position at Vanderbilt University.

Reprinted from NATURE, Vol. 139, page 923, May 29. 1937.)

Structure Types of Protein 'Crystals from Virus-infected Plants

FURTHER work on preparations of proteins from virus-infected plants tends to confirm and extend the conclusions expressed in our previous letter[1]. We had there shown that both concentrated solutions of these proteins and the material prepared from them by drying gave orientated paracrystals. We have now succeeded in preparing the 'crystals' of Stanley and Wyckoff in an orientated form, after trying unsuccessfully a number of methods, such as flow orientation. The method finally used was that of forcing mother liquor out of the 'crystals' by centrifuging in an unglazed porcelain tube, in a saturated atmosphere. The resulting mush was orientated by rolling between glass slides, at 0° to avoid evaporation, becoming completely transparent in the process. It was mounted in a sealed brass case with bubble thin lindeman glass windows and examined by X-ray and optical methods without loss of water. It proved to be practically indistinguishable—in water content (50 per cent), optics and X-ray pattern at low and high angles—from the preparation previously called wet gel. On being allowed to dry, it likewise turns into a horny mass giving similar X-ray photographs to 'dry gel', though fainter and with slightly larger spacings, probably due to internal strain. The X-ray pattern obtained from the unorientated crystals shows all the main lines given by Wyckoff and Corey[2] for their crystals, and we have been able to assign all but four very weak lines on these photographs to definite spots given by the orientated gel. There can therefore remain little doubt that their 'crystals' are substantially identical with ours and with our previously described paracrystals.

As many deductions have already been drawn from the crystalline nature of these proteins, it is important to state precisely in what sense the word 'crystalline' should be used in respect of them on the basis of interpretation of X-ray photographs. Normally, crystallinity presupposes an indefinite repetition of identical units in three-dimensional space. These proteins appear not to possess this degree of regularity. The degree of regularity which seems to be the only one consistent with the X-ray evidence has thus far not been described, though it has been predicted theoretically as one of the types of liquid crystal. The long molecules of the protein are packed with a perfect hexagonal two-dimensional regularity at right angles to their length, but there is no evidence of regularity of molecular arrangement in the direction of their length. None of the reflections in directions other than at right angles to the molecular length is due to the packing of the molecules, but rather to regularities within them, as shown by the

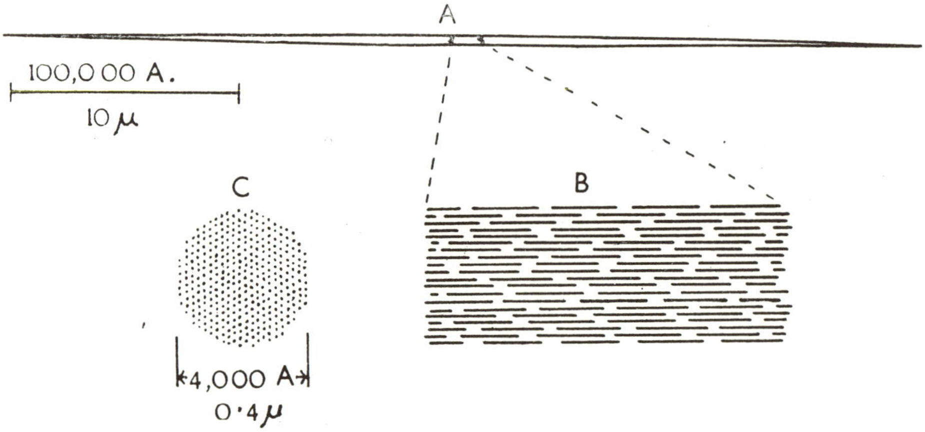

Fig. 1. DIAGRAMMATIC.
A. TYPICAL 'CRYSTAL'. B. LONGITUDINAL SECTION.
C. TRANSVERSE SECTION.

fact that such reflections persist unchanged when the molecules are in dilute solution. Lines observed by Wyckoff refer not to the regular packing of equivalent

RELATIVE INTENSITIES OF INTERMOLECULAR REFLECTIONS OBTAINED FROM VARIOUS VIRUS STRAINS.

$h^2 + k^2 + hk$	*hkil*	Tobacco mosaic	Enation	Aucuba	Cucumber 3	Cucumber 4
1	10$\bar{1}$0	$s+$	$w-m$	m	m	w
3	11$\bar{2}$0	s	s	s	m	$w-m$
4	20$\bar{2}$0	m	m	$w-m$	w	vw
7	21$\bar{3}$0	$m-s$	$m-s$	$m-s$	s	s
9	30$\bar{3}$0	vvw	0	0	0	0
12 13	22$\bar{4}$0 31$\bar{4}$0	0	vvw	0	$m-s$	m

molecules but to regularities within the molecule. They can only be said to be evidence for crystallinity if by that it is understood that each molecule is itself a crystal. The relatively small lateral dimensions of the molecules, 150 A., can only permit of imperfect diffraction effects in these directions as shown by the diffuseness of the spots, while the great length of the molecule is indicated by the extreme sharpness of the reflections from planes perpendicular to its length. A complete interpretation of this intramolecular pattern has not yet been possible; but a rough survey indicates that the molecule is made of piles of submolecules of dimensions 22 A. × 20 A. × 20 A., somewhat smaller than the normal protein molecule, and themselves divided into nearly identical units with half these dimensions.

The microscopic appearance of the Stanley 'crystals' is quite consistent with their liquid crystal character, for although fairly long, about 40 μ, they are extremely thin, about 0·4 μ thick, and have gently tapering ends. This is quite different from the prismatic or pyramidal forms of true protein crystals and more like the 'micro tactoids' observed in colloidal solutions of the fibrous type. On pressing they readily fuse together into a homogeneous mass, emphasizing their partially liquid nature. Their most probable internal arrangement is that shown in Fig. 1. This arrangement would also serve very well to explain the observations of R. J. Best[3] on the formation of mesomorphic fibres from infected sap. A further consequence of this picture of the internal structure of these 'crystals' is that their formation furnishes even less evidence than is provided by ordinary crystallization as to the homogeneity of the material.

Further light on the structure of the protein may be given by comparative study of similar materials. We have so far examined three strains of tobacco virus: mosaic, enation and aucuba, as well as two varieties of cucumber virus prepared by Bawden and Pirie[4]. The intramolecular patterns resemble one another very closely, but interesting and significant differences appear in the intermolecular patterns. All three varieties of tobacco virus show the same intermolecular distance in dry gels prepared in the same way but show consistent differences of relative intensity of the different lines (see Table). The cucumber virus protein shows even more marked intensity differences and also a different intermolecular distance, 146 A., 4 per cent smaller than that of tobacco virus. There was no marked difference in spacing between the two cucumber strains.

The significance of these observations is that all the protein molecules have substantially the same general shape and size and are made of similar subunits, but that these are arranged somewhat differently in the various tobacco virus strains and more markedly so in the cucumber virus. These results indicate the possibility of deriving a system of classification of viruses on the basis of their X-ray patterns.

J. D. BERNAL.
I. FANKUCHEN.

Crystallographic Laboratory,
Cambridge.

[1] Bawden, F. C., Pirie, H. W., Bernal, J. D., and Fankuchen, I., NATURE, **138**, 1051 (1936).

[2] Wyckoff, Ralph W. G., and Corey, Robert B., *J. Biol. Chem.*, **116**, 51 (Nov. 1936).

[3] Best, R. J., NATURE, **139**, 628 (1937).

[4] Bawden, F. C., and Pirie, N. W., NATURE, **139**, 546 (March 27, 1937).

Printed in Great Britain by FISHER, KNIGHT & CO., LTD., St. Albans.

[Reprinted from THE JOURNAL OF GENERAL PHYSIOLOGY, September 20, 1941, Vol. 25, No. 1, pp. 147–165]

X-RAY AND CRYSTALLOGRAPHIC STUDIES OF PLANT VIRUS PREPARATIONS. III

BY J. D. BERNAL AND I. FANKUCHEN*

(*From the Department of Physics, Birkbeck College, University of London*)

(Received for publication, March 14, 1941)

1. The Structure of Particles

Almost the whole of our evidence for the internal structure of the particles is derived from x-ray observations, particularly those of the high angle or intramolecular photographs. In contradistinction to the low angle photographs, the most striking feature is that the general appearance of the pattern is independent of the state of hydration.

Most of the photographs have been taken from oriented specimens. This is because on unoriented specimens the wealth of lines makes direct interpretation impossible, though with a knowledge of the oriented pattern they may all be identified. Table IV shows the close correspondence between the spacings of the spots on our oriented photographs (Plate 4, Fig. 25)[1] with the rings on the unoriented photographs of Wyckoff and Corey (1936). The oriented specimens gave typical fibre photographs. In the case of the large two dimensional crystals of the dry gel, which possess secondary orientation relative to the water surface, a careful search was made to see if the large-angle scattering also showed such orientation, but no trace of it was found (Plate 4, Fig. 23). It would therefore appear that the only mutual orientation which affects the inner portions of the particles is that of parallelism of their long axes. But this negative result may be due to the disorientation of the dry gel on shrinking.

The large-angle scattering of the various virus preparations shows characteristic similarities and differences. Only four distinct patterns have been observed. The first, which has been most carefully studied, is common to all the strains of tobacco mosaic virus, a second barely distinguishable from it is given by the cucumber virus (Plate 3, Fig. 20). A third is given by both strains of the potato virus X (Plate 3, Fig. 21). The fourth pattern, which is difficult to compare with the others owing to the absence of orientation, is given by the isotropic bushy stunt virus of tomato (Bernal, Fankuchen, and Riley, 1938).

* Now National Research Fellow in Protein Chemistry, Physics Department, Massachusetts Institute of Technology, Cambridge, Massachusetts.

[1] The plates appear at the end of Parts I and II.

The Intramolecular Pattern

Of these only the pattern of the tobacco mosaic strains has been examined, in any detail, though enough is known of the others to show their general relationships. Early photographs of the tobacco mosaic virus, as either oriented gel or as liquid, gave the pattern shown in Plate 3, Fig. 19, containing four main spots (0003), (0006), $(44\bar{8}2)$, and $(30\bar{3}7)$. Of these the $(44\bar{8}2)$, (0006), and $(30\bar{3}7)$, lay approximately on a Debye circle of 11 Å, the region of strong reflection for proteins. This and indications of another ring of reflections at about 4.5 Å were sufficient to show that the picture was of the same type as those of proteins. The sharpness of the spots was, however, definitely greater than for the natural fibre proteins, except perhaps feather keratin. They were not, however, at this stage by any means as sharp as those from crystalline proteins (Bernal, Fankuchen, and Perutz, 1938). Further studies, however, showed that this was largely due to lack of orientation. It was found that the clearest photographs could be obtained from solutions oriented in capillary tubes and after many trials remarkably perfect photographs were obtained. The best (Plate 4, Figs. 24 and 25) which required an exposure of nearly 400 hours, showed hundreds of distinct spots some of which, particularly near the meridian line, were as sharp as those from crystalline proteins.

Interpretation

The interpretation of the intramolecular pattern seemed at first sight fairly straightforward. It corresponded to that of a crystal with a hexagonal lattice of $a = 87$ Å, $c = 68$ Å (Bernal, 1938). As progressively better photographs were obtained several spots appeared which could not be explained by such a small cell. It was hoped that by obtaining still better photographs this difficulty could be cleared up, but the best obtained merely confirmed the impossibility of the simple interpretation. Orthorhombic and monoclinic pseudo-hexagonal cells were next tried, but they also failed to account satisfactorily for the observed reflections, while calling for many others that were not observed. It appeared that it was quite impossible to explain the pattern on the existing theory of x-ray diffraction from a crystal, for any cell large enough to give the observed spots was found to be larger than the size of the particle as inferred from all the intermolecular measurements.

The key to what may be the true explanation of this paradox, and what is in any case a plausible one, was found in a detailed comparison of the intramolecular photographs of the dry gels, wet gels, and solutions. The equatorial line offers the greatest interest, for here the intermolecular pattern of the dry gel is found to pass almost continuously into the near intramolecular pattern of the solutions, as shown in Tables I and IV. Particularly significant are the two reflections at ξ values of approximately 0.031 and 0.043,[2] (corre-

[2] ξ is a coordinate of a reflecting plane in reciprocal space. For details see Bernal (1927).

TABLE I *a*

Intermolecular Data from Air Dried Anisotropic Gels. Relative Intensities, Spacings, and Interparticle Distances in Angstrom Units

$h^2 + k^2 + hk$	*hikl* Hexagonal indices	Spacing		Tobacco mosaic virus			Cucumber virus		Potato virus X
		Observed	Calculated	Ordinary	Enation	Aucuba	Cucumber 3	Cucumber 4	
1	10$\bar{1}$0	131.8	131.8	*s* +	*w* − *m*	*m*	*m*	*w*	
3	11$\bar{2}$0	75.8	76.0	*s*	*s*	*s*	*m*	*w* − *m*	
4	20$\bar{2}$0	65.9	65.9	*m*	*m*	*w* − *m*	*w*	*vw*	
7	21$\bar{3}$0	49.8		*m* − *s*	*m* − *s*	*m* − *s*	*s*	*s*	No intermolecular reflections observed
9	30$\bar{3}$0	43.5	43.9	*vvw*	0	0	0	0	
12	22$\bar{4}$0		37.3	0	*vvw*	0	*m* − *s*	*m*	
13	31$\bar{4}$0		36.7						
Interparticle distance........				152	152	152	146	146	

The intensities are very sensitive to slight variations of water content. It has been observed for tobacco mosaic virus that the presence of moisture tends to weaken the (1010) reflection and to enhance the (22$\bar{4}$0) and (31$\bar{4}$0).

TABLE I *b*

X-Ray Data from Bushy Stunt Virus

	Wet crystals	Dry crystals
Side of unit cell....	394 Å	316 Å
Cell molecular weight...........	48,000,000	25,600,000

Reflections observed

hkl	Observed		*hkl*	Observed	
	Spacing	Intensity		Spacing	Intensity
110	279	*vs*	110	224	*w*
112	160	*w* − *m*			
220	140	*vw*			
310	129	*vw*			
440} 600}	70–67	*vw*			
660	47	*vw*			
880	35	*vw*			
888	28.5	*vw*			

The unit cell of the wet crystal is body centered cubic and contains two molecules. Despite the fact that only one reflection was observed from the dry crystals, there are good reasons for assigning to it the same type unit cell.

TABLE II

Data Illustrating the Relation between Spacing and Concentration in Virus Preparations

N = No. of grams vacuum-dried protein in 100 cc. of specimen.
T.M.V. = tobacco mosaic virus.

Virus used	N	Spacing observed $d(10\bar{1}0)$	Interparticle distance	$\frac{1650}{\sqrt{N}}$
T.M.V. vacuum dried	137	129	149	143
T.M.V. enation, and aucuba, air dried	131	132	152	144
T.M.V. wet gel	57.5	184–192	213–222	217
Aucuba	34	252	291	281
Enation	26	259	300	324
Enation	25	287	332	330
T.M.V.	23	297	343	340
T.M.V.	13	397	458	460

TABLE III

The Dependence of Interparticle Distance in Equilibrium Gels on Salt Concentration and pH

(*a*) Gels in ammonium sulfate solution

Normality of $(NH_4)_2SO_4$	$d(10\bar{1}0)$	Interparticle distance
8.25	150	173
6.2	150	173
4.12	158	182
2.06	161	186
1.03	182	210
0.52	192	223
0.33	198	229
0.26	216–261	251–302

(*b*) Gels at varying pH

pH	$d(10\bar{1}0)$	Interparticle distance
2.5	Specimen dispersed	
3.0	171	198
3.4	158	182
3.4	160	185
3.4	167	192
4.0	174	201
5.0	206	237
6.0	252	289
7.0	~263	~305

sponding to spacings of 50 and 36 Å respectively). In the dry gel they correspond to the reflections of $(12\bar{3}0)$ and the doublet $(22\bar{4}0$ and $13\bar{4}0)$; in the 50 per cent wet gel to $(13\bar{4}0)$ and $(50\bar{5}0)$. Beyond this they cannot be clearly identified with any intermolecular spacing, but their position has hardly changed. There can be no reasonable doubt that they correspond to impor-

TABLE IV

Intramolecular data on tobacco mosaic virus

Powder data (Wyckoff and Corey)					Outstanding intramolecular reflections from best oriented specimen (enation)				
Ordinary strain		Aucuba strain		Ultra-violet irradiated spacing	Spacing in Å	Layer line	Letter for reflection Text-fig. 16	Indices in terms of hexagonal unit cell $a = 88$ Å	Intensity
Spacing	Intensity	Spacing	Intensity						
80	*s*	80	*s*		93	0	a	Fresnel scattering	*s*
53	*m*	55	*m*		47	0	b	$(4/3, 1/3, \overline{5}/3, 0)$	*s*
37	*w*	37	*w*		36	0	c	$(5/3, 2/3, \overline{7}/3, 0)$	*s*
28	*vw*	27	*vw*		29	0	d	$(7/3, 1/3, \overline{8}/3, 0)$	*m*
					19.7	0	e	$(8/3, 5/3, \overline{13}/3, 0)$	*m* — *s*
20.8	*m*	21.0	*m*	21.2	22.0	2	a	$(2, 1, \overline{3}, 2)$	*m*
					23.0	3	a	$(0, 0, 0, 3)$	*vs*
					22.3	3	b	$(1/3, 1/3, \overline{2}/3, 3)$	*m* — *s*
16.2, 14.2	*vw*	14.2	*vw*	14.7	Too many reflections of this spacing to list (see Plate 4, Fig. 25, and Text-fig. 16).				
11.0	*s*	11.0	*s*	11.3	11.4	6	a	(0006)	*vs*
10.2	*w*	10.2	*w*		10.3	2	g	$(44\overline{8}2)$	*m* — *s*
9.2	*m*	9.2	*m*	9.2	9.1	7	a	$(30\overline{3}7)$	*s*
7.44	*m*	7.43	*m*	7.5	For each reflection shorter than 20.8 Å there are too many intramolecular reflections corresponding to it in spacing for the correlations to have any special significance. The intense reflections corresponding to 11.0, 10.2, 9.2 Å, are given merely because these are strong lines and the reflections principally responsible for them can be determined.				
6.5	*w*	6.5	*w*						
5.7	*vw*								
5.44	*w*	5.41	*w*						
4.95	*m*	4.95	*m*						
4.71	*m*	4.68	*m*						
4.44	*m*	4.45	*m*	4.46					
4.08	*w*	4.09	*w*	4.10					
3.88	*m*	3.88	*m*	3.88					
3.70	*vw*								
3.54	*vw*								
3.39	*w*	3.39	*w*						

tant intramolecular reflections which serve to enhance the intensity of the intermolecular reflections lying nearest them. The spot on the third layer line which corresponds to them is a single one at $\xi = 0.036$, which would be $(30\overline{3}3)$ for the dry gel and remains unchanged in position, within errors of measurement, for the other forms.

Now the fact that these intramolecular values though arrived at quite independently fit with intermolecular ones for the dry gel points to the not unexpected conclusion that its intermolecular periodicity is closely linked to the intramolecular one. We cannot, however, attempt to index the intramolecular pattern on this basis for that would bring us back to the paradox that the cell

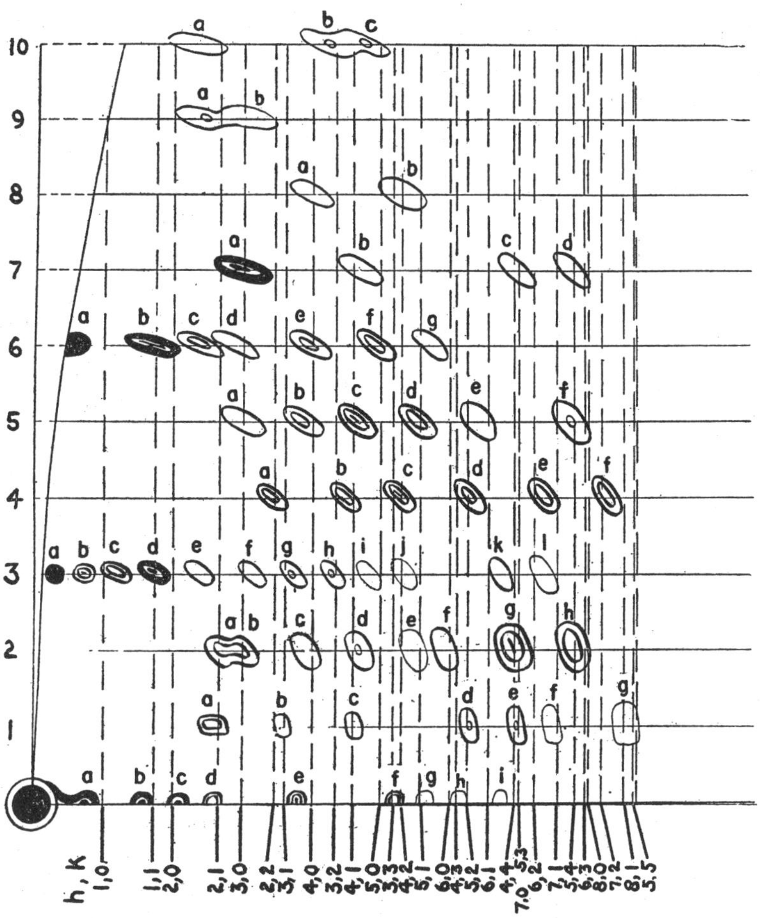

TEXT-FIG. 16. Projection of reciprocal net of hexagonal unit cell to show indexing of reflections of Plate 4, Fig. 24. $a = 87$ Å, $c = 68$ Å. Region between y axis and curved line is a forbidden zone (Bernal, 1927).

size would be larger than the particle, but we can make the complementary assumption *that for the scattering of the isolated virus particle referred to a true repeat unit inside the particle, fractional as well as integral indices are admissible for the reflecting planes.* The assumption of fractional indices, daring as it seems, might but for its novelty have been logically deduced from the scattering conditions of the isolated particle. The Bragg theory of diffraction of x-rays by crystals is based on the assumption that the repeat unit is repeated an indefinite number of times. If this is not the case the subsidiary diffraction maxima can no longer be neglected. The theory is here necessarily complicated and has not up till now been attempted largely because no known experimental examples of it existed. It is, however, in many respects similar to that of electron diffraction from very thin crystals. No attempt will be made here to develop it formally, merely to indicate the general manner in which the anomalous x-ray pattern of the virus particles may be explained.

The particle is short only in two dimensions. In the direction of its length there are many repeat units and therefore the diffraction effects corresponding to this direction are normal. In all other directions, however, fractional (*hki*) indices are observed. Table IV and Text-fig. 16[3] show the allocation of these. Owing to the great number of possible planes for which $\xi > 0.1$ and the intrinsic lack of sharpness in this region only the inner reflections can be indexed with certainty, and most of these occur on the equator and on the third and sixth layer lines. Some characteristic regularities can be noticed but have not as yet been explained. Thus, on the equator no reflections with integral indices appear. Reflections both with integral indices and fractional ones of the type $(hh\overline{2h}l)$ occur on the third and sixth layer lines. The integral reflections here are strong and the fractional weak. It would appear as if each true reflection was forbidden on the equator and replaced by a doublet of fractional indices, while on the layer lines the true reflections appeared with two non-integral satellites forming a triplet. One equatorial reflection in particular, the strong reflection at $\xi = 0.0165$ (93 Å) should be noticed as *it corresponds to the Fresnel scattering of the particle as a whole.* The second order interferences are found only for small values of ξ and here the uncertainty of measurement makes it difficult to identify them with certainty.

Cell Size and Molecular Weight

If the explanation of the observed reflections given above be accepted, the apparent great complexity of the pattern is due to the diffraction effects not conforming with Bragg's law, owing to the finite size of the particles. It follows that in the discussion of the actual structure these effects can be disregarded in the first place, and an ideal repeat unit chosen which corresponds to the disposition of scattering matter inside each particle. This proves to be the cell originally chosen, a = 87 Å, c = 68 Å, which would provide exactly three cells in a hexagonal particle cross-section (Text-fig. 17). This is in excellent agreement with the particle diameter of 150 Å = $87\sqrt{3}$ derived from the intermolecular measurements on the dry gel. For such a cell assuming a density of 1.34 (Bawden and Pirie, 1937 *a*) the molecular weight would be 370,000. On account of the strength of the sixth layer line and the marked pseudo-rhombohedral character of the reflections, the true repeat units may well be one-sixth of this or 62,000. Assuming 6 per cent of this is nucleic acid this

[3] The text-figures for Parts I, II, and III are numbered consecutively.

gives a molecular weight of 58,000, which is of the same order as that of a protein molecule, though it does not fit well into any of the Svedberg classes (Svedberg, 1939). It must be emphasized that this is a cell molecular weight and not the "molecular weight" of the gross virus particle which is, of course, very much greater.

Internal Structure

It is difficult on the basis of the complicated x-ray evidence to go much deeper into the structure of the molecule. Certain deductions may, however, legitimately be made. The two most noticeable features of the pattern are first, the absence of strong reflections on the equator which shows that the particles

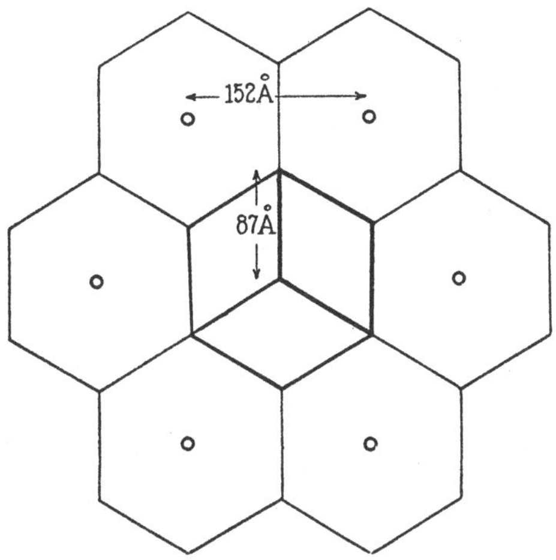

TEXT-FIG. 17. Diagrammatic cross-section of dry gel showing the relation of the unit cell of the particle with that of the interparticle packing.

cannot be composed of bundles of fibrous molecules and, second, the presence of extremely strong reflections perpendicular to the axes, indicating some form of layer structure. Further, the pattern is clearly differentiated into that of the zero, third, and sixth layer lines, and that of the other layer lines. This together with the absence of $(10\bar{1}l)$ and $(20\bar{2}l)$, while $(0003l)$ and $(11\bar{2}l)$ are strongly represented, suggests a rhombohedral structure, a = 55 Å, $\alpha = 104\frac{1}{2}°$.

The presence of the planes $(30\bar{3}2)$ and $(31\bar{4}3)$ would appear to contradict this, but these may be anomalous reflections, or the structure may be only pseudorhombohedral with a triad screw axis.

Strong reflecting planes would appear to indicate sub-units; the larger sub-unit corresponding to the planes (0003) and $(11\bar{2}l)$ would be a flat platelet of dimensions 44 × 44 × 22 Å, the smaller corresponding to the strongest planes of all, (0006), and $(44\bar{8}2)$ and the ring of planes at spacings of about 11 Å would be a roughly equidimensional sub-unit of diameter 11 Å. The larger

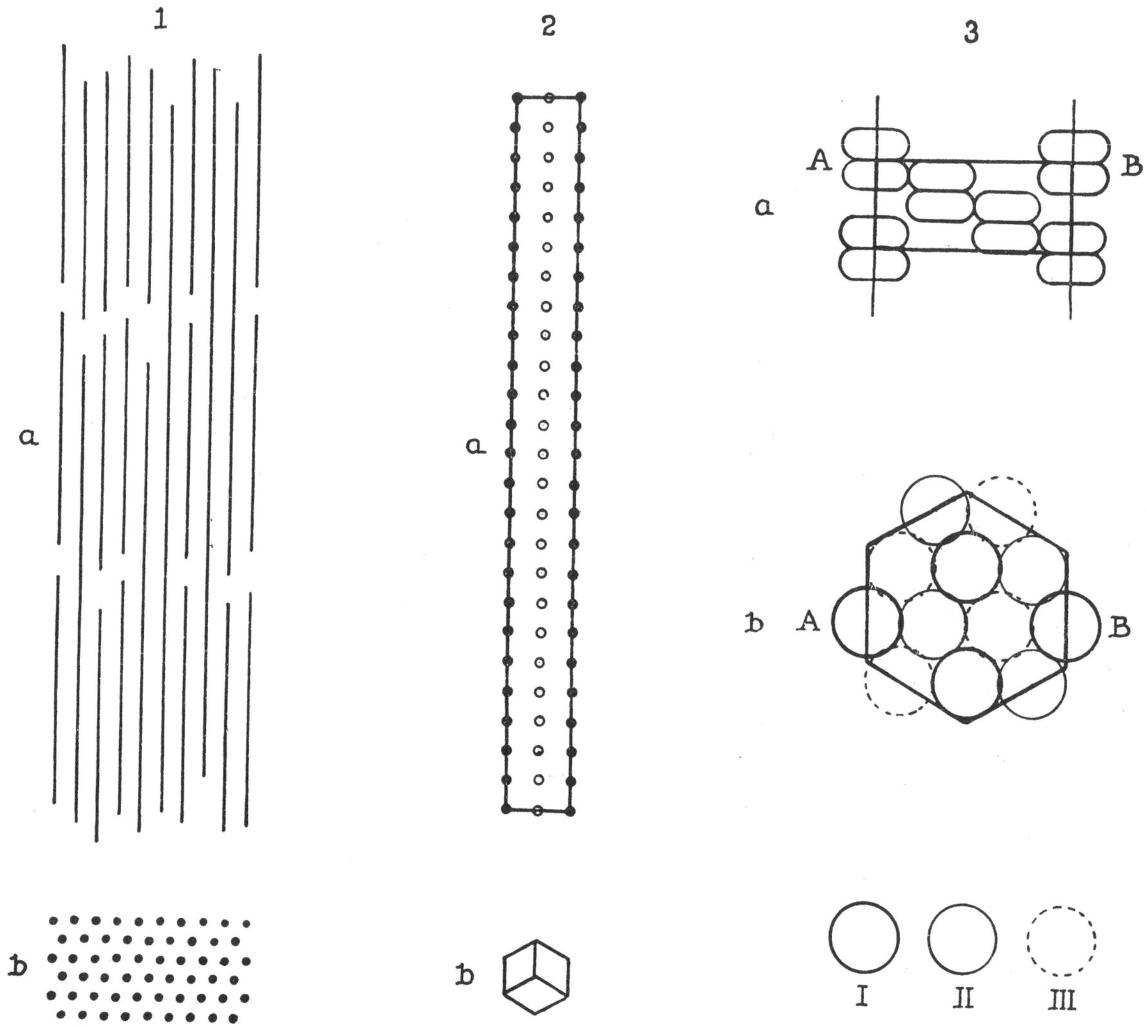

TEXT-FIG. 18. Diagrams to illustrate tobacco mosaic virus structure seen at different magnifications. (In each case the view *above* is normal to the particle length, *below* along the particle length.)

1. Arrangement of particles in dry gel. *a* side view; *b* end view. 1 cm. = 600 Å.

2. Single virus particle showing subdivisions into crystal unit cells. ●—●—●—● indicates *c* axis repeat unit of 68 Å. *a* side view; *b* end view. 1 cm. = 200 Å.

3. Cubic close packing of large sub-units as seen in the three virus unit cells making up one particle. *a* side view; *b* end view. 1 cm. = 67 Å. *a* shows a section at AB of the view below (*b*). I, sub-unit at bottom of cell; II, sub-unit at ⅔ of unit cell height; III, sub-unit at ⅓ of unit cell height.

sub-units could be arranged in cubic close packing, as shown in Text-fig. 18. Further confirmation is furnished by the photographs of potato virus X (Plate 3, Fig. 21). Here we have the same strong reflection at 11 Å, but there are also reflections at 33 and 16.5 Å, suggesting that here also the 11 Å sub-units are present but that the larger sub-units are 44 × 33 Å and are arranged in hexagonal rather than cubic close packing.

This is as far as it is safe to go on the basis of the experimental material available. There is no doubt, however, that with the help of more detailed photographs and comparisons with further virus types, it should be possible to obtain much more detailed information on the structure.

DISCUSSION

The evidence so far supports the view that the virus particle is in some sense intermediate between a molecule and a crystal of a normal crystalline protein. Its lateral dimensions are of the order of six times those of a protein molecule, while in the direction of its length, it contains a multiple of 24 true or 144 pseudocells, almost as many as a small protein crystal (Text-fig. 18 *b*). In detail the intramolecular pattern shows interesting resemblances and differences from those of the proteins that have hitherto been studied. Like them it shows the dominant 11 and 4.5 Å spacings, indicating the same basic internal structure (Bernal, 1939). It differs, however, from them by the fact that one of the 11 Å reflections, that perpendicular to the particle length predominates over all the others. This indicates either sheets or sets of parallel fibres with this lateral spacing, or, what comes to much the same thing, a high degree of alignment of the smallest sub-units. To a certain degree it resembles the fibrous and denatured proteins in this repeat and may like them have a simpler structure than that of the crystalline proteins.[4] Such a deduction would fit with the view that the virus was in some way simpler than other proteins, but whether because it was more primitive than the crystalline, or is a degradation product of them it is of course impossible to say at present. Astbury and Bell (1938) and Astbury (1938 *a*, *b*) have suggested that the hypothetical virus sub-units of 22 Å cube are each linked with one nucleotide of the nucleic acid chain or ring. This fits reasonably with the analytical figures for phosphorus content but as yet there is no direct evidence for the view. The answer to these questions must be found in a far wider study of proteins and nucleoproteins by x-ray methods.

2. *Biological Implications*

So far the discussion has been confined to an elucidation of the modes of aggregation and the internal structure of virus particles, based primarily on an intensive study of tobacco mosaic virus, with its two selected strains, enation and aucuba which are except in the respects mentioned below indis-

[4] More recently the structure of horse serum albumin has been studied by one of us (Fankuchen, 1940) and this seems to offer some analogy to that of tobacco mosaic virus. It is hexagonal, $a = 96$ Å, $c = 144$ Å the main reflection being the (0006) at 24 Å.

tinguishable. The other viruses, cucumber virus strains, potato virus X strains, and bushy stunt, were studied less intensively and can only be discussed here in a comparative way.

Physical Properties

In physical properties the different virus preparations vary widely (see Bawden and Pirie, 1937 *a* and *b*; 1938 *a*, *b* and *c*). The differences that we observed refer only to the preparations of wet and dry gel. In this respect the behaviour of the three strains of tobacco mosaic virus was indistinguishable, all gave gels of similar texture and birefringence, but it was observable that these were more easily obtained in a well oriented form from the ordinary strain.

Cucumber virus strains gave gels of lower birefringence for wet and for dry gel, and these were definitely not so well oriented. Potato virus X strains gave even worse gels of very low birefringence and poor orientation. No orientation phenomena were observable for bushy stunt virus, and the gels it gave were amorphous.

Intramolecular Patterns

All the strains of tobacco mosaic virus gave indistinguishable intramolecular patterns, as far as the position and intensities of the four principal spots, (0003), (0006), (44$\bar{8}$2), and (30$\bar{3}$7), are concerned (see p. 148). Enation and the ordinary strain in solution also gave the same full detailed patterns. The aucuba strain and the cucumber virus were not examined in such detail. We may, however, conclude that they contain the very similar sub-units and particle structure. Potato virus X showed a different intramolecular pattern, (Plate 3, Fig. 21), which, however, was clearly related to that of the tobacco virus. The reflection corresponding to (0006) in tobacco mosaic virus remains as the strongest at 11 Å, but (0003) is replaced by two spots of spacing 33 and 16.5 Å representing the first and second orders of which (0006) is the third. (31$\bar{4}$0) and (44$\bar{8}$2) are replaced by arcs of approximately the same spacing, but it is impossible to determine their precise position owing to the poor orientation of the specimens. No difference was detectable between the ordinary and the masked strains of potato virus X. The most obvious interpretation at this stage is that the virus contains similar sub-units, but that these are arranged in a different way to those of tobacco and cucumber virus. In the case of the bushy stunt virus the distinction established between intermolecular and intramolecular patterns breaks down. As the crystals here are true crystals all x-ray reflections must conform to Bragg's law for the crystal unit, which here corresponds to the intermolecular pattern observed for the other viruses. Powder photographs at short distances, however, do show definite concentrations of scattering at about 11 and 4.5 Å, of which no more can be said than that they are typical of proteins in general and not unlike those given by the other viruses. It may be possible to get more information on this point by examining single crystals, though so far these have proved too small for the purpose.

Intermolecular Patterns

The different varieties of virus showed distinct differences in their intermolecular patterns. These differences are of two kinds, namely, in the spacing of the pattern and in the relative intensities of the different spots. Within the accuracy of measurement of about one-fifth per cent, all strains of tobacco mosaic virus give the same dry gel spacing. It must be emphasized that this reproducibility is one of the most notable characteristics of the virus preparations. Centrifuged sap, dried top and bottom solutions of different concentrations, and dried crystals prepared from different samples and at different times, give identical spacings to this degree of accuracy. Both varieties of cucumber virus give a different spacing though the difference is small, 128 Å against 132 Å for the tobacco mosaic strains.

The patterns given by the bushy stunt virus are necessarily of a very different kind. The virus was prepared in the form of small crystals which on microscopic examination appeared to be regular dodecahedra, which are completely isotropic (Bawden and Pirie, 1938 *b*). This indicates the probability of a spherical rather than an elongated particle. For the x-ray examination powder methods were used, the virus being examined first of all in its mother liquor and subsequently air dried (Bernal, Fankuchen, and Riley, 1938). The results of the measurements are shown in Table I. They correspond to those of a body-centred cubic structure. This is in perfect agreement with the habit of the crystals, for according to Federov's law the rhombohedral habit corresponds to body-centred packing. The size of the cell in the wet case is 394 Å and in the dry case 318 Å. Their ratio corresponds within errors of measurement to a shrinkage observed on the crystals themselves. Using this cell size and the observed density of 1.286 for the wet crystals, the wet particle molecular weight would appear to be 24,000,000. Bawden and Pirie (1938 *b*) estimated the water content of these crystals to be 55 per cent and if this is the case the dry particle molecular weight would be 10,800,000 compared to 8,800,000 the molecular weight measured by McFarlane and Kekwick (1938). More recently, Neurath and Cooper (1940) have reported a molecular weight of 10,600,000, based on sedimentation and diffusion measurements. It has so far proved impossible to measure the density of the dry crystals, and thus to measure the dry molecular weight directly. The value given above would indicate a density of only 1.12 for the dry crystals, which would appear to show that there is considerable zeolitic loss of water. The diameter of the dry particle would appear to be 276 Å or much greater than that of the tobacco mosaic type.

The other viruses did not provide any recognisable intermolecular pattern, though, particularly in the case of potato virus X, exhaustive search was made for such a pattern. This may be due to the larger size of the virus particles or to their irregular shape. Further work is obviously needed here.

The other type of difference between intermolecular patterns of the other virus strains is that of intensity. Plate III, Fig. 18 and Table I show the different reflections of tobacco and cucumber viruses and their estimated intensities. All these were taken from specimens prepared in an identical way and examined in the same orientation. The three strains of the tobacco virus do show definite differences of intensity. In view of the discovery of the secondary orientation of the particles it might seem that the differences are due to this cause, and, indeed, to a certain extent this may be so, the tendency to orientation being possibly more marked in the normal tobacco mosaic strains than in the aucuba and enation strains. This, however, cannot account for the increased intensity of the reflections from the $(12\bar{3}0)$ planes, which, occurring at intervals of approximately 30°, are comparatively independent of secondary orientation. The cucumber virus shows this difference in a far more marked form, the unresolved doublets of planes $(22\bar{4}0)$ and $(31\bar{4}0)$, which is only visible in tobacco after heavier exposures, are here prominent. The specimens of tobacco mosaic virus prepared in 1938 were found to give a pattern which in intensity was very different from that of the earlier preparations. Further this pattern did not itself remain constant (Table I). What is notable is the increase in intensity of the $(22\bar{4}0)$ and the variability of $(12\bar{3}0)$. It was thought that these changes might be due to ageing and some of the specimens were sent back to Mr. Bawden for infectivity tests but were found to behave quite normally. It is possible that these differences are not due to changes in the particle itself, but merely because of its association with water. The line intensity is in fact very sensitive to the amount of water present. An old specimen of tobacco mosaic virus which when dried *in vacuo* gave an excellent four line pattern with $(1\bar{1}00)$ most intense and interparticle distance of 150 Å, gave on exposure to air on a damp day an extremely poor pattern with $(12\bar{3}0)$ strong and $(1\bar{1}00)$ weaker and interparticle distance of 153 Å. Attempts were made to estimate these differences quantitatively by photometer records of the film densities. Owing to the orientation effect, however, the relative intensities can only be used for comparison purposes.

It is difficult to assess the significance of the differences in the intermolecular patterns, but they are probably due to slight differences in the shape of the cross-section and of its association with water. Their full elucidation requires further work in conjunction with a closer study of their intramolecular patterns. They have some value, however, for comparative purposes.

Classification

The comparison between the intramolecular structure of tobacco, cucumber, and potato viruses points clearly to a common pattern. Already, on the basis of the x-ray analysis alone, it is possible to arrange the specimens we have examined on a scale of similarity and differences which is shown in Table I. The three varieties of tobacco mosaic virus which only show slight differences in intensity in their intermolecular patterns and none at all in their intramolecular, might be counted as varieties in a species. The related species would be that of cucumber mosaic virus, which shows marked differences in the intermolecular pattern, not only in intensity but also in spacing. A different genus would be represented by the potato virus, where the intramolecular pattern indicates a different arrangement of similar sub-units. Still another would be given by the bushy stunt virus, in which although the intramolecular pattern is roughly the same in character, no orientation phenomena can be observed. All these virus specimens together show patterns which have more similarities amongst each other than with other proteins though it should be pointed out that those examined so far by x-rays are a small and unrepresentative sample.

These similarities and differences are seen to run closely parallel to those obtained by Bawden and Pirie (1937 *a* and *b*; 1938 *a* and *b*; 1939) on clinical and serological grounds, and it may be hoped that the x-ray method may prove of value in explaining the structural characters underlying the differences between strains and species of virus and similar bodies.

General Conclusions

The provisional picture that we can now give of plant virus particles has already modified earlier conclusions in several respects. The virus that has been most thoroughly studied, that of the tobacco mosaic disease, seems to have very elongated particles of variable length, which we now recognise are in part artefacts. Virus particles in the plant are definitely shorter and possibly all of the same length. In bushy stunt virus, which gives cubic crystals, they must be approximately spherical. The anisotropic character of the other viruses, so interesting from the colloid chemical standpoint and so useful for their x-ray examination, seems now to have been somewhat of a happy accident, brought about by the tendency of cylindrical molecules to aggregate along their length. It must be emphasized, however, that if this is the case, all trace of the length of the original particles may have been lost in the resulting elongated particle.

The x-ray evidence points to a virus particle of complex structure. It is more analogous to a protein crystal than a protein molecule. Its internal regularity is very considerable. The crystallinity of the virus goes further than the aggregation of similar particles that produce the visible crystals or liquid crystals, it is a property of the isolated particles of the virus.

The arrangement seems to be one of small sub-units, 11 × 11 × 11 Å definitely of a smaller order than those of crystalline protein molecules. There are indications that the nucleoprotein of the particle has a different internal structure from that of normal crystalline proteins and that it is in some ways simpler.

Biological Significance

This is not the place to discuss the biological significance of these observations. Until more is known of the structure and biochemical functions of systems such as bacteria, whose chemical activities are usually taken as criteria of life, the crystalline nature of the particles that we have studied cannot in itself be given a biological significance nor give an answer to the question as to whether they are or are not the infective agent (Bawden and Pirie, 1937 *a*), nor to the far more metaphysical question as to whether they are to be considered living organisms (Pirie, 1937). We would moreover like to emphasize that the only fully reliable facts that are here established are the physical and x-ray measurements. The hypotheses used to explain them are more or less provisional and it would be premature to base elaborate or detailed interpretations of structure and function on the evidence available at this stage of the work.

The chief utility of this work up to the present has probably lain in the extension to the biological field of two new methods, the technique of the study of oriented aggregates and the x-ray examination of complex structures of very long spacings. We have shown that this latter technique is by no means difficult and we may therefore hope that many other such structures will be discovered by its means. Effectively by the use of ordinary x-rays at small angles we can measure any regularities of the order of 1000 Å or 1/10 μ and this is close to the lower resolvable limit of the ultraviolet microscope and has already been reached by the electron microscope. It should further be possible to push the examination of fine structures from that obtainable by the microscope right down to atomic dimensions.

The study of the virus preparations by these methods has brought to light hitherto unsuspected or at least undemonstrated properties of colloidal aggregates. It has been demonstrated that linear colloidal particles are arranged, even in dilute solution, in aggregates as regular as those of a crystal. The distance between them can be varied with the ionic character of the medium and this proves the existence of long range interparticle forces and indeed provides quantitative confirmation of the theories of these forces. Further by combining the x-ray studies with optical observations it has been possible to build up a theory of gel formation and thixotropy that may prove of value. The viruses of the tobacco mosaic type owing to their extreme uniformity and stability are indeed an ideal material for the controlled studies of colloids. Finally, it must be emphasized that with such a bewildering set of new phenomena, the work here recorded represents only a preliminary and rough survey and that many more man-years of work will be required before exact and reliable interpretations can be expected.

We are greatly indebted to Mr. N. W. Pirie and Mr. F. C. Bawden for supplying us with the virus preparations used in this work and to Mr. C. Chapman for the construction of the apparatus. Miss F. Bell took some of the earliest intramolecular photographs of the virus preparations. Dr. Crowfoot was of the greatest assistance during the final revision of the paper and was responsible for the preparation of many of the illustrations. A grant from the Royal Society made possible the purchase of the x-ray apparatus at Birkbeck College, and one of us (I. Fankuchen) has been assisted by a grant from the Department of Scientific and Industrial Research.

SUMMARY

These papers give an account of an optical and x-ray examination of preparations of plant virus substances isolated by Bawden and Pirie, in particular of those of tobacco mosaic disease. They open with a historical survey of the work, indicating the order in which new phenomena were discovered. The subsequent treatment is divided into three parts:

I. Introduction and preparation of specimens.

II. Modes of aggregation of virus particles.

III. (1) The structure of the particles.
(2) Biological implications.

Part I, after an historical introduction, describes the method of preparation, from solutions of the virus, of optically oriented specimens of different concentrations. For their examination special x-ray apparatus was developed, in particular cameras working with very low angles and capable of indicating spacings up to 1000 Å.

Part II is concerned with the phenomena of aggregation of the virus particles in solutions. Two modes of aggregation are observed, a disoriented sol at low concentrations and an oriented sol passing continuously over to an oriented gel at higher concentrations. Solutions of medium strength divide into two layers on standing. The top layer, which is more dilute, *i.e.* of concentrations below 1.6 to 4 per cent is an isotropic liquid which shows flow birefringence and other phenomena indicating the presence of long thin particles. The bottom layer in equilibrium with it, and also the higher concentrates prepared by centrifugation, are spontaneously doubly refractive and consist of regions in which the particles are parallel. X-ray investigation shows that the particles are also equidistant and that the sol is a new kind of liquid crystal with a regular hexagonal arrangement in cross-section. The distances between the particles depend simply on the concentration, indicating the homogeneity of their distribution. The equilibria between top—and bottom—layer solutions and the special forms which the latter can exhibit—tactoids and reversed spirals—are also discussed. Preparations with concentrations of 30 per cent and over have gel-like properties and become stiffer as the water content decreases, but there are no abrupt transitions. At pH 7 in the absence of salt these gels swell until a concentration in equilibrium with the top layer is reached, that is down to concentrations of 4 per cent–1.6 per cent. Nearer the isoelectric point, at pH 3.4, or in ammonium sulfate solution, the degree of swelling is much more limited and an equilibrium is reached in which the distance between the particles depends on the pH or the salt concentration. The minimum interparticle distance at the isoelectric point or for high salt concentrations is about 175 Å. The original "crystals" observed by Stanley are shown to be gels of this type in the form of small elongated tactoids. On drying, further shrinkage takes place and the resulting air dry gel has an interparticle distance of only 152 Å. In these gels the arrangement of particles in the cross-section is so perfect that each specimen is a two-dimensional single crystal. The conclusion of this study is that the virus preparations consist of approximately cylindrical particles about 150 Å in diameter. Their length cannot easily be estimated by x-ray methods, but the minimum value of 1,500 Å found by Kausche, Pfankuch, and Ruska (1939) is compatible with the x-ray evidence. Much longer particles formed by aggregation of these also occur. The forces maintaining the particles equidistant and parallel in these gels are probably due to the ionic atmospheres surrounding them. It is shown that for long particles the mutual energy due to the ionic atmospheres is considerably greater than their temperature motion at ordinary temperatures, and consequently they must arrange themselves parallel and in a regular two-dimensional lattice. Further consideration of these phenomena leads to an explanation for the other colloid phenomena shown by the viruses, particularly the formation of tactoids and gels. The spindle-shaped form of the tactoids is shown to be due to anisotropic surface tension due to the parallel arrangement of the long particles on the surface. Dilute gels appear to be composed of tactoids of a multiply branched character showing thixotropic behaviour. This theory may have wide extensions to other colloid systems.

In Part III, Section 1 deals with the x-ray evidence on the internal structure of the particles. Even in solution, they have an inner regularity like that of a crystal. Virus preparations are thus in a sense doubly crystalline. Closer analysis reveals that the x-ray patterns are not directly comparable to those of a crystal as many of the reflections do not obey Bragg's law, but can be understood on the theory of gratings of limited size. The structure seems to consist of sub-units of the dimensions of approximately 11 Å cube, fitted together in a hexagonal or pseudohexagonal lattice of dimensions—a = 87 Å, c = 68 Å. Contrary to what earlier observations seemed to indicate, the particle seems to be virtually unchanged by drying and must therefore contain little water. There are marked resemblances with the structure of both crystalline and fibrous protein, but the virus structure does not belong to any of the classes hitherto studied. There are indications that the inner structure is of a simpler character than that of the molecules of crystalline proteins.

Part III, Section 2 contains a comparative study of the optical and x-ray examinations of three strains of tobacco mosaic virus, two of cucumber disease virus, two of potato virus X, and the virus of bushy stunt disease of tomato. In the last case x-ray measurement confirmed the deduction from its cubic crystal habit that it was composed of spherical rather than long particles, and showed that these had a diameter when dry of 276 Å and were arranged in a body-centred cubic close packing. This single example is sufficient to show that the elongated particle form which gives rise to all the anomalous physical properties of the other viruses studied is of no essential biological importance. The similarity and differences observed between the physical properties of these preparations run closely parallel to their clinical and serological classification. Finally, the biological implications of these results are discussed together with possible applications of the new methods of examination to the study of colloid and biological problems.

REFERENCES

Astbury, W. T., *Proc. Roy. Soc. London, Series B*, 1938*a*, **127,** 31; Cold Spring Harbor symposia on quantitative biology, Cold Spring Harbor, Long Island Biological Association, 1938*b*, **6,** 359. Astbury, W. T., in Luck, J. M., Annual review of biochemistry, Stanford University, Annual Reviews, Inc., 1939, **8,** 125.

Astbury, W. T., and Bell, F., Cold Spring Harbor symposia on quantitative biology, Cold Spring Harbor, Long Island Biological Association, 1938, **6,** 109.

Bawden, F. C., and Pirie, N. W., *Proc. Roy. Soc. London, Series B*, 1937*a*, **123,** 274. *Brit. J. Exp. Path.*, 1937*b*, **18,** 275; 1938*a*, **19,** 66; 1938*b*, **19,** 251. *Tabulae Biologicae*, 1938*c*, 355. *Brit. J. Exp. Path.*, 1939, **20,** 322.

Bawden, F. C., Pirie, N. W., Bernal, J. D., and Fankuchen, I., *Nature*, 1936, **138,** 1051.

Bawden, F. C., and Sheffield, F. M. L., *Ann. Appl. Biol.*, 1939, **26,** 102.

Beale, H. P., *Contrib. Boyce Thompson Inst.*, 1937, **8,** No. 5, 413.

Bechhold, H., and Schlesinger, M., *Phytopath. Z.*, 1933, **6,** 627.

Bernal, J. D., *Proc. Roy. Soc. London, Series A*, 1927, **113,** 117. *J. Sc. Inst.*, 1928, **5,** 241. Perspectives in biochemistry, Cambridge University Press, 1937, 45. *Proc. Roy. Soc. London, Series B*, 1938, **125,** 299. *Nature*, 1939, **143,** 663.

Bernal, J. D., and Fankuchen, I., *Nature*, 1937, **139,** 923.

Bernal, J. D., Fankuchen, I., and Perutz, M., *Nature*, 1938, **141,** 522.

Bernal, J. D., Fankuchen, I., and Riley, D. P., *Nature*, 1938, **142,** 1075.

Best, R. J., *Nature*, 1937*a*, **139,** 628; 1937*b*, **140,** 547.

Eirich, F., and Simha, R., *Monatsh. Chem.*, 1937, **71,** 2, 67.

Eriksson-Quensel, I. B., and Svedberg, T., *J. Am. Chem. Soc.*, 1936, **58,** 1863.

Fankuchen, I., *Nature*, 1937, **139,** 193; 1940, unpublished data.

Freedericksz, V., and Zolina, V., *Tr. Faraday Soc.*, 1933, 919.
Friedel, G., and Friedel, E., *Z. Krist.*, 1931, **79**, 1.
Hermann, C., *Z. Krist.*, 1931, **79**, 186.
Hills, and Vinson, S. G. *Mississippi Agric. Exp. Station Research Bull.*, 1938, **286**, 18.
Van Iterson, G., *Proc. Akad. Wetensch. Amsterdam*, 1934, **37**, 367.
Kausche, G. A., Pfankuch, E., and Ruska, H., *Naturwissenschaften*, 1939, **27**, 292.
Kausche, G. A., *Naturwissenschaften*, 1939, **27**, 77.
Langmuir, I., *J. Chem. Physics*, 1938, **6**, 873.
Lauffer, M., *J. Physic. Chem.*, 1938*a*, **42**, 935. *J. Biol. Chem.*, 1938*b*, **126**, 443. *Science*, 1938*c*, **87**, 469. *J. Am. Chem. Soc.*, 1939, **61**, 2412.
Lauffer, M., and Stanley, W. M., *J. Biol. Chem.*, 1938, **123**, 507. *Chem. Rev.*, 1939, **24**, 303.
Levine, S., *Proc. Roy. Soc. London, Series A*, 1939*a*, **170**, 145; 1939*b*, **170**, 165. 1939*c*, unpublished work.
Loring, H. S., Lauffer, M. A., and Stanley, W. M., *Nature*, 1938, **142**, 841.
Marton, L., *J. Bact.*, 1941, **41**, 408.
McFarlane, A. S., and Kekwick, R. A., *Biochem. J.*, London, 1938, **32**, 1607.
Mehl, J. W., Cold Spring Harbor symposia on quantitative biology, Cold Spring Harbor, Long Island Biological Association, 1938, **6**, 218.
Neurath, H., Cold Spring Harbor symposia on quantitative biology, Cold Spring Harbor, Long Island Biological Association, 1938, **6**, 196.
Neurath, H., and Cooper, G. P., *J. Biol. Chem.*, 1940, **135**, 455.
Neurath, H., and Saum, A. M., *J. Biol. Chem.*, 1938, **126**, 435.
Pirie, N. W., Perspectives in biochemistry, Cambridge University Press, 1937, 11.
Robinson, J. R., Dissertation, Cambridge, 1938. *Nature*, 1939*a*, **143**, 923. *Proc. Roy. Soc. London, Series A*, 1939*b*, **170**, 519.
Staudinger, H., *Helv. Chem. Acta*, 1932, **15**, 213.
Stanley, W. M., *Science*, 1935, **81**, 644. *Am. J. Bot.*, 1937, **24**, 59. *J. Physic. Chem.*, 1938, **42**, 55.
Svedberg, T., *Proc. Roy. Soc. London, Series B*, 1939, **127**, 1.
Takahashi, W. N., and Rawlins, T. E., *Proc. Soc. Exp. Biol. and Med.*, 1932, **30**, 155. *Science*, 1933*a*, **77**, 26; 1933*b*, **77**, 284; 1937, **85**, 103.
Wyckoff, R. W. G., *J. Biol. Chem.*, 1937*a*, **121**, 219; 1937*b*, **122**, 239.
Wyckoff, R. W. G., and Corey, R. B., *J. Biol. Chem.*, 1936, **116**, 51.
Wyckoff, R. W. G., Biscoe, J., and Stanley, W. M., *J. Biol. Chem.*, 1937, **117**, 57.
Zocher, H., *Z. anorg. Chem.*, 1925, **147**, 91.
Zocher, H., and Jacobsohn, K., *Kolloid Beihefte*, 1929, **28**, 167.

Plate 3

Inter- and intramolecular x-ray photographs of various virus preparations.

Intermolecular photographs—Figs. 15–18.

FIG. 15. Showing shift of intermolecular reflections on passing from (lowest film) dry gel to (middle film) wet gel and (uppermost film) 13 per cent solution.

FIG. 16. Effect of concentration of virus on spacing.

(*a*) 13 per cent solution. Unfiltered Cu radiation

(*b*) 23 per cent solution. Unfiltered Cu radiation

(*c*) Top half, 34 per cent solution. Unfiltered Cu radiation
Bottom half, 34 per cent solution. Filtered Cu radiation

(*d*) 34 per cent solution. Monochromatic Cu K_α radiation

(*e*) 50 per cent gel. Unfiltered Cu radiation

FIG. 17. Enation mosaic virus solution photographed with monochromatic Cu K_α radiation showing extremely sharp lines due to perfect orientation and arrangement.

FIG. 18. Intermolecular pattern of dry gels of different virus proteins. (*a*) tobacco mosaic virus. (*b*) enation. (*c*) aucuba. (*d*) cucumber.

Plate distances: Figs. 15, 16, 17—40 cm.; Fig. 18—16 cm.

Intramolecular photographs—Figs. 19–21.

FIG. 19. Typical early dry gel photograph of tobacco mosaic virus.

FIG. 20. Cucumber mosaic virus.

FIG. 21. Potato virus X (shorter plate distance). Note the two lines 2 and 4 inside the strong meridian line 6 instead of the single line 3 on tobacco and cucumber virus photographs.

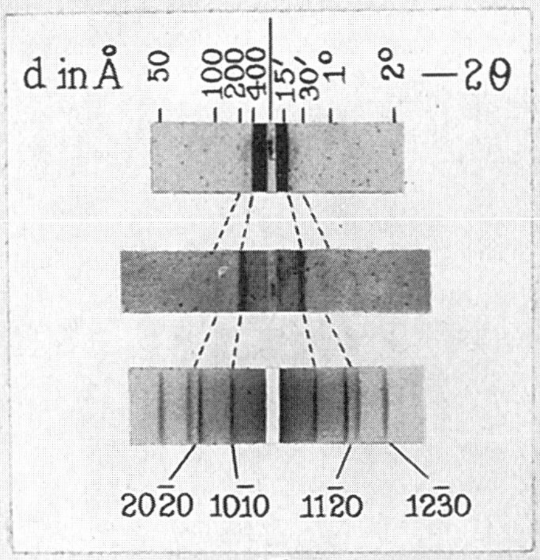

15

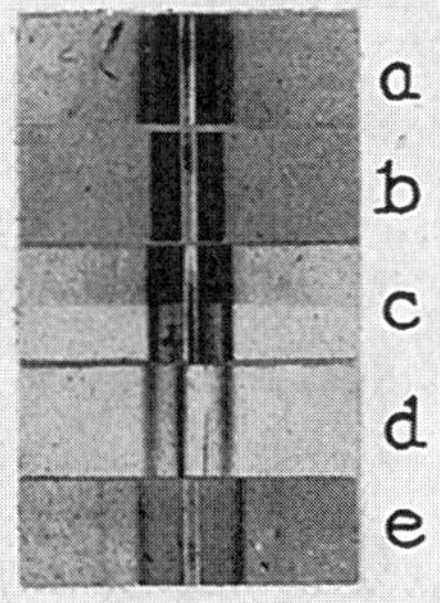

16

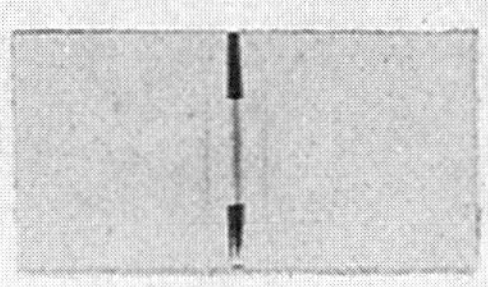

17

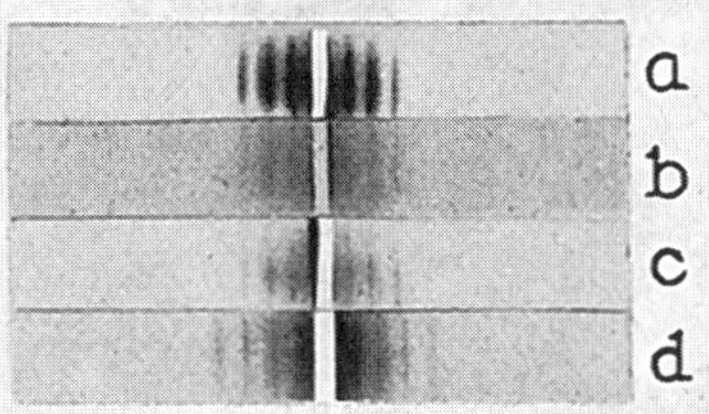

18

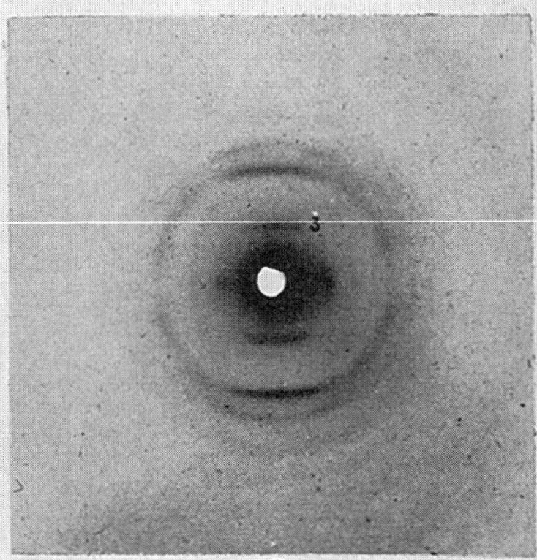

19

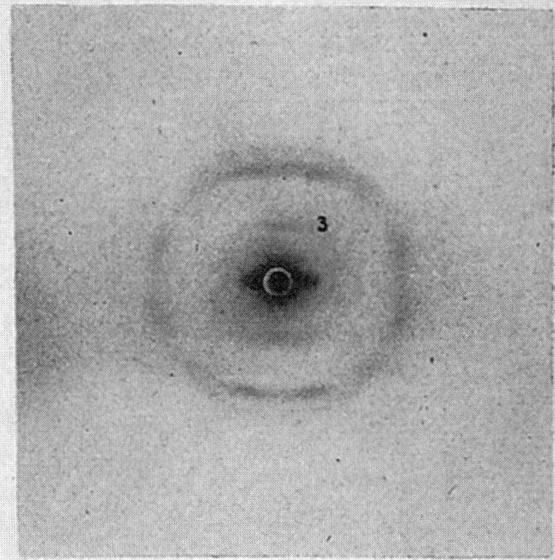

20

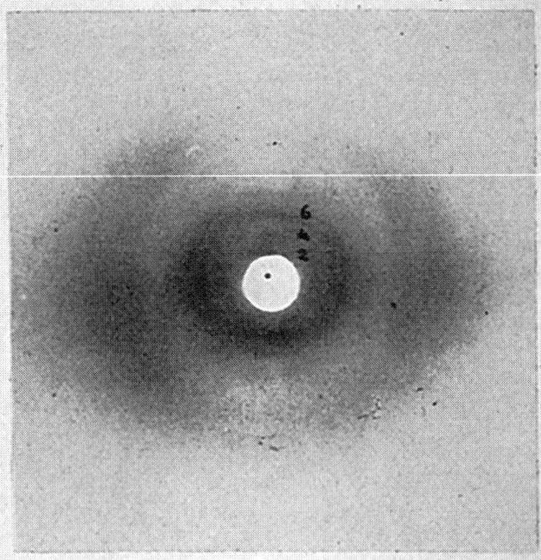

21

(Bernal and Fankuchen: X-ray and crystallographic studies of plant virus. I, II)

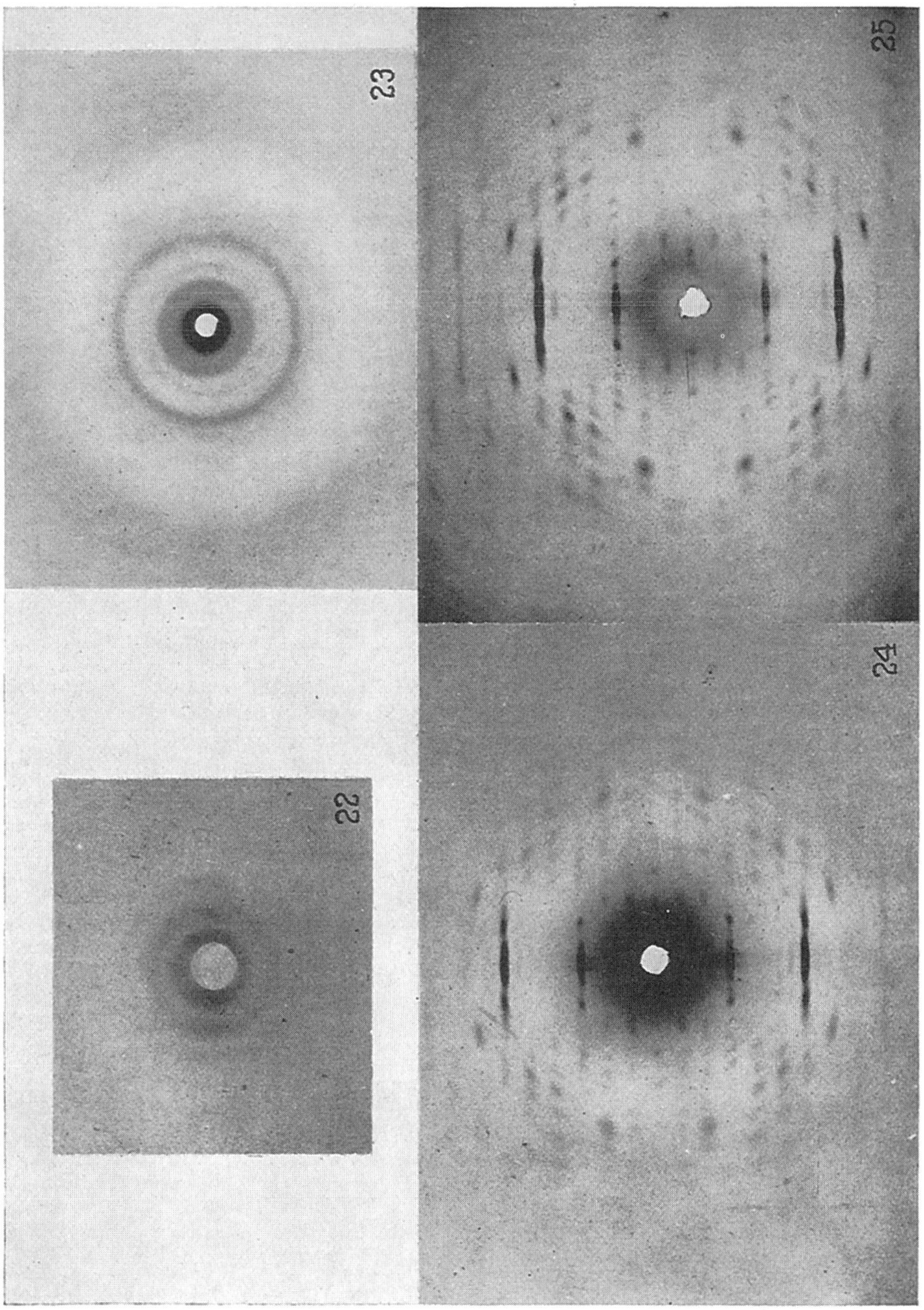

(Bernal and Fankuchen: X-ray and crystallographic studies of plant virus. I, II)

PLATE 4

X-ray photographs of tobacco mosaic virus.

FIG. 22. Intermolecular pin-hole photograph (plate distance 30 cm.) of small piece of dry gel mounted with x-rays along particle direction showing orientation relative to surface (vertical in photograph) and hexagonal single crystal character. It can be seen that the $(11\bar{2}0)$ and not the $(1\bar{1}00)$ planes lie parallel to the surface (air-water interface in drying cell (Text-fig 2).

FIG. 23. Intramolecular pin-hole photograph (plate distance 6 cm.) of the same piece showing completely random orientation of individual particles.

FIG. 24. Intramolecular photograph of 36 per cent enation solution in capillary tube (direct print). (Plate distance 14.8 cm.)

FIG. 25. Same as Fig. 24, but printed through a rotating sector disc to show all important reflections of comparable intensity.

The Electron Microscopy of Tobacco Mosaic Virus

Robert W. Horne
Department of Biological Sciences, University of East Anglia, Norwich, U.K.

One of the interesting observations the science historian will note is the link between the construction of the earliest experimental electron microscopes and the structure of tobacco mosaic virus (TMV).

The first images from Ernst Ruska's experimental microscope showing a resolution better than the light microscope were published in a series of papers from 1933 to 1936. The first attempts to apply the electron microscope to biological material were discussed by L. Marton in 1934. For a detailed historical account of these early experiments, the reader is referred to the excellent publication 'The Early Development of Electron Lenses and Electron Microscopy' by Ernst Ruska (1980; translated by T. Mulvey).

Following Marton's paper in 1934, there was much discussion between the biologists and non-biologists concerning whether or not biological material would be reduced to incinerated remains as a result of electron bombardment. Clearly, whole cells were too thick for penetration at electron beam voltages of about 60 to 70 kv. The search for small biological specimens began, including bacteria and possibly viruses.

The early prototypes for a commercial version of Ruska's transmission EM were well under way by 1938-39. For obvious commercial reasons the need to demonstrate the instrument's use in biology became of importance. It was fortunate that Helmut Ruska (Ernst Ruska's brother), together with his colleagues, decided to develop the instrument's application to biology and medicine. The group subsequently published some twenty papers dealing with EM studies of biological objects. These are fully listed in the book by E. Ruska (1980).

One of the interesting publications included the work of Kausche *et al.* (1939), showing electron micrographs of TMV mixed with colloidal gold. The latter was essential to have some small particulate material of high contrast to focus the image. The micrographs, which the writer understands are still in existence, clearly showed short and long rod-like structures of constant width consistent with the diameter of TMV which was established later. However, Kausche *et al.* (1939), described these rod-like or filamentous objects as 'forms of tobacco mosaic virus-molecules', of 15 mμ across and about 150 mμ long. It is of interest to note here that the electron micrographs of TMV rods reproduced in the book by Ruska (1980; Fig. 38), and recorded on one of the first Siemens prototype EM instruments, used by Kausche *et al.* (1939), are quite remarkable.

The very low contrast from the TMV particles, together with other small biological objects, clearly presented serious problems, although Marton (1934) had discussed how improvement could be made in specimen preparation by the addition of osmium tetroxide and fixatives used in optical microscopy (see Ruska, 1980).

It should be mentioned that many biologists at the time were highly skeptical about EM images recorded from biological materials. They dismissed them as artifacts; moreover, any particulate material or structure that could not be seen with the aid of the optical microscope was regarded as being unimportant. But the significance of the publication by Kausche *et al.* (1939) was recognized by a number of post-war physicists and those working in what was then the new field of 'molecular biology'. The direct observations and measurement from the micrographs by Kausche *et al.* (1939) were discussed in their original paper concerning the data derived from sedimentation studies, X-ray diffraction patterns and other physico-chemical methods. Such a reproducible 'artifact' in the form of a precise structure which fitted within reasonable agreement to other independent data was indeed difficult to dismiss as an artifact.

The immediate post-war period saw a continued increase in the use of TMV particles as a sort of test object in biological electron microscopy, as new commercial electron microscopes rapidly became available from Germany, USA, United Kingdom, and several other countries. Added to this was the development of shadow-casting and freeze-drying by Williams and Wyckoff (1946), which allowed the width and length of TMV rods to be measured with greater accuracy. Moreover, frozen-dried TMV, when sprayed as droplets and shadowed, showed the virus rods standing vertically on the support films.

However, the periodicity of 23 Å along the TMV axis deduced from the X-ray diffraction studies eluded the electron microscopists, although the new post-war instruments were capable of a resolving power approaching 15 to 10 Å. In addition, the X-ray data also revealed the presence of an axial hole of 40 Å diameter. This structural feature was not seen in the EM until Huxley (1956) reported the outlining of TMV rods and fragments and penetration of the axial hole by chemical stains.

In 1954 a new breed of high resolution electron microscopes was in production with a routine performance of 10 Å or better. The first Elmiskop I from Siemens was installed in the

Cavendish Laboratory, Cambridge. The defects and limitations in the preparation of biological specimens using existing methods were clearly apparent when studied at very high resolution.

The installation of the first Siemens high resolution microscope during 1954-55 coincided with the arrival of Sydney Brenner, who joined the Molecular Biology Group in the Cavendish Laboratory headed by Max Perutz. Brenner started work on the separation and purification of the T2 bacteriophage protein components. Electron microscopy was to provide an ideal direct visual means of monitoring the methods and final purity of the T2 fractions.

The defects and limitations of the available preparative methods for high resolution EM have been mentioned above and were clearly unsuitable for the phage work. Electron dense staining was one possibility, but this was normally carried out at acid pH which would further degrade the phage components, making interpretation of the structural components impossible. Horne and Brenner carried out a series of experiments involving raising the pH of phosphotungstic acid positive stain to neutrality using potassium hydroxide. The results from examining the intact T2 phage were totally unexpected. The first negatively stained images of the intact phages were recorded on the day the Russians launched Sputnik-I (Friday, October 4th, 1957), showing the particles to be surrounded by or embedded in electron dense stain, rather than the protein positively stained. Moreover, the level of preservation was remarkable and allowed the gradual disruption of the virus and subsequent separation of its components to be monitored in detail (Horne and Brenner, 1958). The applications of negative staining to a wide range of bacterial, plant and animal viruses followed, with TMV as one of the first plant viruses reported by Nixon and Woods (1960), who showed the regular periodicity along the virus axis. Moreover, the negative stain had penetrated the 40 Å central axial hole.

The detail which could be resolved from negatively stained viruses, and other particles possessing periodic structure, provided objects which could be analyzed more accurately and quantitatively by optical diffraction methods. The particles of TMV again presented a test of object structure consisting of precise assemblies which allowed the 23 Å periodicity of the protein helix and the 69 Å repeat along the rod axis to be resolved (Beeston *et al.*, 1973; Fig. 4.8).

References

Beeston, B. E. P., Horne, R. W., and R. Markham. (1973). Electron and optical diffraction techniques. *In*: Practical Methods in Electron Microscopy. Part II. A. M. Glauert, ed. North Holland/Elsevier, Amsterdam.

Horne, R. W. and S. Brenner. (1958). A negative staining technique for high resolution of viruses. IVth Int. Conf. Electron Microscopy, Berlin. Vol. II. p. 625-627. Springer-Verlag, Berlin. (Published 1960).

Huxley, H. E. (1956). Some observations on the structure of tobacco mosaic virus. Proc. Stockholm Conf., Electron Microscopy. p. 260-261.

Kausche, G. A., Pfankuch, E., and H. Ruska. (1939). Die Sichbarmachung von pflanzlichein Virus im Übermikroskop. *Naturwissenschaften* **27**:292299.

Marton, L. (1934). Electron microscopy of biological objects. *Physik. Rev.* 527-528.

Nixon, H. L. and R. D. Woods. (1960). The structure of tobacco mosaic virus protein. *Virology* **10**:157-159.

Ruska, E. (1980). The early development of electron lenses and electron microscopy. (Translated by Thomas Mulvey). S. Hirzel Verlag, Stuttgart.

Williams, R. C. and R. W. G. Wyckoff. (1946). A shadow-casting technique for electron microscopy. *J. Appl. Phys.* **17**:23-28.

Biographical Sketch

Robert Horne was born at Ascot, England in 1923 and educated at Westmount, Montreal, Canada. From 1941-1946, he served in the Royal Air Force in a branch of Intelligence Countermeasures, including working with a joint US/RAF high energy UHF generator system built at Harvard and M.I.T. from 1943-1946. In 1947 he was appointed to work on the development of electron microscopes at the Cavendish Laboratory, University of Cambridge, under Vernon Ellis Cosslett FRS and received the M.A. Degree in 1956. In 1963 he was appointed head of the Electron Microscope Laboratory at the Agricultural Research Council Institute of Animal Physiology, near Cambridge and received the D.Sc. degree in 1965. For his contribution to the structure of viruses he was made an Honorary Fellow of the Royal Microscopical Society. Subsequently he was appointed Head of the Department of Ultrastructure Research and Professor at the John Innes Institute in 1968. Currently he is a Senior Research Fellow at the University of East Anglia, Biological Sciences.

völligen „Durchsichtigkeit" und Strukturlosigkeit von den nach der gegebenen Vorschrift hergestellten Filmen erfüllt werden.

5. Verfahren zur Scharfeinstellung kontrastarmer Objekte.

Die klare Darstellung feinster organischer Teilchen, um die es sich bei der Aufnahme kleiner Virusarten handelt, verlangt außer einem Fehlen jeglicher Struktur der Trägerhaut eine so weit getriebene Dünne derselben über ein genügend großes Gesichtsfeld, daß ihre Massendicke nicht groß ist gegenüber der Massendicke der feinsten Objekte.

II 378/38.

Fig. 9. Kolloidales Gold zur Scharfeinstellung der Aggregate des Tabakmosaikvirus. el.opt. 20000 : 1. Gesamtvergr. 50000 : 1.

Nur dann kann der Elektronenstrahl, der ein Bild der Massenverteilung des Objekts gibt, diese Objekte mit Kontrast gegenüber dem Untergrund des Trägerfilms abbilden. Die auf dem Leuchtschirm scharf eingestellten Bilder sind bei der sehr großen Tiefenschärfe auf der um nur wenige Millimeter tiefer liegenden photographischen Platte ebenfalls scharf. Der Leuchtschirm gibt aber die im Strahlenbild bestehenden Intensitätsunterschiede weniger kontrastreich wieder als die Photoplatte. Diese Tatsache erschwert die genaue Scharfeinstellung, die auf bequeme Weise nur durch Beobachtung des sichtbaren Bildes mit dem Auge oder durch eine Lupe erfolgen kann. Die Helligkeit des Leuchtschirms ist für die laufende Betrachtung und für die Scharfeinstellung durchaus genügend.

Ein einfaches Verfahren für die Scharfeinstellung des Bildes auf dem Leuchtschirm bei ungenügendem Kontrast des Objekts ist der Zusatz von Metallkolloidteilchen, die infolge ihrer größeren Dichte sehr viel besser zu sehen sind als gleichgroße Teilchen organischer Natur. Die dazu verwendeten Lösungen von kolloidalem Metall dürfen keine Schutzkolloide und keine Salze enthalten. Fig. 9 zeigt einen Film mit kolloidalen Goldteilchen von 20—40 mμ Durchmesser*; daneben erkennt man zarte, fadenförmige Gebilde, welche, wie in der folgenden Mitteilung ausführlich auseinandergesetzt wird, Tabakmosaikvirus darstellen. Damit hat sich die früher von uns ausgesprochene Vermutung, daß das Übermikroskop die bildliche Darstellung von lichtoptisch unsichtbarem Virus ermöglichen wird, in vollem Umfang bestätigt[14].

Die übermikroskopische Untersuchung erscheint gleichzeitig als das empfindlichste Verfahren, um die Einheitlichkeit und den Reinheitsgrad von Viruspräparaten zu prüfen, da es bei einem Auflösungsvermögen von 10 mμ (Einzelteilgröße 5 mμ) noch Teilchen von der Größenordnung von 10^{-19} g darstellt, was vergleichsweise 5000 Kohlenstoffatomen entspricht.

Zusammenfassung.

Die Möglichkeiten der übermikroskopischen Betrachtung beliebiger Objekte werden aufgezeigt und die Forderungen besprochen, die man an Objektträger in der Übermikroskopie stellen muß. Die praktische Herstellung solcher Objektträger und ihre Behandlung beim Beschicken mit Objekten und beim Mikroskopieren wird beschrieben. Schließlich wird ein Verfahren angegeben, um eine Scharfeinstellung des Bildes bei kontrastarmen Objekten zu ermöglichen.

* Die verwendeten Goldsole waren Präparate der Firma Imhausen & Co., Witten a. d. Ruhr. Sie enthielten 66 mg metallisches Gold und 125 mg Elektrolyt im Liter, dagegen *kein* Schutzkolloid. G. A. KAUSCHE hat durch Dialyse bzw. Elektrodialyse gegen reines Wasser die Leitfähigkeit von $\varkappa = 2{,}5 \cdot 10^{-3}$ Siemens/cm auf $\varkappa = 1{,}5 \cdot 10^{-6}$ Siemens/cm gebracht

[14] B. v. BORRIES, E. RUSKA u. H. RUSKA, Klin. Wschr. **1938**, 921.

Die Sichtbarmachung von pflanzlichem Virus im Übermikroskop.*

Von G. A. KAUSCHE, E. PFANKUCH u. H. RUSKA.

Ziel der Arbeit.

Der überraschende Befund STANLEYS[1,2], daß es sich bei dem „Erreger" der Tabakmosaikkrankheit um einen hochmolekularen Eiweißkörper handelt, hat in den letzten Jahren zu Untersuchungen über die vermutliche Form, die Teilchengröße der letzten infektiösen Einheit und das Molekulargewicht dieses Virusproteins geführt. An diesen Arbeiten haben entsprechend der bisherigen Beteiligung an der physiko-chemischen Seite des Problems der phytopathogenen Viren bis auf BECHHOLD und SCHLESINGER[3,4] hauptsächlich

* Mitteilung aus der Biologischen Reichsanstalt für Land- und Forstwirtschaft und dem Laboratorium für Elektronenoptik der Siemens & Halske A.G.

[1] W. M. STANLEY, Science (N. Y.) **81**, 644 (1935).

[2] W. M. STANLEY, J. of biol. Chem. **115**, 673 (1936).

[3] H. BECHHOLD u. M. SCHLESINGER, Biochem. Z. **236**, 392 (1931).

[4] H. BECHHOLD, Phytopath. Z. **6**, 342 (1933).

amerikanische und englische Autoren mitgewirkt. Die Untersuchungen wurden mit dem röntgenographischen Verfahren nach DEBYE-SCHERRER, der Viskosimetrie und den Bestimmungen der Diffusionskonstanten, der Sedimentationskonstanten bzw. der Sedimentationsgeschwindigkeit im Schwerefeld hochtouriger Zentrifugen nach dem Verfahren von THE SVEDBERG durchgeführt. Die Ergebnisse dieser Arbeiten, obwohl in einigen Punkten noch widerspruchsvoll, rechtfertigen die Annahme, daß es sich bei diesen Viren um langgestreckte Moleküle von sehr hohem Molekulargewicht handeln muß.

Aus dem Wunsche heraus, möglichst voraussetzungsfrei zu arbeiten, haben wir geglaubt, in der übermikroskopischen Aufnahme von Makromolekülen ein Verfahren zu besitzen, das über die Form, Größe und Eigenschaften dieser Makromoleküle hypothesenfreie Aussagen liefern kann. Wir haben daher das Verfahren der Übermikroskopie[5, 6] auf die Sichtbarmachung von Tabakmosaik- und Kartoffel-X-Virus angewandt, um mit den Ergebnissen unserer Arbeit mit in die Lösung des Problems der Form und Teilchengröße einzugreifen.

Bisheriger Stand der Kenntnis von Größe und Form des Tabakmosaikvirus-Moleküls.

Nach den Untersuchungen von WYCKOFF und COREY[7] ergibt das Tabakmosaikvirus sowohl als Feuchtgel wie auch als kristallines Ultrazentrifugensediment Debye-Scherrer-Diagramme. Dieser Befund ist von BERNAL und FANKUCHEN[8] bestätigt worden. Die Identitätsperiode beträgt nach diesen Autoren $22 \times 20 \times 20$ Å. Die Reflexe lassen sich den Prismenebenen eines hexagonalen Gitters zuordnen, dessen Nebenachse eine Länge von 152 Å aufweist.

Die experimentell gemessenen intermolekularen Abstände lassen den Schluß zu, daß in einem Trockengel von Tabakmosaikvirus stäbchenförmige Moleküle parallel gelagert sind. Ihr Querschnitt würde normal eine hexagonale Anordnung zur Längsachse darstellen, bei der der Achsenabstand der benachbarten Moleküle 152 Å betragen würde. Ließ sich auf diesem Wege das maximale Ausmaß der kleineren Achse berechnen, so konnte im röntgenographischen Verfahren noch nichts über die Länge der Moleküle festgestellt werden. KRATKY und MARK[9] deuten das bezüglich der Längsachse negative Ergebnis von BERNAL und FANKUCHEN[8] so, daß es beim völligen Fehlen von durchlaufenden Schichtlinien und dem Fehlen von charakteristischen Gruppen, welche die durch die Moleküllänge bedingte Periodizität hervorheben, möglich sei, daß die angewendeten Expositionszeiten nicht ausreichend gewesen seien, um die Schichtlinien zu erfassen. Immerhin kann als recht sicheres Ergebnis der röntgenographischen Untersuchungen festgestellt werden: das Virusmolekül ist ein regulär sechseckiges Stäbchen von 15 mμ Durchmesser.

Weitere Aussagen über die Form, insbesondere die Länge des Virusmoleküls sind also erst dann möglich, wenn das Molekulargewicht bekannt ist. Versuche, das Molekulargewicht aus der in einer hochtourigen Zentrifuge bestimmten *Sedimentationskonstanten* zu berechnen, sind öfters unternommen worden, wir wollen hier aber nur die neuesten Versuche von WYCKOFF[10, 11] und von NEURATH und SAUM[12] berücksichtigen. Die Berechnung gründet sich auf die Formel:

$$M = \frac{R \cdot T \cdot s}{(1 - V \cdot d) \cdot D}.$$

M = Molekulargewicht; R = Gaskonstante in Erg; T = absol. Temperatur; s = Sedimentationskonstante; V = spez. Vol. des gelösten Teilchens; d = Dichte des Lösungsmittels; D = Diffusionskonstante.

Im Gegensatz zur alten STOCKESschen Gleichung ist diese Formel nicht nur auf kugelförmige Teilchen anwendbar, da der Einfluß der Teilchenform in der Diffusionskonstanten zum Ausdruck kommt. Die Sedimentationskonstante s von homodispersem, nicht aggregiertem Tabakmosaikvirus ist von WYCKOFF zu $174 \cdot 10^{-13}$, die Diffusionskonstante D von NEURATH und SAUM zu $3 \cdot 10^{-18}$ bestimmt worden; aus diesen Werten ist ein Molekulargewicht von 54 Millionen zu berechnen. Das Virusmolekül wäre danach 15 mμ dick (s. o.) und 330 mμ lang.

Leider kann dieses Resultat nicht als durchaus gesichert gelten; zwei Einwände können nicht übergangen werden. Zunächst ein leicht zu beseitigender, experimenteller Natur: die Diffusionskonstante ist von NEURATH und SAUM[12] an einem 2 Monate alten teilweise aggregierten Virusmaterial gemessen worden und deshalb nicht völlig einwandfrei. Sodann ein weiterer Einwand gegen die theoretischen Grundlagen der Berechnung, dessen Bedeutung zunächst nicht zu klären ist. Sedimentationskonstante und Diffusionskonstante sind in ganz verschiedenen Schwerefeldern bestimmt worden (70000 g bzw. 1 g), ihre Vereinigung in *einer* Formel ist deshalb nur dann zulässig, wenn man Moleküldeformationen bzw. Molekülschwingungen in dem hohen Schwerefeld der Ultrazentrifuge ausschließt, eine Annahme, die zwar wahrscheinlich, aber nicht gesichert ist.

Eine weitere, zunächst sehr elegant erscheinende Methode, um zu Daten über Form und Größe des Virusmoleküls zu kommen, ist von FRAMPTON und

[5] E. RUSKA, Z. Physik **87**, 580 (1934).
[6] B. v. BORRIES u. E. RUSKA, Wiss. Veröff. Siemens **17**, 99 (1938).
[7] RALPH WYCKOFF u. R. B. COREY, J. of biol. Chem. **116**, 51 (1936).
[8] J. S. BERNAL u. FANKUCHEN, Nature (Lond.) **139**, 923 (1937).
[9] O. KRATKY u. H. MARK, Anwendung physikalischer Methoden zur Erforschung von Naturstoffen, in Fortschr. Chemie-organ. Naturstoffe. Wien: Julius Springer 1938.
[10] R. WYCKOFF, J. of biol. Chem. **121**, 219 (1937).
[11] R. WYCKOFF, J. of biol. Chem. **124**, 585 (1938).
[12] NEURATH u. SAUM, J. of biol. Chem. **126**, 435 (1938).

NEURATH[13] und von LAUFFER[14, 15] benutzt worden. Aus Messungen der *Viskosität* von sehr verdünnten Viruslösungen ist mit Hilfe der KUHNschen Formel

$$\frac{\eta}{\eta_0} - 1 = g \cdot \left(2{,}5 + \frac{1}{16}\left(\frac{b}{a}\right)^2\right)$$

(η = Viscosität der Lösung; η_0 = Viscosität des Lösungsmittels; g = Konzentration der Lösung in Vol.-%; b = lange Achse; a = kurze Achse des gelösten Teilchens)

das Verhältnis von langer zu kurzer Achse des angenommenen Stäbchenmoleküls berechnet worden. Das unmittelbare Ergebnis dieser Berechnungsweise war ein Achsenverhältnis von 35 : 1. Schließt man diesen Wert an den bisher sichersten röntgenographisch gemessenen Wert von 152 Å für die kurze Achse an, so erhält das Virusmolekül Ausmaße von 15 mμ mal 530 mμ und ein Molekulargewicht von rund 83 Millionen.

Es bestehen aber doch recht große theoretische Bedenken gegen eine Berechnung der Teilchenform aus Viskositätsmessungen mit Hilfe der KUHNschen Gleichung. Der Einfluß, den die Ladung und die Hydratation der Teilchen auf die Viskosität haben, wird in dieser Formel überhaupt nicht berücksichtigt, eine Fehlerquelle, die bei einem lyophilen Kolloid wie dem Tabakmosaikvirus sicher erheblich ist. Man kann nun natürlich, wie dies LAUFFER notgedrungen tut, diese Fehlerquelle als kompensiert betrachten durch eine andere, in entgegengesetzter Richtung wirkende Fehlerquelle, nämlich die Orientierung der Teilchen während der Viskositätsmessungen. Diese Annahme ist aber willkürlich gemacht und bedeutet den Verzicht auf eine wirklich quantitative Auswertung der experimentell gefundenen Werte. Wir können daher die aus Viskositätsmessungen abgeleiteten Befunde über die Form der Virusteilchen nicht für zuverlässig halten. Das gleiche gilt besonders, wenn es sich wie hier um fadenförmige Teilchen handelt, von den Größenbestimmungen durch Ultrafilter nach ELFORD VON THORNBERRY[16].

Angesichts dieser widersprechenden Angaben über Form und Molekulargewicht haben wir uns entschlossen, über die Ergebnisse zu berichten, die die Untersuchung des Tabakmosaikvirus im Übermikroskop bisher gebracht hat.

Die Virusgelstrukturen.

Die theoretischen und praktischen Grundlagen der Übermikroskopie sind von B. v. BORRIES und E. RUSKA bereits abgehandelt worden. Literaturangaben befinden sich in der vorhergehenden Arbeit dieses Heftes[17]. H. RUSKA[18] hat im gleichen Heft das Wesentlichste über die Untersuchungs- und Präparationstechnik für die Übermikroskopie zusammengestellt. Hinsichtlich der Reindarstellungsverfahren von Tabakmosaik und Kartoffel-X-Virus verweisen wir auf die Arbeit von PFANKUCH und KAUSCHE[19]. Auf den Kollodium- oder Zaponlackfilmen trägt man mit einer sterilen Platinöse einen Tropfen der wässerigen Viruslösung auf (s. a. H. RUSKA[18], Fig. 4, ds. Heft), der dann darauf eintrocknet. Nach dem Auftrocknen ergibt sich die Situation von Fig. 1e (bei H. RUSKA[18] ds. Heft), d. h. die Virus- und evtl. beigemischten Goldteilchen haften am Trägerfilm. Bei hochkonzentrierten Viruslösungen bildet sich nach dem Eintrocknen ein mehr oder weniger zusammenhängender Viruseiweißfilm, bei niedrigen Konzentrationen haften dagegen die Einzelteilchen isoliert an dem Kollodiumträgerfilm an.

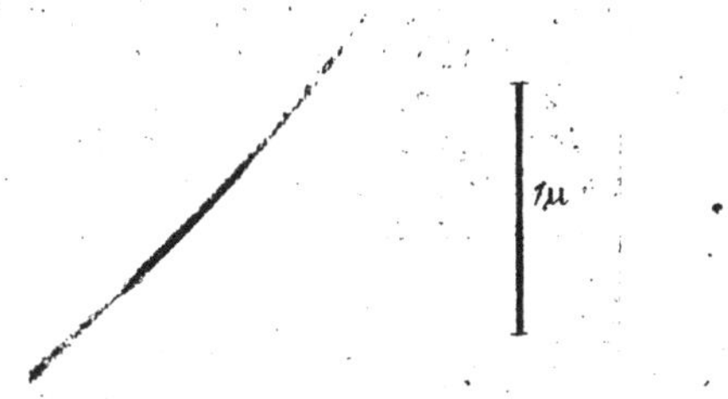

II 261/38.

Fig. 1. Kollodiumfilm mit destilliertem Wasser beschickt. el.opt. 20000:1.

II 258/38.

Fig. 2. Kollodiumfilm mit niedermolekularem Eiweiß beschickt. el.opt. 23000:1.

Fig. 1 zeigt einen solchen reinen Kollodiumfilm mit reinem Wasser beschickt. Es ist keinerlei Eigenstruktur erkennbar. An dem sichtbaren Rand links oben auf dem Bild ist der Film eingerissen und aufgerollt. In Fig. 2 bringen wir ein Bild von einem Film, auf dem nach Beschickung mit reinem Wasser und Kontrollaufnahme im Übermikroskop zusätzlich niedermolekulares Eiweiß

[13] C. FRAMPTON u. J. NEURATH, Science (N. Y.) 87, 468 (1938).
[14] M. LAUFFER, Science (N. Y.) 87, 469 (1938).
[15] M. LAUFFER, J. of biol. Chem. 126, 443 (1938).
[16] H. H. THORNBERRY, Phytopathol. 25, 601 (1935).
[17] v. BORRIES u. E. RUSKA, Naturwiss. 27, 281 (1939).
[18] H. RUSKA, Naturwiss. 27, 287 (1939).
[19] E. PFANKUCH u. G. A. KAUSCHE, Biochem. Z. 299, 334 (1938).

(Pseudoglobulin und Albumin) aus Pferdeserum aufgetragen wurde. Auch hier ist keine Einzelstruktur bei der angewendeten Vergrößerung von 23000fach sichtbar geworden. Man sieht höchstens eine schwachwolkige Masse, die den Eiweißfilm auf dem Kollodiumfilm darstellt. Trägt man nun auf einen solchen mit niedermolekularem Eiweiß zuvor beschickten Film noch nachträglich TM-Virus auf, dann wird eine Struktur von ganz bestimmter Anordnung sichtbar (Fig. 3). Ähnliche, aber im einzelnen doch wieder abweichende Strukturbilder erhält man, wenn Viruslösungen von TM- und Kartoffel-X-Virus auf reine, d. h. vorher nicht mit Wasser oder Eiweiß beschickte Filme in genügend hoher Konzentration zum Eintrocknen gebracht werden (Fig. 4—7).

Hinsichtlich der präparativen Ausgangssituation ist folgendes zu beachten: Die von uns benutzten hochgereinigten Viruslösungen waren sicherlich völlig frei von niedermolekularem Eiweiß. Der nach der Dialyse noch vorhandene Salzrest betrug, bezogen auf Trockensubstanz, noch nicht 0,01%. Die Leitfähigkeiten der Lösungen lagen bei $\varkappa = 1,5 \times 10^{-5}$ S/cm. Im Sinne der Phasenlehre stellen die verwendeten Präparate also dispersoide Systeme dar, in denen das Dispersionsmittel reines Wasser, die disperse Phase reines Viruseiweiß ist. Dieses Suspensoid müßte vom *kolloid*chemischen Standpunkt aus als inhomogen bezeichnet werden, weil das System im ganzen von der Inhomogenität der Grenzflächen der fadenförmigen Moleküle beherrscht wird. Da innerhalb bestimmter Konzentrationsgrenzen in den Viruslösungen ein molekularer Verteilungszustand herrscht, bei dem allerdings den einzelnen Molekülen kolloider Charakter zugesprochen werden muß, resultiert, daß die disperse Phase unserer Lösungen aus kolloiden Makromolekülen besteht. Solange eine molekulare Lösung des Virusproteins vorliegt, können wir also, wie bei jeder anderen molekularen Lösung vom rein *chemischen* Standpunkt aus mit einer völligen Homogenität der Viruslösung rechnen.

Bei der Überführung solcher Lösungen in den Gelzustand müssen die Makromoleküle dann eine bestimmte Lage zueinander einnehmen, wenn ihre Form stäbchen- oder fadenartig ist. Die Lage kann nach den Regeln der Strukturlehre bei Trockengelen mit Fadenmolekülen nur die eines Gerüst- oder Netzwerkes mit vorzüglich linearer Aggregation der Moleküle sein.

Wenn eine Virussollösung über den hochkonzentrierten Zustand in eine Gellösung und daraus über den Grenzkonzentrationszustand in ein Gel durch Entziehung des Lösungsmittels und Hydratationswassers übergeführt wird, dann könnten sich nach Maßgabe der Schnelligkeit und des Ausmaßes der Dehydratation allmählich die physikalischen Eigenschaften der Moleküle beim Überschreiten der Grenzkonzentration ändern, indem Störungen in der Struktur der langen Moleküle auftreten. Indessen haben BAWDEN, PIRIE, BERNAL und FANKUCHEN[20] mit indirekten Methoden sowohl an feuchten als auch trocknen Gelen Faserstruktur nachweisen können. TAKAHASHI und

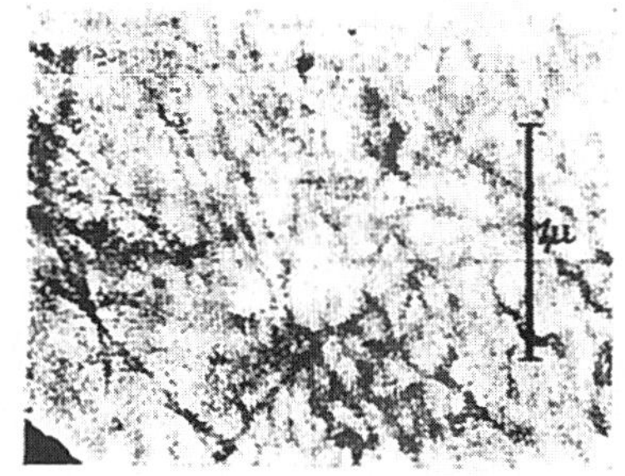

II 263/38.

Fig. 3. Kollodiumfilm mit niedermolarem Eiweiß und Tabakmosaikvirus beschickt. el.opt. 19000 : 1.

II 678/38.

Fig. 4. Kartoffel-X-Virus (Gelstruktur, Fäden und breite Balken). el.opt. 25000:1.

II 671/38.

Fig. 5. Tabakmosaikvirus mit großen Anhäufungen von kolloidalem Gold. (Gelstruktur, Fäden und Balken.) el.opt. 25000:1.

RAWLINS[21] haben auch in Viruslösungen mittels der Feststellung der Strömungsdoppelbrechung stäbchenförmige Gestalt der gelösten Teilchen

[20] F. C. BAWDEN, N. W. PIRIE, J. D. BERNAL u. I. FANKUCHEN, Nature (Lond.) **138**, 1051 (1936).

[21] W. N. TAKAHASHI u. T. E. RAWLINS, Science (N. Y.) **85**, 103 (1937).

wahrscheinlich gemacht. Da auch Lösungen, Feucht- und Trockengele und die ausgefällten Kristalle in ihrem röntgenographischen Verhalten sich übereinstimmend zeigten[7, 8], müssen wir annehmen, daß beim Eintrocknungsprozeß auf dem Trägerfilm eine Veränderung der physikalischen Eigenschaften der Moleküle nicht in dem Sinne eintritt, daß die grundsätzliche Struktur und damit die elementaren Dimensionen sich ändern. Die

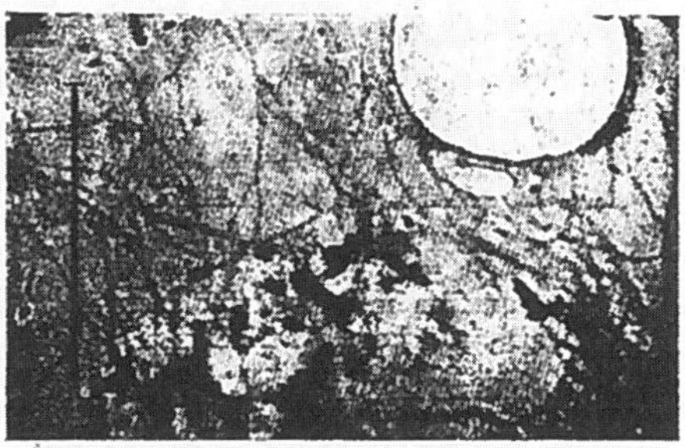

II 672/38.

Fig. 6. Tabakmosaikvirus (Gelstruktur, vorwiegend Einzelfäden). el.opt. 25000:1.

II 378/38.

Fig. 7. Tabakmosaikvirus (links unten zwei Einzelfäden). el.opt. 20000:1.

sichtbargemachte submikroskopische Struktur der Viruseiweißgele mußte also der Form der Moleküle entsprechen.

Bei hinreichend hoch konzentrierten Viruslösungen entsteht demnach während der aus untersuchungstechnischen Gründen notwendigen Eintrocknung ein formbeständiges Trockengel, das die Fig. 4—7 als festes Gelgerüst zeigen. Dieses Gel ist im physikalisch-chemischen und im mikroskopischen Sinne als homogen zu bezeichnen, weil die einzelnen Faserstränge keine Diskontinuitäten mehr erkennen lassen. Das Ausmaß der Aneinanderlagerung der langgestreckten Moleküle scheint von der Schnelligkeit des Eintrocknungsprozesses abhängig zu sein. Die Bänder und Balken sind längenmäßig schwer zu messen, weil sie gegenseitig miteinander verflochten sind. Man kann dagegen Breiten von 16 und 45, 90 und 120 bis zu 160 mμ feststellen. Die Differenzen in den Schwärzungsintensitäten an einzelnen Stellen sind aus ihrem sich kreuzenden Verlauf als räumliche Übereinanderlagerungen zu erkennen (Fig. 4, Kartoffel-X-Virus). Auch dort, wo Parallellagerung besteht, ist mit Übereinanderlagerungen zu rechnen. Die Breiten der aus lateral aggregierten Molekülfäden bestehenden Bänder brauchen deshalb nicht unbedingt Vielfache der Fadenbreite des Moleküls zu sein.

Hier taucht das Problem auf, ob die abgebildeten Gelgerüste einphasige oder zweiphasige Gele darstellen, bei welch letzteren die Fadenmoleküle in bestimmten Gebieten zu Bündeln und geordneten Gitterbereichen zusammentreten. Nach GERNGROSS, HERRMANN und Mitarbeitern[22, 23] und FREY-WYSSLING[24] ist die Wahrscheinlichkeit der Entstehung kristalliner Bereiche um so größer, je länger die Molekülketten parallel miteinander verlaufen. Dieser Hinweis hat seinen Grund darin, daß wir uns bei den hier erstmalig sichtbar gemachten Strukturen fragen müssen, ob wir es noch mit Aggregationen von molekularen Ausmaßen oder bereits mit kristallinen Bildungen zu tun haben, wie sie im Prinzip für den mikroskopischen Bereich beim TM-Virus erstmalig von STANLEY[1, 2] und später von BAWDEN und PIRIE[25, 26] bzw. PFANKUCH und KAUSCHE[19] nachgewiesen und gefärbt (KAUSCHE[27, 28]) werden konnten.

Bei einem Hydrogel werden seit NÄGELI und SCHWENDENER (1877) und später AMBRONN und FREY (1926) und SCHMIDT (1934) die kristallinen Bereiche als Micelle bezeichnet. Nach der Auffassung von NÄGELI können die Micelle eines Gels bei dessen Lösung den micellaren Charakter behalten. Nach STAUDINGER (1932) gehen dagegen die Hochpolymeren nicht als kristalline Teilchen, sondern als einzelne Fadenmoleküle in Lösung. Die Micelle sind nach SIGNER (1933) an sich optisch anisotrop, und diese Eigenschaft bleibt auch bei den isolierten Molekülen erhalten. Da feststeht, daß Lösungen von TM-Virus Strömungsdoppelbrechung zeigen, erhebt sich die Frage, ob das gelöste Virus in einem übermolekularen kristallinen Zustand vorkommt oder ob der monomolekulare Verteilungszustand der primäre ist. Wir glauben bei den von uns verwendeten Lösungen das letztere annehmen zu können und hätten an denjenigen Stellen auf unseren Figuren, wo die gemessenen Dimensionen den theoretisch zu fordernden Molekularbreiten entsprechen, von einem Molekulargerüst zu spre-

[22] GERNGROSS, HERRMANN u. ABITZ, Biochem. Z. **228**, 409 (1930).

[23] GERNGROSS, HERRMANN u. LINDEMANN, Kolloid-Z. **60**, 276 (1932).

[24] A. FREY-WYSSLING, Submikroskopische Morphologie des Protoplasmas und seiner Derivate. Berlin: Bornträger-Verlag 1938.

[25] F. C. BAWDEN u. N. W. PIRIE, Brit. J. exper. Path. **18**, 275 (1937).

[26] F. C. BAWDEN u. N. W. PIRIE, Proc. roy. Soc. Lond. B **274** (1937).

[27] G. A. KAUSCHE, Naturwiss. **26**, 741 (1938).

[28] G. A. KAUSCHE, Mitt. d. biol. Reichsanst. **1939**, im Druck.

chen (Fig. 6). Diejenigen Bezirke, in denen es zu einer mächtigen Aggregation parallel gelagerter Moleküle gekommen ist, tragen offensichtlich übermolekularen Charakter (Fig. 4). Wir haben also die Gerüstbildungen als ein Gemisch von molekularen und kristallinen Bereichen anzusehen, wobei wir die Frage nach den Ausmaßen eines Moleküls hier noch offen lassen. Der auffallende Umstand, daß gleich breite Virusaggregate nadelartig starr, oder auch bandartig geschwungen sein können, ist möglicherweise so zu deuten, daß in einem Falle die Molekülenden der nebeneinander liegenden Virusteile an der gleichen Stelle des Querschnittes liegen und im anderen Falle gegeneinander versetzt sind.

In Fig. 7 sind Aggregationen in der Längs- und in der Seitenrichtung sichtbar und daneben noch einzelne stäbchenförmige Gebilde, deren Dimensionen bei 330×15 mμ liegen. Die Breite der Bänder schwankt zwischen 100 und 15 mμ.

Das Virusmolekül.

Bei Verwendung von Viruskonzentrationen von 10^{-5}·g Virusprotein je Kubikzentimeter gelingt es, beim Eintrocknen die Virusteilchen einzeln fixiert zu erhalten. In Fig. 8 ist der Trägerfilm in der einen Richtung stark gefaltet. Senkrecht bzw. schräg zu dieser Faltenbildung sind Virusfäden zu sehen, die durch die Fältelung der Folie zerknittert und in ihrem Verlauf unregelmäßig geworden sind. Die Länge der Einzelfäden ist dadurch nicht genau bestimmbar. Ihre Breite dürfte bei 10—25 mμ liegen. Die Fäden weisen deutlich knotenartige Verdickungen auf, die dadurch zustande kommen, daß die aufgetrockneten Fäden eine spätere Faltung des Trägerfilms mitmachen. In Fig. 9 wurden 14 der fadenartigen Gebilde einzeln vermessen und ihre Längen maximal zu 1800 mμ und minimal zu 300 mμ bestimmt. Die Schwankungen in der Länge betragen also etwa das 6fache, die Durchmesser liegen zwischen 10 und 20 mμ. Alle Einzelteilchen liegen mehr oder weniger getrennt voneinander. Es fällt auf, daß sie in der Länge zwar verschieden, im Durchmesser jedoch fast einheitlich sind. Die Dimension der eingestreuten kolloidalen Goldteilchen liegt bei 20—30 mμ. Fig. 10 zeigt bei hoher Verdünnung mehrere Stäbchen, deren Ausdehnungen bei 1200 bzw. 300 mμ in der Längsrichtung und 20 bzw. 15 mμ in der Querrichtung liegen. Das gabelförmige Stäbchen im Bild dürfte aus zwei Einzelstäbchen bestehen, die halb schräg zu- und übereinander liegen. Fig. 11 weist ebenfalls fadenartige langgestreckte Einzelteilchen auf, deren Ausmaße sich zwischen maximal 900 mμ und minimal 300 mμ für die lange Achse, bei 15—20 mμ für die kleine Achse halten. Alle Stäbchen liegen getrennt voneinander. In Fig. 12 sind bei sehr hoher Virusverdünnung nur mehr 2 Fäden zu sehen, deren einer völlig freiliegt und etwa 850×15 mμ mißt.

Die Fig. 7—12 haben also gemeinsam, daß sie faden- bzw. stäbchenförmige Einzelteilchen aufweisen, die in ihrer Länge stark variieren, hinsichtlich ihrer Breiten aber recht gleichmäßig sind. Angesichts dieser Tatsache ergab sich die Frage, ob diese distinkten Einzelteilchen als die Moleküle des TM-Virus bzw. als vorwiegend in der Längsrichtung aggregierte Moleküle oder aber schon als

II 366/38.

Fig. 8. Tabakmosaikvirus (geknickte Fäden auf gefaltetem Film). el.opt. 20000:1.

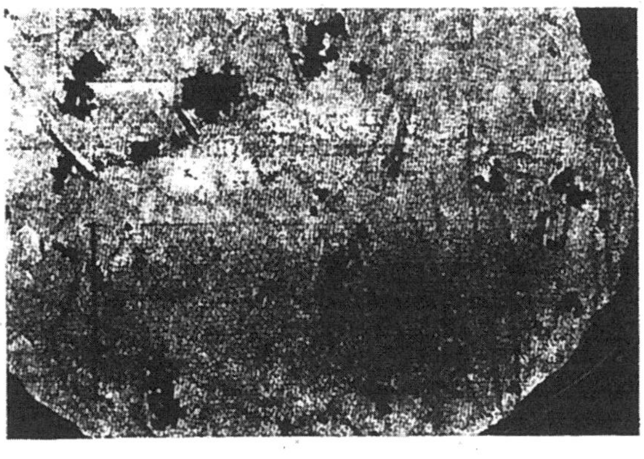

II 374/38.

Fig. 9. Tabakmosaikvirus (Einzelfäden, Parallel- und Queraggregate). el.opt. 20000:1.

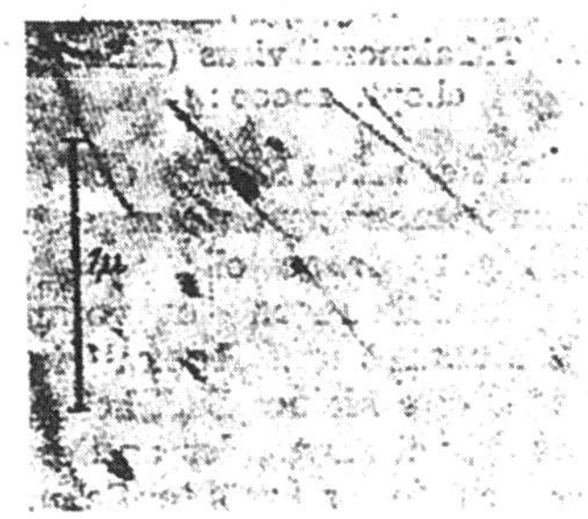

II 221/38.

Fig. 10. Tabakmosaikvirus (gabelförmig gelagerte Stäbchen). el.opt. 20000:1.

übermolekulare Gebilde anzusprechen sind. Die Deutung der Befunde hat aus dem Vergleich derjenigen Zahlen, die aus den indirekten Methoden gewonnen sind, mit den auf diesen Bildern meßbaren Größen unter Berücksichtigung des Auflösungsvermögens und der Abbildungstreue des Übermikroskopes zu erfolgen.

Die Breite der Teilchen haben wir aus der schärfsten der Aufnahmen, der Fig. 7, zu messen versucht. Das Auflösungsvermögen für diese Aufnahmen beträgt etwa 7—8 mμ. Man kann also von vornherein nicht hoffen, die Breite der Teilchen auf einige wenige mμ genau bestimmen zu können. Dennoch haben wir den an der linken

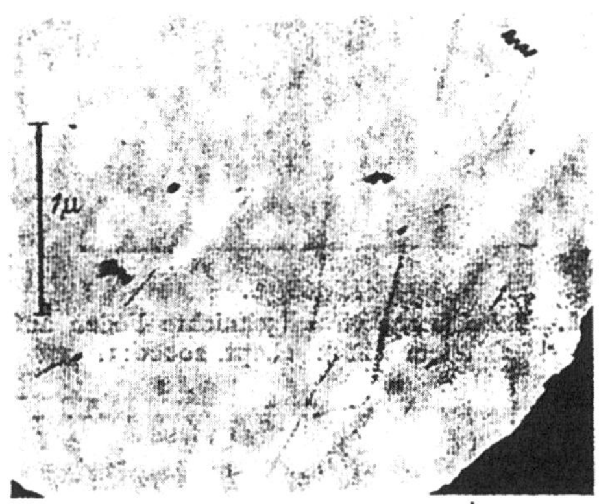

II 371/38.
Fig. 11. Tabakmosaikvirus (Einzelfäden). el.opt. 20000 : 1.

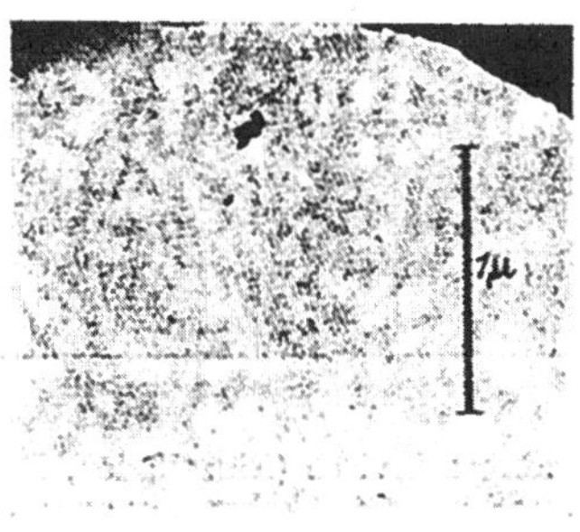

II 372/38.
Fig. 12. Tabakmosaikvirus (Einzelfäden). el.opt. 20000 : 1.

Seite des Bildes liegenden Einzelfaden an 3 Stellen einer Schwärzungsmessung unterzogen*. Das Ergebnis ist in Fig. 13 dargestellt.

Nach den Ausführungen der voranstehenden Arbeit von v. BORRIES und RUSKA[17] ist eine Bestimmung der Breite kleiner Partikel möglich aus der Breite des Schwärzungstales an der Stelle, wo sich zwischen kleinster und größter Schwärzung die mittlere Schwärzung befindet. Man erkennt, daß die Breite des Teilchens bei etwa 15 mμ liegen muß.

Die Längenabmessungen der Einzelfäden wurden unmittelbar durch Anlegen eines Meßstabes bestimmt. Besonders häufig treten Längen von etwa 300 mμ auf. Kürzere Stücke wurden nur vereinzelt gefunden (Abb. 9 u. 11). Die längeren Fäden sind meistens Vielfache von etwa 300 mμ, gelegentlich aber auch Vielfache von etwa 150 mμ. Es ist uns daher nicht möglich, endgültiges darüber auszusagen, ob das Tabakmosaikvirusmolekül etwa 150 oder etwa 300 mμ lang ist. Aus dem plötzlichen Absturz der Kurve der Häufigkeitsverteilung bei der Länge von etwa 300 mμ möchten wir aber diesem Wert den Vorzug geben. Die Klärung dieser Frage soll einer weiteren Arbeit vorbehalten bleiben.

Wir glauben, daß wir mit den genannten Werten von 15 mμ Dicke und 300 bzw. 150 mμ Länge an

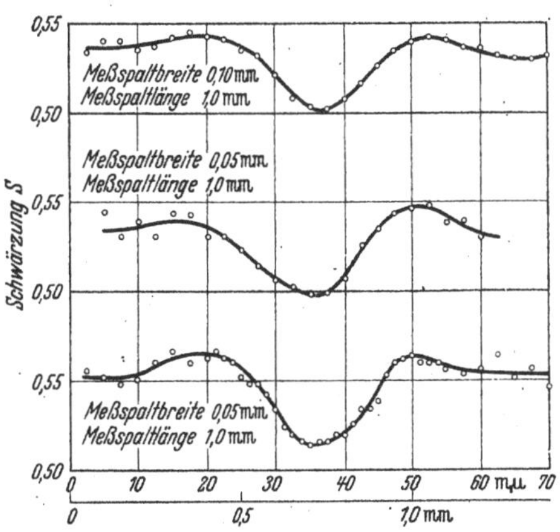

Fig. 13. Schwärzungsverlauf der Originalplatte von Fig. 7 quer zum einzelnen Virusmolekül links unten an drei verschiedenen Stellen gemessen. An der Abszissenachse sind die Entfernungen auf der Platte (mm) und die mit der Vergrößerung von etwa 20000 auf das Objekt umgerechneten Strecken (mμ) eingetragen.

der unteren Grenze der Ausmaße der intakten Elementarteilchen angekommen sind. Die ausgemessenen Dimensionen an den kleinsten dieser Gebilde entsprechen bezüglich der Dicke den röntgenographisch erzielten Ergebnissen; bezüglich der Länge entspricht das Ergebnis von 300 mμ etwa dem mit der Ultrazentrifuge ermittelten. Bei einem spez. Gewicht von 1,35 und der plausiblen Annahme, nämlich, daß es sich um Stäbchen mit sechseckigem Querschnitt handelt, errechnen wir bei der gemessenen Längsachse von rund 300 bzw. 150 mμ und einem Durchmesser von etwa 15 mμ ein Molekulargewicht von 48 bzw. 24 Millionen, während sich aus den zu Anfang genannten Zentrifugenversuchen ein Molekulargewicht von 54 Millionen ergab.

Wir müssen die stäbchen- oder fadenförmigen Gebilde des genannten Dimensionsbereiches für die Moleküle des Tabakmosaikvirus halten und sehen die Vielfachen davon als lineare bzw. laterale Aggregationen an. Es ist durchaus wahrscheinlich, daß die beobachteten Teilchen von 150 mμ Länge dem Tabakmosaikvirusmolekül mit der Sedimentationskonstante $174 \cdot 10^{-13}$ und die Teilchen von 300 mμ Länge dem Virus mit Sedimentationskonstanten $202 \cdot 10^{-13}$ zuzuschreiben sind. Bei

* Herrn Dr. H. BODE sind wir für die liebenswürdige Durchführung dieser Messung sehr zu Dank verpflichtet.

entsprechenden Unterschieden in den Diffusionskonstanten wären die angeführten Differenzen in den Sedimentationskonstanten groß genug, um sie als Folge einer linearen Dimerisation des Virusmoleküls von 15×150 mμ ansehen zu können. Daß dieses Molekül sehr leicht in das Aggregat mit der höheren Sedimentationskonstante übergeht, ist von WYCKOFF[29] mit Hilfe der Ultrazentrifugentechnik gezeigt worden, ebenso daß dieses Aggregat eine ganz besonders bevorzugte Stellung gegenüber den möglichen höheren Aggregationsstufen einnimmt.

Unter Zugrundelegung eines Molekulargewichtes von 48 bzw. 24 Millionen sind die hin und wieder zu beobachtenden kleineren kugel- oder kurzstäbchenförmigen Gebilde als Bruchstücke der Virusmoleküle anzusehen. Wir können unsere Annahme, daß es sich bei den Stäbchen von rund 300 bzw. 150×15 mμ tatsächlich *um die Moleküle des TM-Virus handelt,* noch durch andere Argumente stützen. Das mit Hilfe der Sedimentationskonstanten bestimmte Molekulargewicht resultiert aus Ultrazentrifugenversuchen mit scharfer Sedimentationsgrenze. Die aussedimentierten Virusteilchen müssen also homodispers sein, oder mindestens eine scharfe *untere* Grenze der Teilchengröße besitzen. Es gelingt, mit chemischen Agentien diese Molekülgrößen zu zerschlagen; jedoch hat sich gezeigt, daß die Bruchstücke dann nicht mehr pathogen aktiv sind. *Damit ist erwiesen, daß die Infektiosität an ein Mindestmaß der Einzelteilchen gebunden ist.* Wir haben im übrigen Grund zur Annahme, daß zwischen der pathogenen Aktivität und den Ausmaßen der infektiösen Einheiten bestimmte rechnerisch definierbare Beziehungen bestehen. Diese weisen darauf hin, daß homodisperse Viruslösungen mit dem Molekulargewicht von etwa 48 bzw. 24 Millionen und unseren Beobachtungen nach von der Teilchengröße 300 bzw. 150×15 mμ das Höchstmaß an pathogener Aktivität aufweisen.

[29] R. W. G. WYCKOFF, J. of biol. chem. **121**, 219; **122**, 239 (1937).

Zusammenfassung.

1. Es wird die Möglichkeit aufgezeigt, im Übermikroskop Tabakmosaik- und Kartoffel-X-Virus sichtbar und abbildbar zu machen.

2. Die präparative Ausgangssituation wird geschildert. Durch das Austrocknen des Virussols entsteht auf dem Trägerfilm ein Trockengel mit gerichteter Struktur.

3. Bei hinreichend hoch verdünnten Virussolen gelingt es, die laterale Aggregation zu Bändern zu verhindern, so daß die Stäbchen oder Fäden isoliert voneinander auftrocken.

4. Die ausgemessenen Dimensionen an den kleinsten der Gebilde entsprechen denjenigen, die auf Grund der Röntgenographie und Ultrazentrifugation ermittelt worden sind.

5. Wir halten die stäbchen- oder fadenförmigen Gebilde des Dimensionsbereiches von rund 300 mμ bzw. rund 150 mμ Länge und rund 15 mμ Durchmesser für die Moleküle des TM-Virus und sehen die Vielfachen davon als lineare bzw. laterale Aggregationen an.

6. Im Sinne der submikroskopischen Morphologie konnte zum erstenmal an einem Eiweißtrockengel die Faserstruktur für den micellaren bis molekularen Bereich sichtbar gemacht und abgebildet werden.

7. Wir haben die bisher hypothetische und nur mittels indirekter Methoden nachgewiesene Aggregation der TM-Virusmoleküle direkt sichtbar gemacht und sie als eine überraschend leicht, zunächst in linearer Verkettung eintretende Reaktion festgestellt, die sich kontinuierlich fortsetzt bis zur Bildung von lichtmikroskopisch und dann weiter von makroskopisch sichtbaren Kristallen.

Commentary on the Two Papers by Caspar and Franklin

Aaron Klug
MRC Laboratory of Molecular Biology, Cambridge CB2 2QH, England

As described in earlier chapters in this volume, the X-ray diffraction study of TMV (Bawden *et al* 1936), initiated by Bernal at Cambridge and subsequently moved to Birkbeck College, London, was brought to a stop by the outbreak of the Second World War; and the full paper was published later (Bernal and Fankuchen, 1941). The subject was taken up again by J. D. Watson in 1952, in a brief interlude from his pursuit of the DNA structure, and by Rosalind Franklin at Birkbeck in 1953, after she had moved there from King's College. Watson showed that the arrangement of protein subunits in the virus particle was helical, and Franklin determined the parameters of the helical array.

Franklin set out to determine the internal structure of the virus particle by X-ray diffraction and realized that the way ahead was to use the method of isomorphous replacement, just established by Perutz and his colleagues in their work on single protein crystals. To this end she began searching for suitable heavy atom "derivatives" of the virus.

Independently, Donald Caspar had chosen to study TMV for his Ph.D. project under Professor Pollard in the Biophysics department at Yale, and to this end had resuscitated the X-ray scattering apparatus used many years before by A. H. Compton (in his work which led to the discovery of the Compton effect, establishing the particle nature of X-rays). He too had resorted to isomorphous replacement to study the radial density distribution in the rod shaped particle by analyzing the equatorial layer line of the X-ray diffraction pattern. He accomplished this by preparing a lead-substituted TMV. Lead acetate was added to concentrated TMV solutions up to about one lead atom per subunit, and from the changes produced in the equatorial X-ray scattering he was able to deduce that equal amounts of lead were bound at radial distances of 25 and 84Å from the particle axis. The determination of the two positions at which lead binds, regardless of the chemical nature of the binding, made possible a definite determination of the signs of those equatorial diffraction maxima, and these were used to calculate the cylindrically averaged radial density distribution in the particle.

This was the subject of Caspar's Ph.D. thesis (1955) and of his *Nature* paper (1956) which is a subject of this commentary. Caspar's allocation of signs was confirmed by Franklin, by a second successful application of the method of isomorphous replacement to TMV, using a mercury-substituted TMV prepared by Fraenkel-Conrat (1955) at Berkeley. The result was published in Franklin's accompanying paper in *Nature*. One could, therefore, have complete confidence in the correctness of the radial density distribution obtained in the two papers. The small fluctuations which occur near the axis are of no significance since in this region the effect of experimental error is large.

The most significant features of the radial density distribution are as follows: (1) the virus particle has a hole of diameter 35-40Å extending along the axis; (2) the maximum radius of the particle is close to 90Å; (3) there are regions of high density at radii of 25, 40, 66, and 78Å, with intervening minima at radii of 31, 60, and 73Å.

The central hole came as a surprise, since it was then generally believed that the RNA occupied the central core of the particle (see below), and indeed was strongly questioned by Linus Pauling at a conference on viruses in 1957. The entrance of the hole was soon directly confirmed by electron microscopy. The axial hole of TMV is occupied by water when the virus is in solution, and the inside surface of the virus is as accessible to small molecules as the outside. When TMV is dried from solutions containing salt molecules, these molecules fill in or line the hole, and under these conditions Huxley (1957) showed that the hole is directly revealed in electron micrographs. (This was an early illustration of negative staining in electron microscopy, before the introduction of phosphotungstate). The hole in the virus is a rather natural consequence of the fact that the subunits cannot taper down to a wedge narrower than molecular dimensions that would fill the space near the axis of the helical array.

Electron density carries no chemical label, so the location of the RNA cannot be determined from the density distribution of TMV alone, but studies on nucleic acid-free particles by Franklin demonstrated that the very high density region about a radius of 40Å is due to the RNA. Franklin (1955) had earlier shown the overall similarity of the X-ray patterns of rods made of reaggregated protein alone (originally supplied by Schramm from Tübingen) to those of the intact virus. She made a mercury-substituted derivative of the protein rod and determined the cylindrically averaged radial density distribution. In her 1956 *Nature* paper this is compared with that obtained for the intact virus rod, and the difference interpreted.

The location of the RNA at 40Å radius in the virus particle is indicated by the replacement of the prominent maximum in the radial density map of the virus by a density minimum in the RNA-free protein. This conclusion was borne out by measurement of the relative mass difference in the 40Å region which corresponds in amount to the 5% contribution of RNA to the mass of the virus. Additional confirmation that the structural difference between the virus and the RNA-free protein lies mainly at this radius came from comparison of the nonequatorial parts of the two diffraction diagrams.

Franklin's results on the location of the RNA required a revision of the apparently obvious earlier interpretation of electron micrographs (Hart, 1955; Schramm *et al.*, 1955) of partially degraded virus particles in which the exposed RNA appears to form a central core, like the wick of a candle. The X-ray results refer to the intact virus in wet gel specimens which have not undergone any special treatment, and what was observed in the electron micrographs is a collapsed form of the RNA. A collapse would be expected to take place when the protein is removed, since it is the protein which maintains the RNA configuration deeply embedded in the intact virus.

The location of the RNA at a radius of 40Å does not by itself indicate how the RNA is arranged in three dimensions in the virus, nor how many chains of RNA there are (Franklin *et al.*, 1957). To determine the configuration of the RNA it is necessary to compare the nonequatorial reflections of TMV and of the repolymerized protein rod. This was later initiated by Franklin, and finalized by myself and K. C. Holmes after her untimely death in 1958 (Franklin *et al.*, 1959). From her work we were able to determine with a high degree of certainty the general configuration of the RNA in the virus. The RNA is in the form of a single chain following the basic (or lowest pitch) helix in the array of subunits. There are 49 nucleotides in one turn of the RNA helix.

These two papers of Caspar and Franklin laid the basis of the further X-ray analysis of the three dimensional structure of TMV, described in the commentary by Stubbs in this volume. They were technically remarkable in their own time, for they were among the very first applications of the newly established technique of isomorphous replacement and the first to a non-crystalline specimen.

Perhaps even more remarkable, quite unforeseen at the time, is the crucial role played by the hole in the center in the assembly of the virus from the RNA and protein. As discussed in this volume in the commentary by Richards, nucleation occurs by the insertion of a stem-loop structure of the RNA into the central hole of an intermediate 20 S protein aggregate. This picture, originally developed by the Cambridge group (Butler *et al.*, 1977 and references therein), was confirmed by the spectacular electron micrographs of the Strasbourg group (Lebeurier *et al.*, 1977), showing two tails of RNA present at one end of the growing virus rod.

References

Bawden, F. C., Pirie, N. W., Bernal, J. D., and I. Fankuchen. (1936). Liquid crystalline substances from virus-infected plants. *Nature* **138**:1051-1053.

Bernal, J. D., and I. Fankuchen. (1941). X-ray and crystallographic studies of plant virus preparations. *J. Gen. Physiol.* **25**:111-165.

Butler, P. J. G., Finch, J. T., and D. Zimmern. (1977). Configuration of tobacco mosaic virus RNA during virus assembly. *Nature* **265**:217-219.

Caspar, D. L. D. (1956). Radial density distribution in the tobacco mosaic virus particle. *Nature* **177**:928.

Fraenkel-Conrat, H. (1955). The reaction of tobacco mosaic virus with iodine. *J. Biol. Chem.* **217**:373-381.

Franklin, R. E. (1955). Structural resemblance between Schramm's repolymerised A-protein and tobacco mosaic virus. *Biochim. Biophys. Acta* **18**:313-314.

Franklin, R. E. (1956). Location of the ribonucleic acid in the tobacco mosaic virus particle. *Nature* **177**:928-930.

Franklin, R. E., Klug, A., and K. C. Holmes. (1957). X-ray diffraction studies of the structure and morphology of tobacco mosaic virus. *In* "Ciba Foundation Symposium on the Nature of Viruses," p. 39-52. Churchill, London.

Franklin, R. E., Caspar, D. L. D., and A. Klug. (1959). The structure of viruses as determined by X-ray diffraction. *In* "Plant Pathology: Problems and Progress, 1908-1958" (C.S. Holton, ed.), p. 447-461. University of Wisconsin Press, Madison.

Hart, R. G. (1955). Electron-microscopic evidence for the localization of ribonucleic acid in the particles of tobacco mosaic virus. *Proc. Natl. Acad. Sci. USA* **41**:261-264.

Huxley, H. E. (1957). Some observations on the structure of tobacco mosaic virus. *In* "Proceedings of the Stockholm Conference on Electron Microscopy" (F. Sjöstrand and J. Rhodin, eds.), p. 280. Academic Press, New York.

Lebeurier, G., Nicolaieff, A., and Richards, K. E. (1977). Inside-out model for self-assembly of tobacco mosaic virus. *Proc. Natl. Acad. Sci. USA* **74**:149-153.

Schramm, G., Schumacher, G., and W. Zillig. (1955). An infectious nucleoprotein from tobacco mosaic virus. *Nature* **175**:549-550.

Biographical Sketch

Aaron Klug was educated at the Universities of Witwatersrand, Cape Town and Cambridge. He began as a medical student, transferred to science, and his Ph.D. at the Cavendish Laboratory was in Physics. From 1954 until her untimely death in 1958, he worked with Rosalind Franklin at Birkbeck College, London on the structure of tobacco mosaic virus. There he also began his studies on spherical viruses, including turnip yellow mosaic virus. In 1962, he joined the MRC Laboratory of Molecular Biology in Cambridge, became Joint Head of the Structural Studies Division in 1978 and was the Director of the Laboratory from 1986 to 1996. He now continues as a member of staff, leading a research group on gene expression.

His work has been on the interactions of proteins and nucleic acids and on the analysis of the structures of large biological molecules and assemblies, including simple viruses and chromatin, by X-ray diffraction and electron microscopy, and the development of new methods for their study. In 1982 he was awarded the Nobel Prize in Chemistry for "his development of crystallographic electron microscopy and his elucidation of the structure of protein-nucleic acid complexes of biological importance". His current research is on the structure of DNA and RNA binding proteins which regulate gene expression and in particular on the interaction with DNA of the zinc finger family of transcription factors which he discovered.

He is President of the Royal Society, a member of the Order of Merit, a Foreign Associate of the U. S. National Academy of Sciences, and of the French Academy of Sciences, and has received many honorary degrees.

(*Reprinted from Nature, Vol.* 177, *pp.* 928–930, *May* 19, 1956)

STRUCTURE OF TOBACCO MOSAIC VIRUS

Radial Density Distribution in the Tobacco Mosaic Virus Particle

MEASUREMENTS of the equatorial scattering of X-rays by orientated gels of tobacco mosaic virus[1] have been used for calculating the mean electron density in the virus as a function of distance from the axis of the rod-shaped particle. Using a Geiger-counter diffractometer with monochromatic copper $K\alpha$ radiation, accurate, high-resolution recordings of the equatorial scattering have been obtained for values of $2\sin\theta/\lambda$ up to 0·06, and for $2\sin\theta/\lambda$ between 0·06 and 0·094 intensities have been estimated from photographs.

Over the angular range examined, the amplitudes of the intensity maxima are real since, for a helix such as tobacco mosaic virus[2], this part of the pattern is determined only by the cylindrically averaged density[3]. Sign relations between some of the maxima can be established from their regular periodicity in certain regions. Where this periodicity corresponds to a dimension about that of the particle diameter, it follows that the amplitude changes sign at the zeros separating the maxima[4]. Radial density distributions were calculated for the sign combinations consistent with these sign relations, and all but two sets were found to be improbable since they would require an unreasonably large diameter for the particle and unlikely density values. Unambiguous determination of the correct set of signs was made from the changes in the equatorial X-ray scattering produced by lead bound to the virus.

When tobacco mosaic virus is treated with lead acetate, lead is specifically bound, and the virus (in concentrated solution) remains soluble with up to about one lead atom added per 17,000 molecular weight of virus. The differences in X-ray scattering from orientated gels of normal and lead-substituted tobacco mosaic virus were used to calculate the radii at which the lead is fixed and to establish the signs of the amplitudes. Equal amounts of lead are found to bind at radii of 25 A. and 84 A.

Using the signs so determined, the Fourier–Bessel transform of the equatorial amplitudes has been calculated to give the radial density distribution in the virus particle. This is shown in Fig. 1.

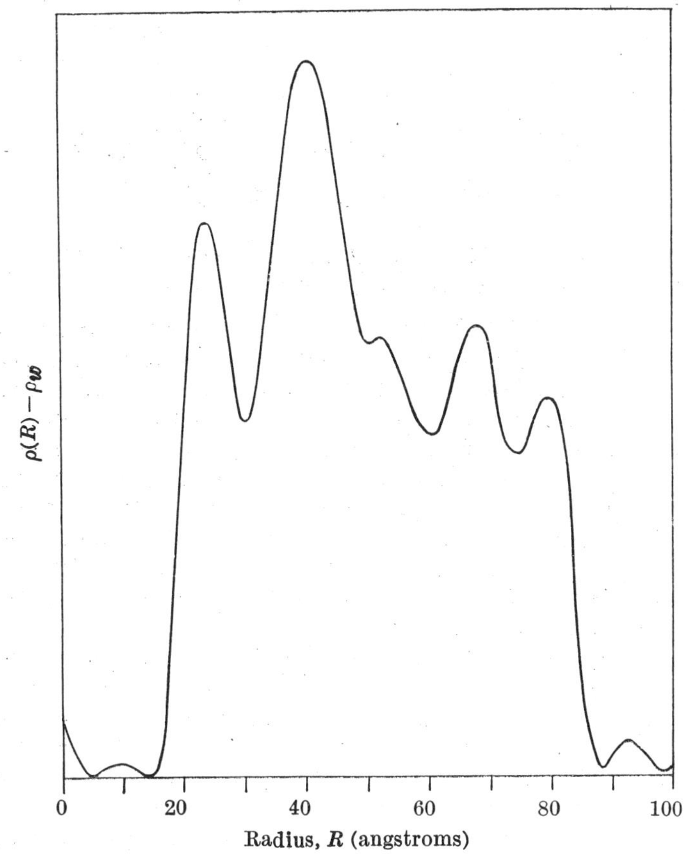

Fig. 1. Radial electron-density distribution in the tobacco mosaic virus particle plotted as a function of distance from the axis of the particle. Ordinate values are proportional to the mean density in excess of that of water

The following features of this transform should be noted: (1) The tobacco mosaic virus particle is hollow. A hole of radius about 19 A. extends along the axis of the particle. (2) The effective radius of the virus in solution is about 84 A., which is appreciably greater than the value 75 A.[1] obtained from the inter-particle distance in dry paracrystals. This difference is consistent with the results of Franklin and Klug[5], who found that there are rather deep indentations in the virus surface which follow the pitch of the helix in such a way that the particles can intermesh in the dry paracrystals. (3) There is a region of high density at a radius of 24 A. and one of higher density at 40 A. The peaks of smaller density at about 67 A. and 79 A. are also significant, but the other minor peaks are diffraction effects from the termination of the Fourier–Bessel integral.

Electron microscope studies[6] have shown that the ribonucleic acid in tobacco mosaic virus is located near the particle axis. From the present work alone it is not possible to decide at what radius the ribo-

2

nucleic acid is located. This question is resolved by comparison of this density distribution for the normal virus with that for the ribonucleic acid – free virus protein, described by Dr. R. E. Franklin in the following communication.

Details of this work, which was supported in part by a grant from the U.S. National Science Foundation, will be described elsewhere.

D. L. D. CASPAR*

Biophysics Department, Yale University,
New Haven, Conn.,
and
Biology Division,
California Institute of Technology,
Pasadena, Calif. Feb. 27.

* Public Health Service Research Fellow of the National Cancer Institute, U.S.A.

[1] Bernal, J. D., and Fankuchen, I., *J. Gen. Physiol.*, **25**, 111 (1941).

[2] Watson, J. D., *Biochim. Biophys. Acta*, **13**, 10 (1954). Franklin, R. E., *Nature*, **175**, 379 (1955).

[3] Cochran, W., Crick, F. H. C., and Vand, V., *Acta Cryst.*, **5**, 581 (1952).

[4] Bragg, Sir Lawrence, and Perutz, M. F., *Proc. Roy. Soc.*, A, **213**, 425 (1952).

[5] Franklin, R. E., and Klug, A., *Biochim. Biophys. Acta*, [**19**, 403 (1956)].

[6] Schramm, G., Schumacher, G., and Zillig, W., *Nature*, **175**, 549 (1955). Hart, R. G., *Proc. U.S. Nat. Acad. Sci.*, **41**, 261 (1955).

Location of the Ribonucleic Acid in the Tobacco Mosaic Virus Particle

IN the rod-shaped particle of tobacco mosaic virus, the protein is in the form of sub-units set in helical array about the long axis of the particle[1]. Electron microscope studies[2] have indicated that the virus nucleic acid may form an axial core. In this communication, however, a brief account is given of an X-ray diffraction investigation which provides strong evidence that the nucleic acid in the intact virus particle does not lie very close to the particle axis. The nucleic acid seen as a core in electron micrographs of partially degraded tobacco mosaic virus is therefore presumably in a collapsed form.

Using monochromatic copper $K\alpha$ radiation and a high-resolution focusing X-ray camera, measurements have been made of the scattered intensities on the equator and layer-lines of X-ray fibre-diagrams of orientated gels of the following substances: (1) tobacco mosaic virus, normal strain, prepared by Dr. H. Fraenkel-Conrat; (2) a mercury-substituted tobacco mosaic virus containing one mercury atom per 20,000 molecular weight of protein, prepared by Dr. H. Fraenkel-Conrat; (3) repolymerized *A*-protein prepared by Prof. G. Schramm[3]. This material is

3

obtained from a normal strain of tobacco mosaic virus by breaking down the virus in weak alkali, removing the nucleic acid by electrophoresis, and repolymerizing the protein by lowering the *p*H.

A few non-monochromatic measurements were also made on a mercury-substituted repolymerized *A*-protein prepared by Dr. H. Fraenkel-Conrat.

The above materials, although prepared in two different laboratories, were all obtained from 'normal' virus strains. All normal strains of tobacco mosaic virus which we have so far examined give X-ray diagrams which are almost indistinguishable from one another. Moreover, our equatorial intensities, obtained photographically, are in excellent agreement with Caspar's data[4] obtained with a Geiger counter and a different virus preparation. Quantitative comparisons of diffraction data obtained from the materials listed above are therefore justified. The main purpose of this communication is to report such a comparison of accurate measurements on the normal tobacco mosaic virus and the repolymerized *A*-protein preparations (1 and 3 above).

It has previously been shown[5] that the X-ray fibre-diagram of repolymerized *A*-protein resembles that of tobacco mosaic virus so closely that there can be no doubt that the structural arrangement of the protein characteristic of the virus is preserved in the repolymerized protein. There are, however, important points of difference in the two diagrams. These are most marked on the equator, but are significant also on all the layer lines which can be accurately measured. It is from a study of these differences that the location of the nucleic acid in the tobacco mosaic virus particle can be determined.

Before discussing the differences between the tobacco mosaic virus and repolymerized *A*-protein diagrams, however, it is necessary to mention some results obtained from a comparison of the normal and the mercury-substituted virus, that is (1) and (2) above (Franklin, R. E., and Holmes, K. C., unpublished results). This has both served to confirm Caspar's allocation of signs[4] to the equatorial intensity maxima of the normal virus, and led to a new determination (Franklin, R. E., unpublished results) of the hitherto somewhat uncertain parameters of its helical structure[1,6]. This makes it possible to make the comparison between tobacco mosaic virus and repolymerized *A*-protein by two independent methods.

In the first method, the intensities recorded on the layer-lines for the two materials were compared. Using the newly determined parameters of the helical structure of the virus it can be shown that,

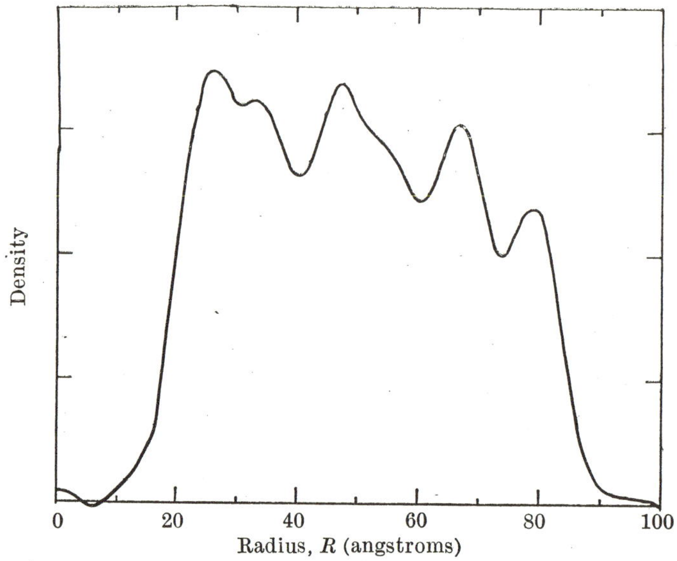

Fig. 1. Radial density distribution in repolymerized, nucleic acid-free tobacco mosaic virus protein. Ordinates are proportional to the excess electron density of the protein over that of water. Note that the density maximum observed at 40 A. for the complete virus (ref. 4) is here replaced by a density minimum

on each of the first eight layer-lines, the principal difference between the X-ray diagrams of tobacco mosaic virus and repolymerized *A*-protein must be attributed to a structural difference at a radial distance of about 40 A. from the particle axis. This strongly suggests that the nucleic acid in tobacco mosaic virus is situated in this position.

The second method is independent of the helical structure of the virus. The radial density distribution in repolymerized *A*-protein has been calculated by means of the Fourier–Bessel transform of the equatorial amplitudes. Amplitudes for values of $2\sin\theta/\lambda$ up to 0·1 were used for this purpose. It was found possible to determine the signs of the amplitudes unambiguously in this range by comparing the intensity data with those for tobacco mosaic virus on one hand, and mercury-substituted repolymerized *A*-protein on the other. To eliminate or reduce spurious peaks in the transform, an 'artificial temperature factor' of $\exp\{-(11{\cdot}8 \times 2\sin\theta/\lambda)^2\}$ was imposed on the amplitude data. The resulting radial distribution function is shown in Fig. 1.

This distribution bears sufficient resemblance to the density distribution of tobacco mosaic virus calculated by Caspar[4] to show once again (cf. ref. 5) that the structural arrangement of the protein in the two materials is very similar indeed. (Some minor points of difference between the two curves may be attributed to the different cut-off position used and to the use of an 'artificial temperature factor' for the repolymerized *A*-protein. For example, the small peak at 53 A. in Caspar's curve is a spurious maximum resulting from the sharp cut-off at $2\sin\theta/\lambda = 0{\cdot}094$.)

5

But there is one striking difference between the two curves. Whereas tobacco mosaic virus has its strongest density *maximum* at a radial distance of 40 A., repolymerized *A*-protein has a pronounced *minimum* at just this position. The radial density distribution in the protein thus provides strong confirmation of the conclusion, reached independently from a study of the layer lines, that the principal structural difference between tobacco mosaic virus and repolymerized *A*-protein occurs at a radius of 40 A. The strong density maximum at 40 A. in tobacco mosaic virus must therefore be due to the presence of nucleic acid at this radius.

In X-ray scattering by nucleic acids, the phosphate group of the nucleotides acts effectively as a 'heavy atom'. X-ray studies of deoxyribonucleic acid have shown[7] that scattering by this compact group can have an effect which predominates over that due to the rest of the nucleotide. It is thus the phosphate back-bone chain of the ribonucleic acid molecule which must lie on a radius of about 40 A. in the tobacco mosaic virus particle.

It will be noted that the first density maximum, at a radius of 25 A., is present in both tobacco mosaic virus and repolymerized *A*-protein, and represents an approximately equal weight of material in the two substances. We must therefore conclude that, whereas the 40-A. density peak in tobacco mosaic virus is due to nucleic acid, the 25-A. peak is due to protein. Thus the nucleic acid in tobacco mosaic virus is deeply embedded in the virus protein. Its molecular structure and chain direction must therefore be closely related to the structural arrangement of the protein. Further study of the X-ray data is in progress with the view of determining how many nucleic acid chains the tobacco mosaic virus particle contains, and in exactly what direction they run.

The work summarized above will be described in detail elsewhere. I am most grateful to Dr. H. Fraenkel-Conrat and to Prof. G. Schramm for supplying me with the tobacco mosaic virus and virus derivatives used in this investigation. The work was supported by the Agricultural Research Council.

ROSALIND E. FRANKLIN

Birkbeck College Crystallography Laboratory,
University of London, W.C.1.

Feb. 27.

[1] Watson, J. D., *Biochim. Biophys. Acta*, **13**, 10 (1955).

[2] Schramm, G., Schumacher, G., and Zillig, W., *Nature*, **175**, 549 (1955). Hart, R. G., *Proc. U.S. Nat. Acad. Sci.*, **41**, 261 (1955).

[3] Schramm, G., *Z. Naturforsch.*, **2b**, 112, 249 (1947).

[4] Caspar, D., thesis, Yale University (1955); and *Nature*, preceding communication.

[5] Franklin, R. E., *Biochim. Biophys. Acta*, **18**, 313 (1955).

[6] Franklin, R. E., *Nature*, **175**, 379 (1955).

[7] Franklin, R. E., and Gosling, R. G., *Nature*, **172**, 156 (1953).

Printed in Great Britain by Fisher, Knight & Co., Ltd., St. Albans.

UV Irradiation of Tobacco Mosaic Virus

Albert Siegel
Dept. of Biological Sciences, Wayne State University, Detroit, MI

I went to UCLA in 1951 supported by a United States Public Health Service postdoctoral fellowship under the sponsorship of John G. Bald to work on the genetics of TMV, if such should exist. The type of experiments I proposed could be more conveniently done in Samuel G. Wildman's laboratory and, thus, I had the advantage of the tutelage and collaboration of both of these experienced investigators. Genetic recombination had recently been demonstrated to occur between bacteriophage strains that differed in at least two easily recognized phenotypes and I had in mind to determine if the same could be shown to occur with TMV. As a first step I set out to collect a group of strains and to characterize them in as many ways as possible (Siegel and Wildman, 1954). I found that, among other things, strains differed considerably in their sensitivity to inactivation with UV. The attempts to observe recombination between TMV strains were not going well so we decided, in the meantime, to try to determine the nature and time course of infection events. Methods were not available at that time for biochemical analysis of cells undergoing synchronous infection, so we cast about for other techniques. It occurred to us that the method of Luria and Latarjet (1947) might be useful to study the events occurring in the primary infected cells of an inoculated hypersensitive host. These investigators had analyzed the events of bacteriophage infection by observing changes in the sensitivity to irradiation of infection at progressive times after inoculation. Preliminary work showed that inoculated *Nicotiana glutinosa* leaves could be detached and maintained in flat trays with their petioles immersed in wet cotton wool for a period long enough for necrotic lesions to develop. We gained confidence in the feasibility of our approach by observing that when leaves were irradiated with UV soon after inoculation, lesions were prevented from forming. Most importantly, the sensitivity of different TMV strains to UV prevention of lesion formation mimicked their UV sensitivity *in vitro*, thereby demonstrating that irradiation was killing the infectious agent and not the leaf cells. Information obtained concerning the time course of early infection events is detailed in the accompanying paper (Siegel and Wildman, 1956).

About this time the discovery was made that TMV RNA was infectious (Fraenkel-Conrat, 1956; Gierer and Schramm, 1956) and William Ginoza became expert in its preparation. We soon discovered that infectious RNA had the same UV sensitivity as the most sensitive of our TMV strains (U2), even though it was derived from a resistant strain (U1, common) (Siegel *et al.*, 1956), indicating that the RNA of the U1 strain was protected from UV damage by its protein coat. Interesting observations were made: one, that resistance was not restored when U1 virus was reconstituted from its RNA and protein, suggesting that rare host proteins found in TMV preparations (Dunigan *et al.*, 1988) might perhaps play a role in the resistance phenomenon; and, two, that UV resistance was lost when the virus was irradiated dry (Siegel, 1957). Irradiation of inoculated leaves revealed that the course of infection proceeded instantaneously with RNA whereas there was about a two hour time lag with virus (Siegel *et al.*, 1957). We interpreted this to signify that, as with bacteriophage, TMV came apart into its components upon inoculation and that this took about two hours at 20°C. In confirmation of this, Wildman (1959) observed that lesions appeared on attached leaves two hours sooner with RNA as inoculum than with virus.

I went on to do other things, but work continued in the Wildman laboratory on the nature of infectible sites, infection centers and different types of host resistance, primarily by Warren Furimoto, Jia-Hsi Wu and Irving Rappaport. I had the opportunity to review these experiments and others dealing with early events of infection at a memorable symposium in Wageningen (Siegel, 1966). Later, after I had moved on to the University of Arizona and then to Wayne State University, I participated in studies of other aspects of the infection process, first in leaf tissue and then in protoplasts (Jackson *et al.*, 1971; Jackson *et. al.*, 1972; Siegel *et al.*, 1973; Siegel, *et. al.*, 1976; Siegel *et al.*, 1978). I have been fortunate in having stimulating and compatible graduate students, fellows and collaborators during my active career, some of whom have contributed much to the work described here.

References

Dunigan, D. D., Dietzgen, R. G., Schoelz, J. E., and M. Zaitlin. (1988). Tobacco mosaic particles contain ubiquitinated coat protein subunits. *Virology* **165**:310-312.

Fraenkel-Conrat, H. (1956). The role of the nucleic acid in the reconstitution of active tobacco mosaic virus. *J. Am. Chem. Soc.* **78**:882-883.

Gierer, A. and G. Schramm. (1956). Infectivity of ribonucleic acid from tobacco mosaic virus. *Nature* **177**:702-703.

Jackson, A. O., Mitchell, D., and A. Siegel. (1971). Replication of tobacco mosaic virus. I. Isolation and characterization of double-stranded forms of ribonucleic acid. *Virology* **45**:182-191.

Jackson, A. O., Zaitlin, M., Siegel, A., and R. I. B. Francki. (1972). Replication of tobacco mosaic virus. III. Viral RNA metabolism in separated leaf cells. *Virology* **48**:655-665.

Luria, S. E. and R. Latarjet. (1947). Ultraviolet irradiation of bacteriophage during intracellular growth. *J. Bacteriol.* **53**:149-163.

Siegel, A. (1957). Ultraviolet irradiation of dry tobacco mosaic virus. *Nature* **180**:1430.

Siegel, A. (1966). The first stages of infection. *In*: Viruses of Plants. A. Beemster and J. Dijkstra, eds. North-Holland Co., Amsterdam, pp. 3-18.

Siegel, A. and S. G. Wildman. (1954). Some natural relationships among strains of tobacco mosaic virus. *Phytopathology* **44**:277-282.

Siegel, A. and S. G. Wildman. (1956). The inactivation of the infectious centers of tobacco mosaic virus by ultraviolet light. *Virology* **2**:69-82.

Siegel, A., Ginoza, W., and S. G. Wildman. (1957). The early events of infection with tobacco mosaic virus nucleic acid. *Virology* **3**:554-559.

Siegel, A., Hari, V., and K. Kolacz. (1978). The effect of tobacco mosaic virus infection on host and virus-specific protein synthesis in protoplasts. *Virology* **85**:494-503.

Siegel, A., Wildman, S. G., and W. Ginoza. (1956). Sensitivity to ultraviolet light of infectious tobacco mosaic virus nucleic acid. *Nature* **178**:1117-1118.

Siegel, A., Zaitlin, M., and C. T. Duda. (1973). Replication of tobacco mosaic virus. IV. Further characterization of viral related RNAs. *Virology* **53**:75-83.

Siegel. A., Hari, V., Montgomery, I., and K. Kolacz. (1976). A messenger RNA for capsid protein isolated from tobacco mosaic virus-infected tissue. *Virology* **73**:363-371.

Wildman, S. G. (1959). Observations on the biological behaviour of tobacco mosaic virus nucleic acid. *Proc. Natl. Acad. Sci. USA* **45**:300-305.

Biographical Sketch

Albert Siegel received a B. A. in 1947 from Cornell University and a Ph.D. in 1951 from the California Institute of Technology. He was a USPH Postdoctoral Fellow at UCLA from 1951-1953 and an Assistant Research Botanist from 1953-1959. He was a Professor of Agricultural Biochemistry at the University of Arizona from 1959-1972. From 1965-1966, he was a Visiting Investigator at the Virus Research Laboratory, Cambridge, England and he served as the Biochemistry Program Director at NSF from 1967-1968. From 1972-1993, he was Professor of Biology, Wayne State University (WSU) and Visiting Professor, University of Michigan, 1978-1979. Since 1994 he has been Professor Emeritus at WSU. In 1978 he was elected a Fellow of the American Phytopathological Society. In addition, he received a Faculty Research Award at the WSU chapter of Sigma Xi in 1983, the Distinguished Graduate Faculty Award in 1984, and was elected to the Academy of Scholars at WSU in 1984. Other awards include the Achievement Award, Michigan Association of Trustees and Governors, 1984, and the Gershenson Distinguished Faculty Fellowship, 1986.

Reprinted from VIROLOGY, Volume 2, No. 1, February 1956
Academic Press Inc. *Printed in U.S.A.*

VIROLOGY **2,** 69–82 (1956)

The Inactivation of the Infectious Centers of Tobacco Mosaic Virus by Ultraviolet Light[1]

ALBERT SIEGEL AND SAM G. WILDMAN

Botany Department, University of California, Los Angeles, California

Accepted October 27, 1955

The early events of tobacco mosaic virus infection were studied by following the radiosensitivity of infection as a function of time after inoculation. At 20°, the survival of infection as a function of ultraviolet light dose is exponential for approximately 5 hours after inoculation, after which the appearance of new intracellular infectious particles is indicated by a change to a multitarget survival curve. During the first 5 hours after inoculation, the infection is characterized by three discernible phases of sensitivity to ultraviolet light (UV): (1) a period during which the infection has essentially the same sensitivity as the virus *in vitro*; (2) a period during which resistance to UV increases; and (3) a period during which a secondary level of resistance is maintained. The timing of these phases is markedly dependent on temperature.

The exponential survival curve during the first phase is independent of inoculum concentration, indicating that an exclusion mechanism is operative which permits only one infectious virus particle to initiate an infection.

INTRODUCTION

Although the extracellular particles of tobacco mosaic virus have been extensively characterized, little is known about the mode of entry of the virus particle into the host cell and the subsequent intracellular events which lead to the production of more virus. This paper presents the results of a series of experiments designed to gain information about the early time sequence of TMV infection. The basic approach is that originated by Luria and Latarjet (1947) for the study of bacteriophage infection, in which radiosensitivity of infection is followed as a function of time after inoculation.

[1] This work was supported in part by contract AT(11-1)-34, Project 8 of the Atomic Energy Commission, and a research grant, G-4124, from the Division of Research Grants, National Institutes of Health, Public Health Service.

MATERIALS AND METHODS

Two strains of TMV were used in these experiments, U1, a common strain of TMV maintained in this laboratory, and U2, a strain which gives rise to a mild, light green-dark green, mosaic in tobacco and which differs from the common strain in a wide range of chemical, physical, and biological properties (Siegel and Wildman, 1954; Ginoza and Atkinson, 1955). Except where otherwise specified, the inoculum contained 10^{-3} mg of virus per ml in pH 6.9 phosphate buffer. All inocula contained 50 mg/ml of Celite.

The local-lesion host *Nicotiana glutinosa* was used in all experiments.

The irradiation source was a Westinghouse Sterilamp, 782L-30 whose output is primarily at 2537 A. Material to be irradiated was placed 25 cm from the lamp. Virus was irradiated in phosphate buffer suspensions 1 to 2 mm thick.

The half-leaf assay method, used for determining the effects of UV on virus *in vitro*, as well as for determining the infectivity dilution curve, has been described (Siegel and Wildman, 1954). This method allows for the infectivity comparison of thirteen samples within an experiment. Each leaf contains a comparison of two samples appearing on the opposite halves of the leaf. Within the experiment every possible comparison of samples is made. Each sample is replicated on 24 half-leaves, and the assay is so designed as to minimize variation due to the use of different plants, different leaf levels on the same plant, and the two halves of the same leaf. In experiments where virus samples were irradiated *in vitro*, the zero dose value comprised three of the thirteen samples, the other ten samples consisting of duplicates of each of five virus aliquots which had received a graded series of UV doses. A similar procedure was followed in determining the infectivity dilution curve; that is, each virus concentration comprised at least two of the thirteen samples to be compared within an experiment. The sum of lesions on the 24 half-leaves for an individual sample seldom differed by more than 10% from its duplicate (or triplicate).

The following procedure was used when inoculated leaves were irradiated. After inoculation the leaves were washed with tap water. For ease of irradiation, the leaves were detached at the base of their petioles and laid flat into trays 12 × 18 × 1 inch containing cotton saturated with water. Between the time of inoculation and irradiation, the leaves were maintained under controlled conditions of light and temperature, either detached from the plants in early experiments, or left attached to the plants in later experiments. Subsequent to irradiation, the trays were placed in moist chambers in the greenhouse for 3 to 5 days until lesions appeared on the leaves and became large enough to count conveniently. The assay consisted of the half-leaf analysis described above, modified so that a group of irradiated half-leaves each receiving the same dose comprised a sample. Each tray contained the six leaves of a single plant. At the time of irradiation the tray was placed 25 cm from the UV lamp; the doses on the individual half-leaves were controlled with the use of aluminum foil shields. Each experiment consisted of thirteen plants, which allowed for twelve replicates of each treatment. As in the *in vitro* irradiation experiments, the zero dose value comprised three of the thirteen treatments, and each of five UV doses, graded in intensity, was duplicated. Again as in the *in vitro* irradiation experiments, duplicates rarely differed by more than 10%. In later experiments, only three leaves were taken from each plant to ensure more uniform results. When this was done, twice the number of plants were used and an assay similar to the one above was employed.

In some experiments, only a single UV dose was tested at different times after inoculation. For each post-inoculation time period, the number of lesions appearing on irradiated half-leaves of three leaves from each of four plants was compared with the number of lesions on the opposite unirradiated half-leaves.

The interpretation of irradiation survival curves obtained in this work closely parallels that of Latarjet (1953).

RESULTS

Effect of UV on the Host

Interpretation of experiments in which the host-virus system is irradiated depends in part on the damaging effects of the radiation to the host. In order that the results be meaningful for the intracellular events of infection, it is of importance that the effect on lesion formation be the action of the radiation on the virus and not on the host. That this is so in the current experiments is shown by Fig. 1, which presents a comparison of the survival curves of strains U1 and U2, irradiated *in vitro*, and soon after inoculation to the host. The two strains differ 5½-fold[2] in

[2] The difference in sensitivity to UV was previously reported to be 7-fold for these two strains, based on uncorrected infectivity data. More precise data corrected to virus survival as measured by the infectivity-dilution curve show the above figure to be more nearly correct.

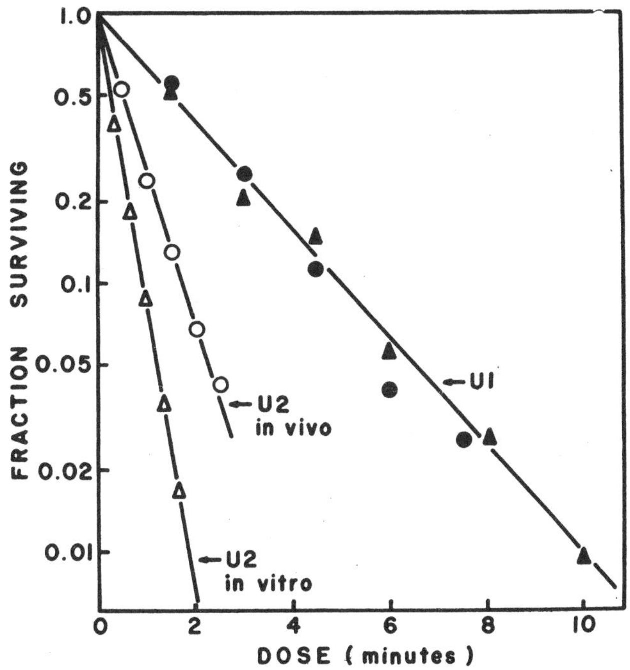

FIG. 1. A comparison of the survival curves obtained when virus suspensions are irradiated *in vitro* and when leaves are irradiated soon after inoculation. Triangles represent fraction of virus surviving when irradiated *in vitro*. Circles represent the number of lesions appearing on irradiated leaves as compared with the number on nonirradiated leaves.

their *in vitro* sensitivity to inactivation by UV (Siegel and Wildman, 1954); nevertheless, the *in vivo* curves closely match the *in vitro* curves for each strain, demonstrating that the further course of infection is largely prevented by the action of UV on the virus and not on the host. The strains differ in so far that U2 is somewhat more resistant when irradiated *in vivo* than *in vitro*, whereas U1 has the same sensitivity under both conditions.

Benda (1955) and Bawden and Kleczkowski (1952) have reported that UV irradiation of bean leaves before inoculation will reduce the number of lesions formed. We have confirmed this effect for *Nicotiana glutinosa* and have found it to be dependent on UV dose, but independent of the strain used as inoculum. This preinoculation irradiation effect, however, has little bearing on experiments in which leaves are irradiated after inoculation. Preinoculation irradiation temporarily prevents the initiation of infection; the leaves recover the ability to become infected on incubation in light. The infection, once initiated, is then eliminated primarily by UV action on the virus and not on the host. In addition to decreasing the ability to become infected, UV will, under the proper conditions, kill the host cells and may thus interfere with postinoculation irradiation experiments. This type of damage can also be largely prevented by incubating the irradiated leaves in the light (Bawden and Kleczkowski, 1952; Benda, 1955). We have also noted that irradiated leaves are more subject to damage by excessive heat and lack of humidity than are nonirradiated leaves and have taken the necessary precautions to eliminate this source of error in the experiments. Finally, in order to obviate the need for excessive doses of UV in these experiments, we have used the U2 strain of TMV which, as previously noted, is several times more sensitive to UV than is the common strain.

Change in the Survival Curve with Time after Inoculation

Figure 2 presents the UV survival curves of infective centers obtained at increasing times after inoculation. In this series of experiments the leaves were maintained at 20°, 1000 foot-candles, between the time of inoculation and irradiation. No change in the survival curve is observed

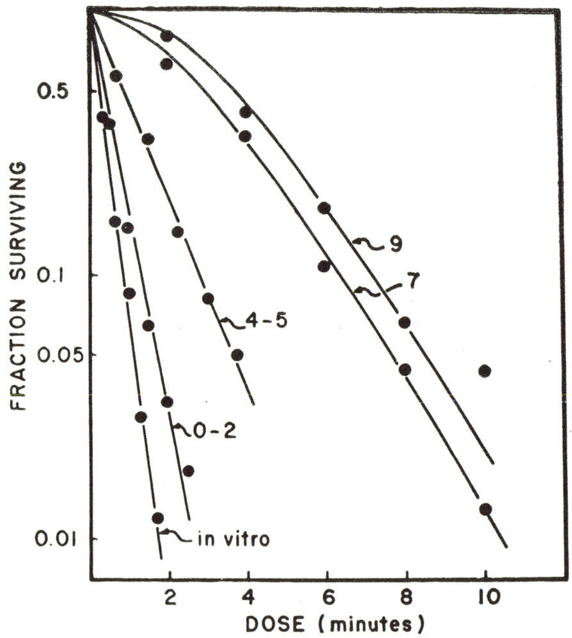

FIG. 2. UV survival curves of strain U2 infective centers at different times after inoculation. The numbers indicate the time in hours between inoculation and irradiation. The points indicate the number of lesions on irradiated leaves compared with the number on nonirradiated leaves. Temperature between inoculation and irradiation, 20°.

for at least 2 hours; precisely the same curve is obtained when the leaves are irradiated immediately after inoculation and 2 hours later. At 4 and 5 hours after inoculation, an exponential curve is obtained whose slope is now a little less than one-half that of the 0- and 2-hour curves. At 7 hours, a multitarget curve is observed whose final slope is approximately the same as that of the 4- and 5-hour curves. Nine hours after inoculation, a multitarget curve is again observed, indicating a greater number of targets than at 7 hours. Although it is difficult to determine the precise number of targets represented by the 7- and 9-hour curves, a reasonable approximation can be made, since the curves are, respectively, theoretical three- and five-hit curves, the points being experimental.

Effect of Environmental Conditions between Inoculation and Irradiation

Figure 3 shows the results of a series of experiments in which the leaves were maintained at 35° between inoculation and irradiation. At this higher temperature, we again observe the same sequence of events, first a change in slope, and then a change from an exponential to a multi-target-type curve. The changes, however, occur more rapidly at 35° than

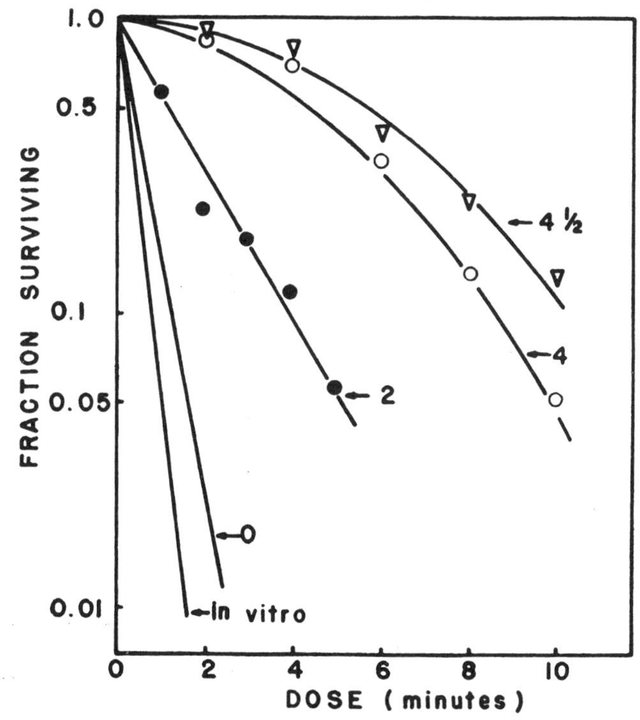

FIG. 3. Same as Fig. 2. Temperature, 35°.

at 20°. The change in slope has already taken place at 2 hours after inoculation and a multitarget curve is observed in 4 hours.

Irradiation with a Constant Dose

The time sequence of change in the survival curve is most readily seen in Fig. 4, which presents the results of experiments at two different temperatures in which inoculated leaves were irradiated with a single dose at different times after inoculation. The early time sequence of infection as determined by changes in sensitivity of infectious centers to UV is divided into four phases:

Phase 1. The sensitivity of the infection to inactivation by UV is essentially that of the *in vitro* virus particle.

Phase 2. The infection increases its resistance to UV with time.

Phase 3. A plateau of resistance is maintained approximately two and one-half times that of phase 1.

Phase 4. A second rise in resistance occurs.

During the first three phases the survival curve is exponential, whereas the secondary rise in resistance during phase 4 is due to the change in the survival curve from exponential to multitarget.

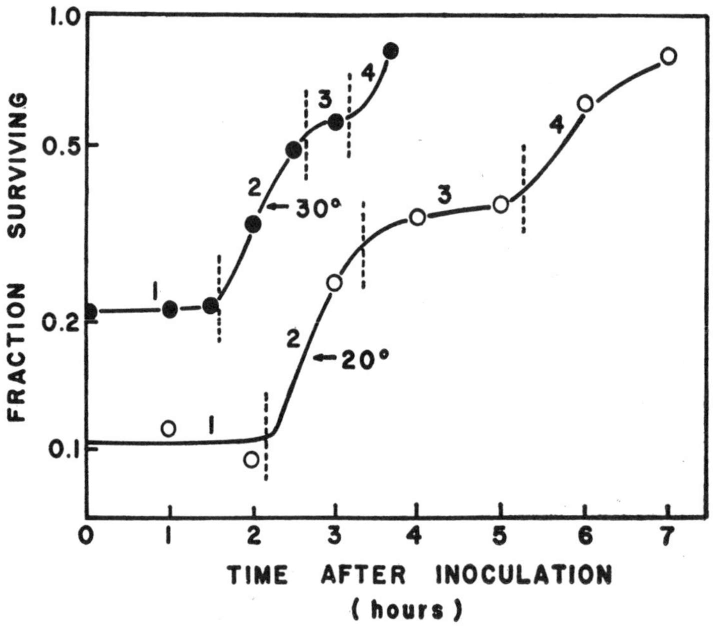

FIG. 4. UV survival of infective centers as a function of time after inoculation. UV dose, 90 seconds. The curves are divided into four phases as described in the text.

The absolute timing of the phases is dependent on the conditions of incubation between inoculation and irradiation. At 20°, the approximate times for the first three phases are, respectively, 2 hours, 1¾ hours and 1½ hours. At 30°, all the phases are of considerably shorter duration, the comparative time intervals being 1½ hours, 1¼ hours, and 1 hour. No significance can be attached to the difference in the level of the curves

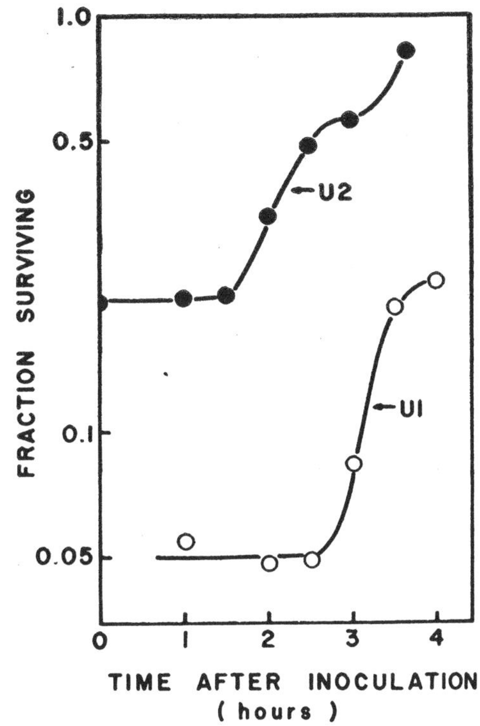

FIG. 5. UV survival of infective centers as a function of time after inoculation. UV dose: U1, 6 minutes, U2, 90 seconds. Temperature, 30°.

obtained at 20° and at 30°; differences of this nature are obtained in experiments performed on different days.

Difference between Strains

Strain U1 has been investigated in a preliminary fashion, and it is evident, as shown in Fig. 5, that the length of time at 30° before the U1 infection changes in sensitivity to UV is about 50% greater than phase 1 of the U2 strain. Thus, the change in the sensitivity of infection to UV is, apparently, due to some change specifically concerned with the infection and cannot be explained in terms of a nonspecific host effect, such as the recovery of the leaves with time from the rubbing damage associated with inoculation.

Effect of Inoculum Concentration on the Survival Curve of Infective Centers

The survival curve of infective centers is independent of the inoculum concentration. Exponential curves were obtained during phase 1 over the tenfold concentration range of applied virus from 5×10^{-4} to 5×10^{-3} mg of virus per milliliter. Following the analysis of the infectivity-dilution curve formulated by Youden *et al.* (1935) and Bald (1937), these concentrations are equivalent to an average of 1.4 to 7 infectious units per infected site. If all the particles which are present at a susceptible site actually participate in the infection, we would have expected that the UV survival curves would be of a multi-target nature and markedly dependent on the inoculum concentration over the range investigated. Since an exponential curve, independent of concentration, was observed, there is a strong indication that only one infectious particle participates in an infection even when several such particles are present.

The infectivity-dilution curve obtained under our assay conditions

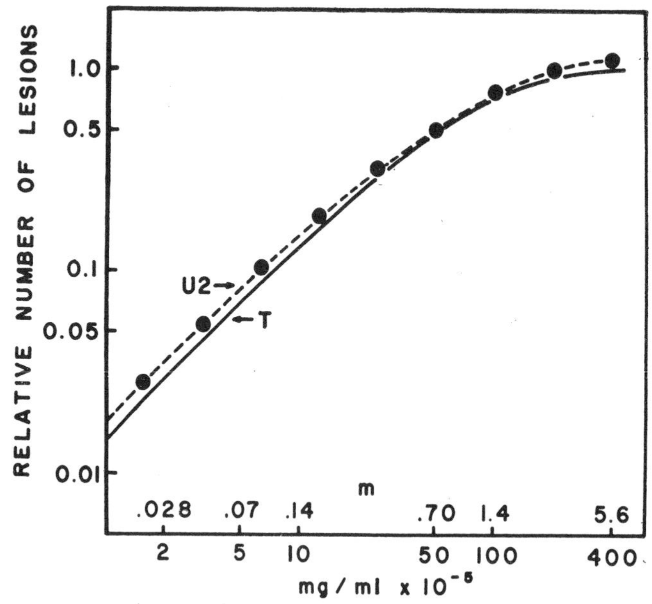

FIG. 6. Relationship between number of lesions and concentration of virus in the inoculum. T is the theoretical curve described in the text; m is the average number of infectious particles per susceptible site calculated from the theoretical curve.

is shown in Fig. 6 along with the theoretical curve for the expression

$$\frac{Y}{N} = 1 - e^{-m}$$

where $\frac{Y}{N}$ = relative number of lesions.

m = average number of infectious units per susceptible site.

The experimental values are very close to the theoretical. The calculated values of m corresponding to the different concentrations of our virus preparation are given on the abscissa.

As an independent check that the higher virus concentrations present several infectious units per susceptible site on the leaf, different concentrations of virus were irradiated in buffer solution and then assayed. The results, as shown in Fig. 7, demonstrate that the shape of the infectivity survival curve is markedly dependent on the concentration of virus irradiated. The curves are of a definite multitarget nature with the

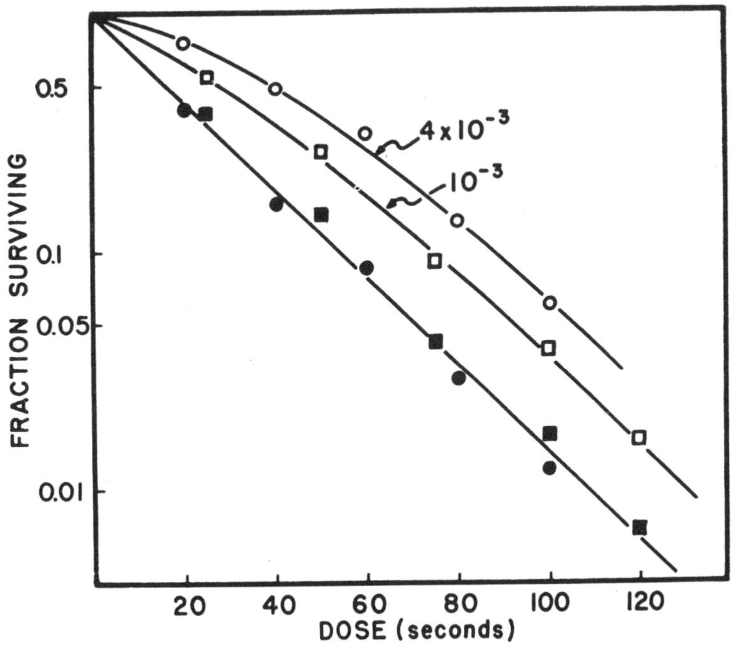

FIG. 7. Infectivity and virus survival when strain U2 is irradiated *in vitro*. Open symbols, infectivity survival; closed symbols, virus survival. The concentration in milligrams per milliliter at which the virus preparation was irradiated and assayed is indicated for the upper two curves. The lowest curve is derived from the upper two curves with the use of the infectivity-dilution curve shown in Fig. 6.

higher concentrations indicating a greater number of targets. This information serves to confirm the inference from the infectivity-dilution curve that at the higher inoculum concentrations the susceptible sites are each presented with several particles which are independently capable of initiating the infection. That the virus particles are inactivated in single-hit fashion in these experiments, in agreement with the results of other workers (Bawden and Kleczkowski, 1953; Oster and McLaren, 1950; Price and Gowen, 1937), can be seen from the exponential curve for virus survival that is obtained by correcting the infectivity survival curve with the standard infectivity-dilution curve.

DISCUSSION

We interpret these experiments to indicate that the first three phases of infection are preparatory to virus multiplication and that during the fourth phase new infectious units appear within the cell which are independently capable of maintaining the infection. At 20°, the first three phases take approximately 5 hours, and when new particles appear in the cell their initial rate of increase is an approximate doubling every 2 hours. At higher temperatures the first three phases are completed in a little over 3 hours, and the rate of appearance of new infectious units is greater. Yarwood (1952) has reported the latent period for TMV to be two to three times as great as the above values at comparable temperatures. The significance of this discrepancy cannot be assessed at the present time, since the strain of virus, as well as the host, were different.

The proper interpretation of the change in sensitivity of the infecting virus particle which occurs during the second phase may depend on the extent to which it becomes shielded from UV on entering a cell. Benda (1955) provides evidence that the epidermis transmits only about 25 to 40% of incident UV. If the rise in resistance results from the infecting particle becoming protected by other UV-absorbing materials in the cell, then we would presume that during the first phase the virus particle is attached to the infectible site, but that it is sitting on the outside of the leaf where it is exposed to the full force of irradiation. During the second phase, the particle enters the cell and becomes screened from the radiation and radiation products by cell wall and cytoplasm; the third phase would represent the time that the virus particle is inside the cell before it has started to multiply. We would predict, on the basis of this model, that during the first phase the further course of infection could be prevented by subjecting the leaf to agents which would non-specifically inactivate or remove virus but leave the leaf undamaged. The infection should be partially resistant to such treatment during the second phase and completely resistant during the third and fourth phases. Yarwood (1955) has demonstrated that lesion formation can be prevented by water treatment of leaves for a limited time after inoculation. This, perhaps, may represent just such a treatment in view of the fact that the time period within which the treatment is effective is markedly dependent on temperature.

Another interpretation would follow the one suggested by Benzer (1952) to explain a similar phenomenon observed for the bacteriophage T2. This hypothesis states that the virus in some manner changes in state once inside the cell so that it is now more resistant to UV damage. Zech (1952) has postulated that the TMV particle passes through a stage where it is infectious *in vivo* but much smaller in size than the *in vitro* virus rod. It would be reasonable to suppose that this infectious entity would be more resistant to UV than is the *in vitro* virus particle. Phase 2, on the basis of such an interpretation, would represent the transition from the extracellular to the intracellular infectious unit. Our current data indicate that the final slopes of the survival curves during phases 3 and 4 are nearly alike. This would suggest that the infection, once initiated by the characteristic virus rod, can be maintained by the smaller intracellular infectious units. This is in contrast to the situation in phage T2, where, after an initial change in slope, comparable to phase 2 in this instance, the final slope returns to the phase 1 slope when new infectious particles appear within the cell. It is possible, however, that a similar increase in slope of the multitarget curve would be observed in the TMV system, were the time increased between inoculation and irradiation.

The conclusion that only one particle may participate in an infection stands in seeming contradiction to the results of Lauffer and Price (1945), who have recovered two different strains of TMV from single lesions when a mixed inoculum had been used. Our data, however, bear only on the number of particles of a single strain that may participate in an infection. The possibility exists that no exclusion mechanism is operative between strains. On the other hand, it is possible that the exclusion mechanism is imperfect and that at high concentrations of applied virus a low level of multiple infection takes place which has gone undetected in our experiments. In support of this notion, Lauffer and Price (1945) found many fewer mixed infections than had been predicted from the concentrations of the strains in their inocula.

The conclusion that an exclusion mechanism is operative for TMV is equally valid if one accepts the interpretation of Kleczkowski (1950) for the infectivity-dilution curve. This formulation is based on the hypothesis that regional susceptibility of the leaf varies in such a way that the logarithm of the minimal effective concentration is normally distributed. If one considers on the basis of this hypothesis that the infectible sites require different numbers of particles to cooperate in the initiation of an infection, then one would expect a multitarget curve for survival of infective centers if all the particles present could enter the site, and a curve with an upward concavity in the case that exclusion was operative. Our current data, therefore, rule out this model. The assumptions are also permissible that only one particle is sufficient to initiate infection but that the sites, for instance, differ in size and are infected with different probabilities when a uniform concentration of virus is applied to the leaf. On the basis of this model, we would again expect a multitarget survival curve in the case that more than one particle could participate in the infection, and an exponential curve, as observed, in the case of exclusion.

ACKNOWLEDGMENTS

The authors should like to thank Miss Maris Allen and Miss Sue Smith for their valuable assistance in performing the experiments.

REFERENCES

Bald, J. G. (1937). The use of numbers of infections for comparing the concentration of plant virus suspensions. *Ann. Appl. Biol.* **24,** 33–55.

Bawden, F. C., and Kleczkowski, A. (1952). Ultraviolet injury to higher plants counteracted by visible light. *Nature* **169,** 90.

Bawden, F. C., and Kleczkowski, A. (1953). The behaviour of some plant viruses after exposure to ultraviolet radiation. *J. Gen. Microbiol.* **8,** 145–156.

Benda, G. T. A. (1955). Some effects of ultraviolet irradiation on leaves of french bean. *Ann. Appl. Biol.* **43,** 71–85.

Benzer, S. (1952). Resistance to ultraviolet light as an index to the reproduction of bacteriophage. *J. Bacteriol.* **63,** 59–72.

Ginoza, W., and Atkinson, D. E. (1955). Comparison of some physical and chemical properties of eight strains of tobacco mosaic virus. *Virology* **1,** 253–260.

Kleczkowski, A. (1950). Interpreting relationships between the concentrations of plant viruses and numbers of local lesions. *J. Gen. Microbiol.* **4,** 53–69.

Latarjet, R. (1953). In *The Nature of Virus Multiplication* (P. Fildes and W. E. Van Heyningen, eds.), pp. 175–193. Cambridge University Press, New York.

Lauffer, M. A., and Price, W. C. (1945). Infection by viruses. *Arch. Biochem.* **8,** 449–468.

Luria, S. E., and Latarjet, R. (1947). Ultraviolet irradiation of bacteriophage during intracellular growth. *J. Bacteriol.* **53,** 149–163.

Oster, G., and McLaren, A. D. (1950). The ultraviolet light and photosensitized inactivation of tobacco mosaic virus. *J. Gen. Physiol.* **33,** 215–228.

Price, W. C., and Gowen, J. W. (1937). Quantitative studies of tobacco mosaic virus inactivation by ultraviolet light. *Phytopathology* **27,** 267–282.

Siegel, A., and Wildman, S. G. (1954). Some natural relationships among strains of tobacco mosaic virus. *Phytopathology* **44,** 277–282.

Yarwood, C. E. (1952). Latent period and generation time for two plant viruses. *Am. J. Botany* **39,** 613–618.

Yarwood, C. E. (1955). Deleterious effects of water in plant virus inoculations. *Virology* **1,** 268–285.

Youden, W. J., Beale, H. P., and Guthrie, J. D. (1935). Relation of virus concentration to the number of lesions produced. *Contribs. Boyce Thompson Inst.* **7,** 37–53.

Zech, H. (1952). Untersuchungen über den Infektionsvorgang und die Wanderung des Tabakmosaikvirus im Pflanzenkörper. *Planta* **40,** 461–514.

In Vitro Assembly of Tobacco Mosaic Virus

George Lomonossoff
John Innes Centre, Norwich, U. K.

The 1955 publication by Heinz Fraenkel-Conrat and Robley Williams constitutes a major landmark not only in the history of virology, but also in the development of molecular biology in general. It was the first instance in which a relatively complex biological entity (tobacco mosaic virus) was shown to be capable of correctly assembling from its components outside the cell. As well as prompting further investigations on the *in vitro* self-assembly of TMV itself, the paper stimulated new fields of research, such as the mechanisms of macromolecular assembly and the study of protein-nucleic acid recognition.

At the time the work in question was performed, it was already known that TMV particles consisted of over two thousand identical protein subunits arranged with helical symmetry and that the viral RNA was contained within the core of the particles. It had also been demonstrated that it was possible to dissociate the virus under alkaline conditions, enabling the coat protein subunits to be isolated. Studies on the properties of the isolated protein showed that while it formed only small aggregates (termed "Alkalischer- or A-protein") under alkaline conditions, under acidic conditions it formed helical rods similar in appearance to virus particles, but of indeterminate length (Schramm and Zillig, 1955).

In view of the ability of the isolated protein to form virus-like structures, it was perhaps the next logical step to investigate whether the protein could interact with the viral RNA to form biologically active virus particles. A prerequisite for such a study was the development of a detergent method for purifying the virion RNA since the alkaline conditions used to isolate the coat protein resulted in highly degraded RNA preparations. The authors also had to go to considerable lengths to demonstrate that each component was effectively free from the other and that neither was contaminated with undissociated TMV particles. In the original experiment to achieve reconstitution, preparations of isolated protein (in the form of A-protein) and RNA were mixed at a ratio of approximately 10:1 by weight at pH 9.0, after which the pH of the solution was reduced to between 7.0 and 6.0. Reconstitution was then allowed to proceed at 3°C for 24 hours, after which the mixture was ultracentrifuged under conditions designed to pellet any virus particles. Formation of infectious virus was monitored by inoculating the resuspended pellet on a local lesion host (*Nicotiana glutinosa*). To the stated surprise of the authors, they found that their samples of reconstituted TMV consistently gave a significant number of lesions when inoculated on test plants at 10-100 μg/ml. The lesions obtained were similar in number and appearance to those obtained when the plants were inoculated with a 0.1 μg/ml suspension of native TMV. Critically, no lesions were obtained when the purified protein or nucleic acid were inoculated in equivalent amounts. However, in a footnote the authors did note than some infectivity was obtained when their RNA preparations were inoculated at high doses, a prelude to the subsequent demonstration that the RNA by itself is infectious. As further evidence that correct assembly had taken place, electron microscopy was used to show the presence of 300 nm rods with the appearance of the TMV particles in the reconstituted material; no such rods were seen when the protein or RNA components used for the reconstitution were similarly analysed.

The ability to reconstitute infectious virus particles from apparently inert components caused a great excitement at the time, leading to newspaper headlines to the effect that life had been created in the test tube. Of course, with the demonstration the following year that the isolated RNA component was, indeed, infectious on its own, such claims were quickly disproved and it became apparent that the effect of reconstitution was only to increase, through protection, the specific infectivity of the labile RNA. Because of the low specific infectivity of the reconstituted virus, there was also, apparently, a degree of skepticism in the scientific community about the true meaning of the results. In hindsight, this low specific infectivity can be attributed to two factors: degradation of the RNA preparation used in reconstitution experiment and the fact the protein to RNA ratio (10:1) used was only half that (20:1) in fully assembled TMV particles. In the light of these two factors, it is perhaps fortunate that any infectivity at all was obtained.

The group at Berkeley followed up their initial discovery with a series of papers refining the conditions required for efficient reconstitution. However, even under their optimum conditions, reconstitution of full-length virus particles still took 3-6 hours (Fraenkel-Conrat and Singer, 1964), rather a long time for the assembly of an entity the size of a virus particle. The explanation for this was provided only after a detailed analysis of the factors (temperature, pH and ionic strength) which affect the aggregation state of the isolated coat protein had been carried out. These studies showed that the predominant aggregate under physiological conditions is neither A-protein nor virus-like helices, but rather a two layer disk structure containing 34 subunits. Disks could be formed from A-protein if a preparation of the latter was incubated at 20°C for several hours. In a crucial experiment Butler and Klug (1971) demonstrated that if reconstitution was performed by adding preformed disks to viral RNA, the rate of formation of virions was greatly enhanced. The interpretation of this result was that the two layer disk forms a preformed "jig" which can interact efficiently with a defined sequence on the RNA to initiate or nucleate virus assembly. The slow rate of assembly in the experiments of the Berkeley group can be ascribed to the time taken for the A-protein in the reconstitution mixture to form disks.

While the role of disks in the initiation of TMV assembly was rapidly accepted, the same was not true for the suggestion that disks (as well as smaller aggregates) also played an important part in the elongation of initiated rods. My own contribution to this controversy involved an analysis of the lengths of viral RNA encapsidated during the initial stages of assembly. The results indicated that at least over the first 2 kb, the RNA was encapsidated in series of steps of 50 or 100 nucleotides, a result compatible with addition of one or both layers of the disk (Butler and Lomonossoff, 1978). Unfortunately, like all other attempts to resolve the role of disks in elongation, the results were open to alternative explanations. In retrospect, the controversy regarding the role of disks in elongation is probably unresolvable as it is impossible to completely separate initiation from elongation and the protein aggregates involved all exist in equilibrium with one another.

Throughout all the studies that have been carried out since 1955 on the *in vitro* reconstitution (or reassembly) of TMV, there has always been one nagging question: how relevant are these *in vitro* studies to the assembly process *in vivo*? For most of this time the only answer has been that the conditions optimal for *in vitro* assembly are at least physiologically plausible. However in the past few years, the development of PCR-based methods has allowed the process of both assembly and disassembly to be analysed in protoplasts (Wu and Shaw, 1995). It is reassuring that the results obtained so far as to the rate and bidirectional nature of *in vivo* assembly are compatible with the results obtained through the years of study of the process *in vitro*.

References

Butler, P. J. G. and A. Klug. (1971). Assembly of the particle of tobacco mosaic virus from RNA and disks of protein. *Nature New Biol.* **229**:47-50.

Butler, P. J. G. and G. P. Lomonossoff. (1978). Quantized incorporation of RNA during assembly of tobacco mosaic virus from protein disks. *J. Mol. Biol.* **126**:877-882.

Fraenkel-Conrat, H. and B. Singer. (1964). Reconstitution of tobacco mosaic virus. IV. Inhibition by enzymes and other proteins, and the use of polynucleotides. *Virology* **23**:354-362.

Schramm, G. and W. Zillig. (1955). Uber die strukture des tabakmosaikvirus. IV. Mitt: Die reaggregation des nucleinsaure-freien proteins. *Zeitscrift fur Naturforschung* **10b**:493-499.

Wu, X. and J. G. Shaw. (1996). Bidirectional uncoating of the genomic RNA of a helical virus. *Proc. Natl. Acad. Sci. USA* **93**:2981-2984.

Biographical Sketch

George Lomonossoff's familiarity with the paper in question stems from his time in the late 1970s as a Ph.D. student in Dr. Jo Butler's laboratory at the MRC Laboratory of Molecular Biology in Cambridge (U. K.) where he was studying some of the finer points of TMV self-assembly. In particular, his project concerned the rate and timing of the encapsidation of two RNA tails. Following the completion of his Ph.D. studies he remained in Cambridge for a further year to work on the determination of the nucleotide sequence of the viral RNA, after which he moved to the John Innes Centre in Norwich where he has continued to work on positive-strand RNA plant viruses.

Reprinted from the Proceedings of the NATIONAL ACADEMY OF SCIENCES,
Vol. 41, No. 10, pp. 690–698. October, 1955

*RECONSTITUTION OF ACTIVE TOBACCO MOSAIC VIRUS FROM ITS INACTIVE PROTEIN AND NUCLEIC ACID COMPONENTS**

BY H. FRAENKEL-CONRAT AND ROBLEY C. WILLIAMS

VIRUS LABORATORY, UNIVERSITY OF CALIFORNIA, BERKELEY

Communicated June 17, 1955

Much recent evidence from chemical, physicochemical, electron microscopical, and X-ray studies has resulted in a definite concept of the structure of the tobacco mosaic virus (TMV) particle.[1–5] It appears that about 2,800 protein subunits of a molecular weight near 18,000 are arranged in a helical manner to form a rod with a hollow core. The nucleic acid is believed to occur as strands in the core. Electron micrographs which support this concept have been obtained of the virus at various stages of disaggregation.[3–5] A protein isolated from infected plants has been found to reaggregate—first to short pieces of the presumed helix lying on end and resembling disks with central holes and then to much longer, but inactive, rods of the diameter of the virus yet free from nucleic acid.[6] It has now been possible to achieve the co-aggregation of inactive virus protein subunits and inactive virus nucleic acid to give nucleoprotein rods which appear to be infective.

Preparation of Protein and Nucleic Acid Components.—TMV was dialyzed against pH 10–10.5 glycine buffer (0.01 M) or pH 10.5 carbonate-bicarbonate (0.1 M) at 3° C. for 48–72 hours. Undegraded virus was separated by cold ultracentrifugation, and the supernatant was brought to 0.4 saturation with ammonium sulfate. The protein alone precipitates (optical density$_{260\ m\mu}$/optical density$_{280\ m\mu}$ = R = 0.65), leaving only nucleic acid (R = 2.0) in the supernatant; if this separation is not clean, longer alkali treatment is necessary. The protein moiety is precipitated twice more with 0.25–0.35 saturated ammonium sulfate, dialyzed, brought to pH 7.0–8.0 with NaOH, and finally again freed from heavy particles, such as undegraded virus, by ultracentrifugation. The protein gives a water-clear solution at pH 7; the masked —SH group is still present. The spectrum resembles that of a mixture of tryptophan, tyrosine, cysteine, and phenylalanine, simulating the composition of the protein, although the minimum (at 250 mμ) is not quite as low (max./min. = 2.4 versus 2.9) (Fig. 1); P analyses (0.01–0.03 per cent) indicate removal of about 95–98 per cent of the nucleic acid. Evidence for the absence of detectable virus particles will be discussed below.

The nucleic acid fractions from such alkali-degraded TMV are not as effective for reconstitution as that obtained by the detergent method.[2] A virus solution (1 per cent) containing 1 per cent sodium dodecyl sulfate is adjusted to pH 8.5 and held at 40° for 16–20 hours. Following this treatment, ammonium sulfate is added to 0.35 saturation, and the protein precipitate is separated by centrifugation. When the supernate is refrigerated, from 60 to 90 per cent of the nucleic acid precipitates (R = 2.0) and is centrifuged off the next day; it is further purified by repeated resolution in ice water and precipitation with two volumes of cold ethanol and a few drops of 3 M pH 5 acetate. The nucleic acid solution is finally subjected to cold ultracentrifugation to remove any traces of virus. The virtual absence of protein is indicated by a minimum in O.D. near 230 mμ (max./min. = 3.0) (Fig. 1).

In a few preliminary ultracentrifuge experiments, kindly performed by Dr. Howard K. Schachman, the nucleic acid preparations exhibited one principal boundary with a sedimentation coefficient of about 8 S, resembling the preparation of Cohen and Stanley.[7] The protein, in pH 9 borate (0.01 M) or in 0.01 M NaCl at that pH, exhibited largely a single boundary with a sedimentation coefficient of 4.5 S. At lower pH values larger components, presumably due to aggregation, were observed.

Reconstitution of Active Virus.—For reconstitution of the virus, 1 ml. of approximately 1 per cent protein solution is mixed with 0.1 ml. of a 1 per cent nucleic acid solution. Opalescence appears after addition of a suitable buffer; 0.01 ml. of pH 6 acetate (3 M) has given the best results, but phosphate (pH 6.3 and 7.0) and pH 6 ammonium acetate have also been used successfully. The samples are held at 3° for at least 24 hours. They may then be directly diluted and assayed. More often they were ultracentrifuged, the pellets redissolved in water, traces of insoluble material separated by centrifugation, and aliquots of the opalescent supernatants diluted for spectrophotometry. Most of the protein and 40–60

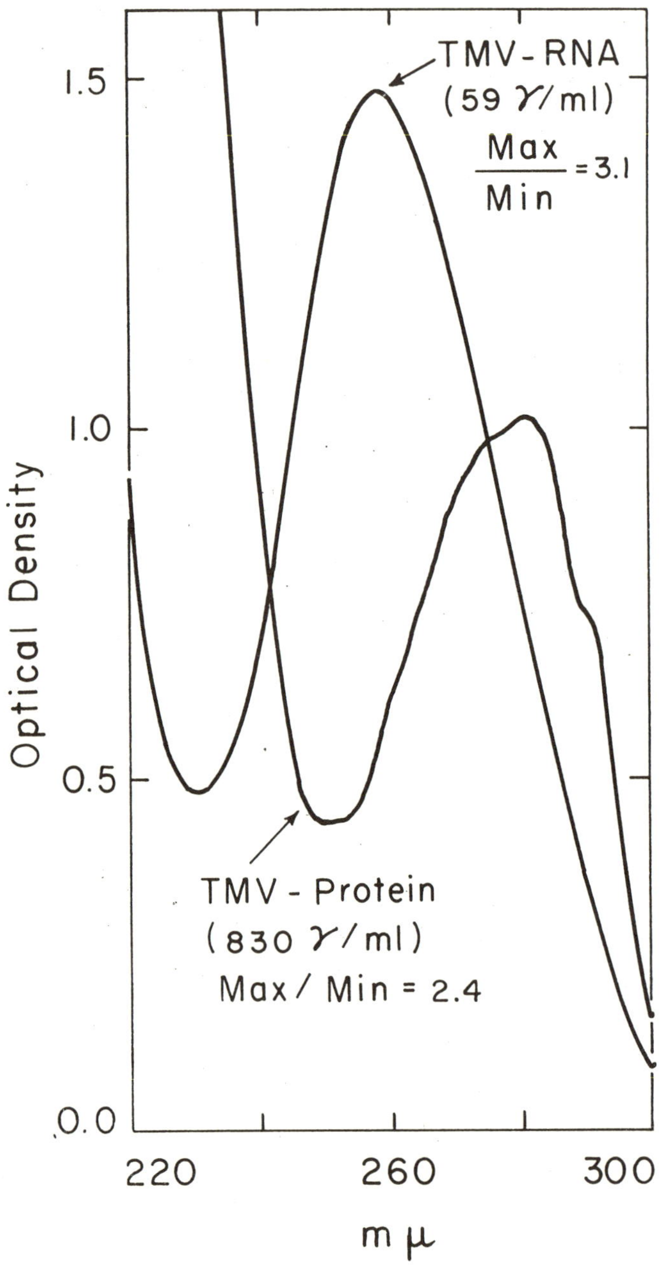

FIG. 1.—Ultraviolet absorption spectrum of purified virus protein and nucleic acid.

per cent of the nucleic acid[8] were in the pellet, and the spectrum was that expected for a nucleoprotein such as TMV (max./min. = about 1.17). On the basis of the O.D. of TMV (0.27 at 260 mμ for an 0.01 per cent solution), approximate concentrations were calculated, and the sample was diluted to a range of 100 to 10 μg/ml. for assay. For pellets composed of protein only (absorption maximum at 280 mμ), calculations were based on the O.D. of the virus proteins (0.13 for an 0.01 per cent solution at that wave length). Assays were performed on groups of 8–12 plants (*Nicotiana glutinosa*), distributing each of an equal number of samples (including at least one standard and often a solvent blank) over 10 equivalent half-leaves. It was indeed quite surprising to find that the reconstituted nucleoprotein preparations produced local lesions (from 2 to 30 per half-leaf), when tested at 10–100 μg/ml., which were indistinguishable in appearance from those of TMV at 0.1 μg/ml. In contrast, no activity was observed when the protein or nucleic acid alone was tested up to 1,800 and 52 μg/ml., respectively.[9] When 0.005 per cent TMV was held for several hours in 5 per cent protein or 0.5 per cent nucleic acid, it was found fully active upon dilution and assay. This strongly suggests that the noninfectivity of the two components cannot be attributed to inhibition phenomena (Table 1).

TABLE 1

ACTIVITY OF THREE PREPARATIONS OF RECONSTITUTED VIRUS AND OF TMV AT DIFFERENT ASSAY LEVELS; ABSENCE OF INHIBITORS FROM COMPONENTS

Material Tested	Assays: μg./Ml.	Assays: Lesions/Half-Leaf*
Reconstituted virus:		
25,000-rpm. pellet	100	27 54 31
		120†
	30	18 — —
	25	— 35 —
	10	9 7 14†
40,000-rpm. pellet	100	2 23 —
TMV stock preparation	0.25	27
		15
	0.1	9 (6.1–12.1)‡
	0.02	5
TMV-protein added at 1000:1	0.1	10
TMV-RNA added at 100:1	0.1	5
		13

* Each figure is the average of 10 equivalent half-leaves. The 3 columns in top half of table refer to separate preparations.

† This is one of the preparations which appeared to increase in activity during storage.

‡ Average and range of 9 assays (10 half-leaves each) of one TMV preparation over 3 months.

When each of the two components was diluted for assay to 0.2 and 0.02 mg./ml., respectively, prior to mixing, no activity was generated. Also, when the complete reaction mixture (5–10 mg./ml.) was diluted to 0.1 mg./ml. one minute after addition of the buffer which triggers the aggregation, no activity was obtained. About one hour at room temperature appears to be required for the formation of any active rods. These experiments represent convincing control data; they also seem to prove that we are dealing with a definite chemical reaction and to exclude most other interpretations. The observation that the nucleic acid gradually loses most of its activity during several weeks of storage at 3° (in aqueous solution, pH 5) also serves as a control experiment, indicating the crucial nature of its physical state. Treatment of the nucleic acid (600 μg.) with ribonuclease (0.2 μg) in 10^{-4} *M* magnesium sulfate rendered it unable to combine with the protein to produce active rods; also, substitution of other nucleic acids (DNA from thymus and RNA from turnip yellow mosaic virus[10]) yielded almost nucleic acid-free pellets which were inactive (Table 2).

A few preliminary experiments on the properties of the reconstituted virus indicated somewhat greater lability to alkali (pH 9) than that of TMV. Ribonuclease (0.2 μg for 5 mg.) also appeared to decrease the activity. A fractionation of the nucleoprotein material by centrifugation at 25,000 rpm. for 30 minutes yielded a pellet of considerably higher infectivity than the residual half of the material, pelleted at 40,000 rpm. (1 hour). Material so purified seemed to increase in activity upon storage in 0.3–0.5 per cent solution at 3°. It also appeared to become more resistant to alkali and to ribonuclease.

Electron Microscopic Studies.—Electron microscopy was used qualitatively to study the shapes and sizes of the components involved in the formation of the virus rods and quantitatively[11] to assess the degree of homogeneity of the suspensions as well as to obtain counts and length distributions of the reconstituted rods. In the protein, the only visible particles, aggregates presumably of the original 4.5 S material, appeared as disks about 5–15 mμ thick and with central holes. The diameter of the disk was 15 mμ and that of the central hole about 4 mμ (Fig. 2).

TABLE 2

EFFECT OF REACTION CONDITIONS ON REGENERATION OF VIRUS ACTIVITY

REACTION CONDITIONS*	ASSAYS μg./Ml.	Lesions/Half-Leaf
Protein (1 per cent) + RNA (0.1 per cent):		
1 minute, pH 6	100	0.6
24 hours	100	10.2, 13.1†
96 hours	100	13.0
Protein (0.01 per cent) + RNA (0.001 per cent):		
24 hours, pH 6	100	0.6, 0.2†
96 hours	100	0.1, 0.2†
Protein + RNase-digested RNA	100	0.0
Protein + TYMV-RNA	100	0.3
Protein‡	500; 625; 1800	0.0; 0.0; 0.0
RNA‡	30; 52; 230	0.0; 0.0; 0.3

* Protein and RNA stand for the preparations isolated from TMV by the methods described in the text. TYMV-RNA is the ribonucleic acid isolated from turnip yellow mosaic virus. RNase is ribonuclease.

† Reassayed dilute solution after one week.

‡ Different preparations, assayed about 10 times, generally on separate plants (see n. 9); in the customary half-leaf assay method, up to 0.5 lesions were occasionally obtained with the same preparations.

The appearance of these perforated disks is identical with that of partially polymerized X-protein.[5] In the nucleic acid solution only occasional poorly defined fibrils could be detected. The reconstituted rods appeared to be identical in shape and size with intact TMV, except for a greater randomness of length (Fig. 3).

Preparations of the purified protein and nucleic acid were examined for any electron microscopic evidence of contamination with intact TMV. In a specific instance the protein was sprayed upon the specimen screens at a concentration of 0.1 per cent and the RNA at 0.007 per cent. These are tenfold and sevenfold greater concentrations of the two components than the highest levels used in the assays of the reconstituted virus (100 μg./ml.). The protein preparation showed no rods of a length even as great as 100 mμ in six droplet patterns each of which had a volume of approximately 3×10^{-9} ml. The RNA material also showed no rods in six droplets of approximately this average volume. From these figures it can be calculated that the constituent protein and RNA solutions contribute fewer than 5×10^{7} typical TMV particles per milliliter to the reconstituted virus in the highest concentration assayed. Additional counts made on the standard TMV used as an infectivity control showed that approximately 7 × 10⁸ particles per milliliter, of typical length, are required to produce the 10 lesions per leaf custom-

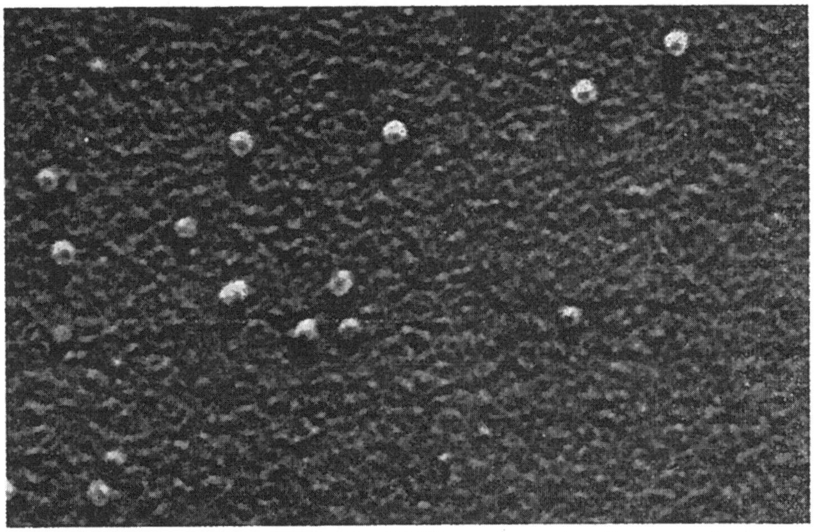

FIG. 2.—Electron micrograph of the TMV protein used in the reconstitution experiments. The particles are characteristically disk-shaped, with central holes. ×120,000.

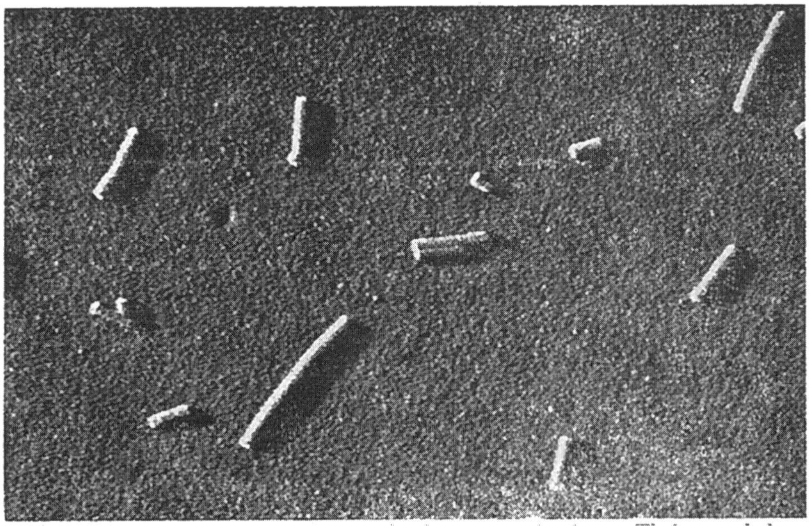

FIG. 3.—Particles of the reconstituted tobacco mosaic virus. Their morphology is identical with that of normal TMV, except for a greater proportion of short particles. The longest rod in this field is about 300 mμ long. ×60,000.

arily obtained. Inasmuch as several preparations of the reconstituted virus gave about twice that number of lesions at the 100 μg./ml. level, we can conclude that the starting materials failed by at least a factor of 30 to contain enough contaminating TMV particles to account for the final infectivity. This conclusion rests upon the supposition that the specific infectivity of any TMV particles that might remain as contaminants in the protein and RNA preparations is the same as that possessed by the untreated control TMV.

Counts and length distributions were secured for the particles of both TMV and the most active reconstituted virus, the 25,000-rpm. pellet described above, in order to ascertain the relative biological efficiency of the two materials in terms of numbers of particles per milliliter per lesions per leaf. The results of the counts and length distributions are given in Table 3. From this table it appears that about one-tenth of the total particles in the reconstituted virus, representing about one-third of the total mass, were of length *ca.* 300 mμ, the length of the monomer of normal TMV. These particles were therefore only about 3 per cent as infective as the particles of similar length in the control TMV. It thus appears that polymerization of the protein by nucleic acid to form ~300-mμ rods is fairly frequent but that only a fraction of the rods is reconstituted with the structural faithfulness necessary for infectivity.

TABLE 3

ELECTRON MICROSCOPE COUNTS OF PARTICLES OF TMV AND RECONSTITUTED VIRUS AT ASSAY CONCENTRATION

Preparation	Assay Concentration (μg./Ml.)	Lesions/ Half-Leaf	Total Particles/Ml.	Particles/Ml. of Length 290–310 Mμ	Particles/Ml./ 10 Lesions
Control TMV	0.1	10	1.6×10^{9}	7.0×10^{8}	7×10^{8}
Reconstituted virus (25,000 rpm.)	10	10	2.2×10^{11}	2.0×10^{10}	2×10^{10}

The length distribution of the rods of this pelleted material exhibited a reasonably random character for lengths less than about 260 mμ, but a highly uniform length between 290 and 310 mμ for the 10 per cent of the rods falling in this range. Only 3 per cent of the particles were of lengths greater than 310 mμ. This is in contrast to aggregates of X-protein, which show a complete randomness of distribution of lengths.

To ascertain whether the nucleic acid was localized in the center of the newly formed virus rods, Dr. R. Hart applied his technique of detergent treatment, followed by electron microscopical analysis.[5] He found the reconstituted rods appreciably more labile to SDS than was standard TMV, but after SDS treatment for 10 seconds many rods were partially degraded and showed a central strand of material protruding from the ends, as does standard TMV after 60 seconds of reaction (Fig. 4). These strands disappeared when ribonuclease was added.

Summary.—The preparation from TMV of protein and RNA fractions which tend to recombine at about pH 6 to form a nucleoprotein carrying virus activity (0.1–1 per cent of that of TMV) is described.

An electron microscopic search revealed no TMV rods in either of the two component solutions at a concentration level thirty fold to three hundred fold greater than those at which the reconstituted virus was assayed. In the latter, on the other hand, up to about one-third of the material consisted of rods of the typical diameter and length of TMV, many, if not all, containing a nucleic acid core.

The concentration, time, and pH dependence of virus regeneration is that of a typical chemical reaction. Freshly prepared RNA is required for appreciable reaction; degraded RNA, or nucleic acids from other sources, are inactive.

No inhibition was observed if know small amounts of TMV were added to concentrated solutions of each of the components and subsequently diluted and assayed.

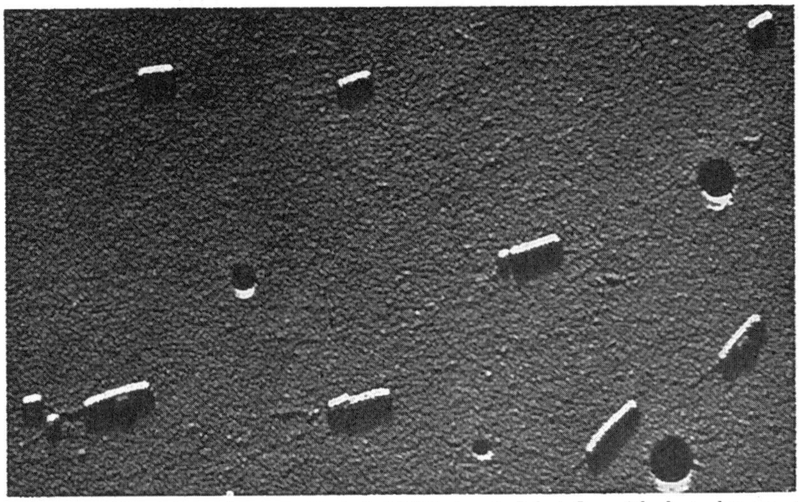

FIG. 4.—Particles of reconstituted TMV treated briefly with hot detergent. Rods are seen with strands of nucleic acid projecting from their ends. This appearance is identical with that found for normal TMV more severely treated. ×55,000.

The evidence thus seems reasonably complete that, under the conditions described, TMV nucleic acid enters into combination with TMV protein subunits and favors aggregation to rods, some of which are of sufficient length and structural integration to carry infectivity.

The capable assistance of Mrs. B. Singer and Mr. J. Toby is gratefully acknowledged.

* Aided by a grant from the National Foundation for Infantile Paralysis and research grant No. C-2245 from the National Cancer Institute of the National Institutes of Health, Public Health Service.

[1] J. I. Harris and C. A. Knight, *Nature*, **170**, 613, 1952, and *J. Biol. Chem.*, **214**, 231, 1 55; G. Schramm, *Z. Naturforsch.*, *2b*, **112**, 249, 1947; J. D. Watson, *Biochim. et biophys. acta*, **13**, 10, 1954; R. Franklin, *Nature*, **175**, 379, 1955.

[2] H. Fraenkel-Conrat and B. Singer, *J. Am. Chem. Soc.*, **76**, 180, 1954.

[3] G. Schramm, G. Schumacher, and W. Zillig, *Nature*, **175**, 549, 1955.

[4] R. V. Rice, P. Kaesberg, and M. A. Stahmann, *Biochim. et biophys. acta*, **11**, 337, 1953.

[5] R. G. Hart, these PROCEEDINGS, **41**, 261, 1955.

[6] W. N. Takahashi and M. Ishii, *Nature*, **169**, 419, 1952; C. C. Delwiche, P. Newmark, W. N. Takahashi, and M. J. Ng, *Biochim. et biophys. acta*, **16**, 127, 1955; B. Commoner, M. Yamada, S. D. Rodenberg, F. Y. Wang, and E. Basler, Jr., *Science*, **118**, 529, 1953.

[7] S. S. Cohen and W. M. Stanley, *J. Biol. Chem.*, **142**, 863, 1942.

[8] When less than 6 per cent of nucleic acid is added to the protein, almost all is incorporated in the pellet, but lower infectivity is the result. Protein alone also forms pellets upon ultracentrifugation at pH 6.

[9] The usual assay method yielded usually 0.0, but at times up to 0.4, lesions per half-leaf when phosphate alone was applied. This result is probably due to accidental contamination by active material applied to other sites of the plant. Therefore, separate plants were used when the absence of active virus was in question. Standard TMV was applied to one bottom leaf

and the unknowns to all others. This gave consistently 0.0 lesions with the components used in virus reconstitution, except when the nucleic acid was applied at 200 μg./ml., twenty times the amount present at the highest assay level of the reconstituted virus (0.3 lesions per half-leaf).

[10] Kindly supplied by Dr. S. S. Cohen and Dr. V. Cosentino.

[11] R. C. Williams, R. C. Backus, and R. L. Steere, *J. Am. Chem. Soc.*, **73**, 2062, 1951.

TMV in Tübingen (1945-1965)

K.-W. Mundry[1]
Biological Institute, Stuttgart University, Stuttgart, Germany

Tübingen on the river Neckar, a relatively small town in southwest Germany with a university founded in 1477, where in 1868-69 Friedrich Miescher discovered his "nuclein" (later called nucleic acid), was for two decades after World War II a center of tobacco mosaic virus (TMV) research. The three publications (Gierer and Schramm, 1956; Gierer and Mundry, 1958; Wittmann, 1960) that are the subjects of this commentary mark highlights of that period. They are the result of 20 years of cooperative research on TMV by a number of people who started that work in Berlin. Impressed by W. M. Stanley's report in 1935 of TMV being a "crystallizable protein" (see this volume) and fascinated by the chance that this "Makromolekül" could serve some day as a model system for replication processes and hereditary mechanisms, the young director of the Kaiser-Wilhelm-Institut für Biochemie, Adolf Butenandt, and two of his colleagues of the Kaiser-Wilhelm-Institut für Biologie, the zoologist Alfred Kühn and the botanist Fritz von Wettstein, began a cooperative investigation. As a first consequence, an "Arbeitsgemeinschaft für Virusforschung" was agreed upon and investigations on TMV began in 1938 in Berlin-Dahlem by the chemist Gerhard Schramm and the botanist Georg Melchers who, initially, were still under the responsibility of their directors.

In 1941, this loose cooperation was set up formally by the founding of the "Arbeitsstätte für Virusforschung der Kaiser-Wilhelm-Institute für Biochemie und Biologie" in Berlin-Dahlem. Gerhard Schramm became head of a chemistry department and Georg Melchers head of a corresponding biology department. Two other departments with the zoologist R. Daneel and the biologist G. Bergold as heads, as well as a steady and close contact with H. Friedrich-Freksa of the Kaiser-Wilhelm-Institut für Biochemie, completed the scientific staff.

The work centered around three basic problems: the structure of the virus, the function(s) of its components, and their possible biological variation. By 1940-42, Schramm had found that chemical modification of the amino groups of the protein did not (!) affect viral infectivity (Schramm and Müller, 1940; 1942). He also discovered that the virus rods, as seen for the first time in an electron microscope by Kausche *et al.* in 1939 (see this volume), not only disintegrated in alkaline media into rather defined classes of protein subunits, but that the subunits could subsequently be reaggregated to form rod-like structures (Schramm, 1943; 1947). Assembling a virus-like structure "does not need the presence of a living organism" (Schramm, 1943); obviously it depends on features of the virus protein because the reaggregated rods lacked the nucleic acid component. The rods were non-infectious, hence the protein was not believed to be the infectious entity.

During the same period, Melchers isolated and biologically characterized some variants of TMV (Melchers, 1942), including a tomato strain of the virus (Melchers *et al.*, 1940), which were found to differ in some but not in other respects. Electrophoresis and serological techniques permitted differentiation of some strains and mutants from others, while analytical ultracentrifugation and electron microscopy demonstrated that they all were members of a unique class of "macromolecules" (Melchers *et al.*, 1940; Friedrich-Freska *et al.*, 1946). It is impressive to see how much the authors of these early papers had been already aware of the potential of this virus to serve as a model system for the study of fundamental genetic mechanisms. Much of their work was aimed in this direction from its very beginning.

Because of the escalation of World War II and the increasing destruction of Berlin by bombing, the Kaiser-Wilhelm-Gesellschaft decided to move its institutes to safer places. Thus, Melchers found shelter in 1943 in the Botanical Institute of the University of Tübingen (taking the glass from the greenhouses in Berlin with him, thus saving it for a new greenhouse in Tübingen!), while the last "trek" with material from the Kaiser-Wilhelm-Institut für Biochemie reached Tübingen as late as March 1945.

Two months later the war was over. While Melchers had only a small office as his own and almost no possibility of doing experimental work, Schramm and some of his colleagues could start again quite soon because Butenandt in 1945 had become Professor of Physiological Chemistry in Tübingen and could house major parts of his Kaiser-Wilhelm-Institut für Biochemie in a spacious building of the university. Thus, they could continue their experiments without much interruption.

In 1950, Melchers moved into "his" institute. Fritz von Wettstein had died unexpectedly in 1945; Melchers became his successor and, as a

[1]Thanks are due to Alfred Gierer, Brigitte Wittmann-Liebold, Hans-Jörg Rheinberger and Holger Jeske for comments on the manuscript, and to Ulrich Kull for help with references.

consequence, got his own building. The "Kaiser-Wilhelm-Gesellschaft" was renamed "Max-Planck-Gesellschaft zur Förderung der Wissenschaften", signalling a new beginning.

Thus, TMV had come to Tübingen. Since 1949, I have had the fortune to witness much of the further development (and many of the stories around the people engaged in it). I will try to report its course chronologically.

My personal relation to TMV research began with a botany student's question in 1949: "Was ist ein Virus?" My professor in Göttingen, Richard Harder (one of the predecessors of the chair I later occupied at the University of Stuttgart), admitted that he didn't know, and he was uncertain if anyone would have the answer to it, "...but if someone in our country would know something about it, this one could be Melchers in Tübingen". Thus I travelled (600 km by bicycle) to Tübingen and got an interview with Melchers. The next day, in the afternoon, he offered me a chance to work with him: "... wenn mein neues Institut fertig ist!"

That was September, 1950. I got a bench in the personal laboratory of "the boss". It was next to his office and he could watch easily if I was doing something, and what. The theme of the proposed Ph.D. thesis was the question of whether TMV can be made to mutate. It was based on a report given by Gowen in 1941 at Cold Spring Harbor (*Symposia Quant. Biol.* **IX**, p.187) that *in vitro* irradiation with X-rays of preparations of a "common strain" of TMV, which produces only faint chlorotic foci on inoculated leaves of *Nicotiana sylvestris*, would lead to an increase in the number of mutant-type localized necrotic lesions. My trials to reproduce these experiments failed as did similar ones with UV instead of X-ray irradiation. However, an increase in the temperature at which Samsun tobacco plants inoculated with TMV "vulgare" (a "common" strain) were kept caused the number of mutant-containing, irregularly shaped, sharp-edged "yellow spots" appearing at the beginning of systemic infection of the plants to increase (first described by McKinney in 1935). At 35°C, and when their number was related to the amount of virus produced, the effect was rather dramatic (for review see Mundry, 1965). In the summer of 1954, I received my Ph.D. degree in Tübingen, and on the 1st of September of the same year, I began to work on sugar beet viruses as head of the virology department of the "Institut für Landwirtschaftliche Technologie und Zuckerindustrie" at the Technische Hochschule in Braunschweig.

Meanwhile, Schramm, Schumacher and Zillig (1955) showed that TMV remains infectious after partial deproteinization, and Schuster, Schramm and Zillig (1956) that RNA can be separated from protein by a procedure used in biochemistry to prepare polysaccharides -- extraction with water-saturated phenol. (This had been brought to their attention in a seminar given in Tübingen by H. Westphal). That was the state of affairs when Alfred Gierer came to Friedrich-Freksa's department in the Max-Planck-Institut für Virusforschung (which had grown out of the "Abteilung für Virusforschung" of Butenandt's Max-Planck-Institut für Biochemie). Gierer became interested in the "RNA problem" of TMV, and Friedrich-Freksa suggested cooperation with Schramm. Gierer's first experiments, intended to determine which component, RNA or protein, defines the length of reconstituted particles, were without clear results. He then turned to the problem of the biological activity of TMV and the role of its components. RNA preparations in those days were obviously labile and their RNA seemingly rather small. A simple trick to avoiding decay of the RNA, direct inoculation of leaves with the diluted RNA-containing aqueous phase immediately following phenol extraction of the protein, resulted in infection of the plants in the very first trial. Following a suggestion by Schramm, the remaining phenol was subsequently extracted with ether and this then removed by a stream of air. This permitted isolation of the infectious, high molecular weight RNA. In their publication, Gierer and Schramm (1956) eliminated the possible contribution of any protein to the phenomenon of infectivity thus distinguishing it from the infectivity of the partially deproteinized, unstable virus particles mentioned above (Schramm *et al.*, 1955). This is the crucial difference between these experiments and those of Fraenkel-Conrat and Williams (see this volume), where it was suggested (in a footnote) that a surplus of RNA left in preparations of reconstituted virus might contribute to the infectivity of their material, and those of Fraenkel-Conrat [*J. Am. Chem. Soc.* (1956) **78**:882-883] on the residual activity of his RNA preparation. Gierer (1957) then demonstrated through inactivation kinetics with RNase that only full-length RNA is infectious and that it is single-stranded.

In the fall of 1957, I gave up my job in Braunschweig and accepted an assistant position offered to me by Georg Melchers at the Max-Planck-Institut für Biologie in Tübingen. I started experiments on the problem of the very first event of infection with TMV: the entrance of the virus into the cells via mechanical inoculation. A first (unpublished) result was the finding that obviously there exist no "receptors" for the virus, and that there is no interference between strains of the virus as long as the interval between the wounding of the leaf (by rubbing its surface with buffer) and dipping of the leaf in concentrated suspensions of the strains did not exceed 10 minutes (after which time the wounds are no longer accessible to the virus).

In the meantime, Schuster and Schramm had taken up experiments to inactivate TMV RNA with nitrous acid [basically the procedure applied by Schramm and Müller (1942) to TMV preparations]. (It has been rumored that these experiments were designed to replace the formaldehyde inactivation procedure for poliovirus, which was used in the production of vaccines, by a more reliable technique because polio infections had occurred after children had been vaccinated with preparations still containing traces of incompletely inactivated polio virus.) Schuster and Schramm (1958) were able to show that the RNA of TMV could be inactivated *in vitro* with nitrous acid and that the amino bases C, A, and G convert into the corresponding oxy bases U, H, and X while the RNA retained its high molecular weight. Inactivation seemed to be a single-hit phenomenon, and the possibility that mutants occur seemed real.

Gierer and I met in a seminar in Melchers' group late in 1957. "Mutants of TMV" was the topic. Gierer had tried several approaches to produce viable mutants of TMV through chemical modification of the hereditary component of TMV with nitrous acid, but all were unsuccessful. We discussed the possibility of detecting mutants by use of the N'-gene host Java tobacco, the leaves of which react like *Nicotiana sylvestris* with faint chlorotic foci upon inoculation with the common strain vulgare, but which might produce local necrotic lesions with some mutants. I had tried this system already, without success, in the X-ray and UV mutagenesis studies of my Ph.D. thesis. All of us were very skeptical, Schramm and Melchers included, but Melchers reacted very positively to our suggestion and gave us all the necessary support (including much space in his greenhouses, which were famous for their sterile conditions -- only very occasionally over a period of years did so-called "spontaneous infections" occur; I know no other greenhouses, with the exception of my own in later years, where such a low level of spontaneous contamination could be maintained). Suddenly, and very unconventionally, Melchers suggested that, in case the two "youngsters" should be successful, they should publish the results without the names of the directors as senior authors, and Schramm agreed. But, on the other hand, Melchers forced us to publish any detailed communication in the *Zeitschrift für Induktive Abstammungs- und Vererbungslehre*, the oldest genetics periodical in the world, because it was on the way down in those days, and Melchers, as chief editor, wanted to draw attention to it. (He had tried that already with my thesis but had obviously not had much success with it!). Over the years, he managed a revival of it; what "the youngsters" used to call without respect "Neckarbote" is now flourishing as *Molecular and General Genetics* ("MGG").

So we tried the experiment. As expected, RNA incubated under control conditions without nitrite caused weak chlorotic foci when inoculated onto Java tobacco leaves, with a very small number of local necrotic lesions in between. However, leaves inoculated with RNA preparations deaminated according to the procedure of Schuster and Schramm were loaded with localized necrotic lesions. When I met Gierer in a concert at night two days after setting up the first experiment I whispered: "There is something coming!" Next morning the result was overwhelming -- hundreds of necrotic lesions on leaves which had received RNA (or virus) that had been treated with nitrous acid up to 4 hours. Never before (and never again) in our life had we experienced such a dramatic effect as result of a "first trial". Another two days later, these leaves were totally necrotic and had collapsed. We were shocked, and our first suspicion was that we had perhaps mixed up the bottles and had taken a preparation of dahlemense-type TMV (a tomato strain) which is of the mutant phenotype we were hoping for. Fortunately, we were not fooled by our excitement, and everything turned out to be reproducible. Mild conditions for the mutagenic procedure permitted a kinetic analysis of the phenomenon, and the new types proved to be genetically stable. This led us to conclude that a single deamination per RNA molecule is sufficient to create a mutant of the local necrotic type, meaning that substitution of a single base on the RNA is sufficient to change its genetic information (Gierer and Mundry, 1958). In addition, kinetics with RNA and complete virus showed no difference between each other, again demonstrating that the protein of the virus does not contribute to the genetics of TMV.

When we reported our results at the Vth International Congress of Biochemistry in 1958 in Vienna, Gierer his characterization of infectious TMV-RNA followed by my presentation of *in vitro* mutagenesis with nitrous acid, a funny episode happened. Requesting the first slide I said: "Now, when you perform the experiment, it looks as follows", and the audience burst into laughter! I had mixed up the buttons -- instead of the projection screen rolling down, the blackboard disappeared and a bright beam of light illuminated something which appeared like Doctor Faustus' laboratory, a room overloaded with glassware, vials, burettes, etc., and all covered with the dust of decades and with the thickest kind of spiders' webs one can imagine.

Otherwise everything was a success. Among the many questions were some by Arthur Knight and Heinz Fraenkel-Conrat, but the very first, if I remember correctly, came from Vernon M. Ingram: "Have you seen any effect on the viral protein?" (He had just demonstrated a single amino acid exchange by which mutant sickle-cell hemoglobin differs from the wild type.) My answer: "Not yet, but this problem is under investigation by Heinz Günter Wittmann." After that session, Fraenkel-Conrat invited me for a cup of coffee, and I wrote for him the recipe for the incubation procedure on a paper napkin. To me the most impressive person at that congress was Erwin Chargaff.

The "nitrous acid story" became accepted among geneticists and molecular biologists rapidly. Some plant pathologists, particularly from Great Britain, however, disputed the validity of the results. They preferred the view of selection of preexisting mutants during the course of the inactivation procedure. They disregarded that we had taken particular precautions to meet this possibility, and they did not accept the quantitative data resulting from the kinetic analyses of the phenomenon. For some years to follow, they remained skeptical in spite of an additional paper on the selection problem (Mundry, 1959; 1965). For two of these years I was at Caltech in Pasadena.

H. G. Wittmann came to Tübingen with a Masters degree in agriculture from the university of Stuttgart-Hohenheim. He got his Ph.D. in Tübingen with a thesis on the mutability of T-phages which he performed in Melchers' laboratory with Wolfhard Weidel as his adviser. After a year as a postdoc with Arthur Knight at the virus laboratory in Berkeley, he returned to Tübingen late in 1957 and started to work on the TMV coat protein. The *in vitro* mutagenesis with nitrous acid was just discovered and immediately had opened up concrete experimental strategies to break the code. Wittmann began this work by learning the techniques of amino acid sequencing in the laboratory of G. Braunitzer of the Max-Planck-Institut für Biochemie, then in Munich. Wittmann was an extremely hard worker and, starting with almost nothing, a protein sequencing laboratory was established in our institute, first with one analyzer from Beckman Instruments, then with several other ones added which were home made in the machine shop of our institute by our excellent mechanic Mr. Böss. There was considerable competition between Wittmann and Anderer [who established the first complete amino acid sequence of the TMV coat protein (Anderer *et al.*, 1960; 1965)] in Tübingen and the group in the virus laboratory in Berkeley headed by Heinz Fraenkel-Conrat. The analyzers ran day and night, and a bell in Wittmann's apartment just across the street rang in case of emergencies, e.g., when a fraction collector was "flooding". It became apparent very soon that Wittmann needed a cooperator for the hundreds of peptide sequence analyses to be performed in order to decipher the genetic code by use of these mutants. He tried to hire Brigitte Liebold from Braunitzer's laboratory in Munich, but she preferred to stay because Munich was a much more attractive place. But then they married, and the attractiveness of Tübingen increased markedly. Thus, cooperation for lifetime began between Heinz-Günter Wittmann and Brigitte Wittmann-Liebold.

The first paper published on the subject by H. G. Wittmann reported a "negative" result: there was no difference between the capsid protein of one of our first nitrous acid mutants (a "very mild" type) and that of wild type TMV (Wittmann, 1959). This, however, was soon followed by the publication in which 9 cases of amino acid replacements resulting from nitrous acid mutagenesis were demonstrated (Wittmann, 1960). The aim and rationale of this work was to establish the existence of the amino acid code, which was still hypothetical at this time, and to reveal its structure. Wittmann's finding, that a single nucleotide substitution results in substitution of one (!) amino acid, if any, was the most clear cut proof that there exists a relation between groups of nucleotides on the one hand and individual amino acids on the other, and that the code is non-overlapping.

The dramatic impact occurred at the International Congress of Biochemistry in 1961 in Moscow when the codon assignments of Nirenberg and Matthaei, obtained with the *in vitro* (prokaryotic) bacterial ribosomal translation system, and the amino acid replacements Wittmann observed with nitrous acid mutants of TMV, translated *in vivo* by the (eukaryotic) infected plant cell, matched perfectly. Through comparison by sequence analysis of induced as well as spontaneous mutants and natural strains of TMV, transition schemes (Wittmann and Wittmann-Liebold, 1963) were experimentally discovered which lent strong support to a degenerate triplet code, with two nucleotides more important than the third, and to the view that the code is probably universal (Wittman and Wittman-Liebold, 1966). Also, the efficient mutagenesis with nitrous acid permitted research resulting in the identification of defective and temperature-sensitive TMV mutants.

One Monday afternoon in the fall of 1965, there was for a moment some excitement of a special kind -- with telephone calls Svenska Dagbladet tried to gather information on several scientists and asked for photographs from the labs. They were rather insistent. Next day in the news there was no name from Tübingen among the "winners".

Later in the 1960s, TMV research in Tübingen diminished. Schramm died in 1969, Melchers began pioneering work on somatic hybridization of plant cells, Gierer switched to developmental biology, the Wittmanns left for Berlin and began impressive work on the structure and function of the ribosome and I accepted an offer to assume the chair of botany at the University of Stuttgart. Work on TMV continued there. Following the discovery of "omega" in 1965 (a G-lacking sequence in the RNA, later found to carry the leader sequence of TMV RNA) there were studies on nucleation during particle assembly, and the detection of an "initiation site" for uncoating the TMV particle. The latter met so perfectly with the requirements for "cotranslational disassembly" as it was suggested by T. M. A. Wilson (see this volume), that we published our results later together with some of his (Mundry *et al.*, 1991). With this and with experiments to locate functionally on the RNA of TMV the gene for triggering (or breaking!) hypersensitivity in leaves of N'-tobacco plants, my experimental career with TMV came to an end. As a "foot-print" of it, the department of botany of the University of Stuttgart was changed to the "Department of Molecular Biology and Virology of Plants" with one of the finest plant experiment stations in our country. Although the place of TMV has been taken by DNA plant viruses, the young crew under my successor, Professor Holger Jeske, is actively and successfully keeping alive the tradition of basic plant virus research in Germany, which was initiated 60 years ago in Berlin with TMV.

The castle of Tübingen (16th century). It marks the beginning of nucleic acid research: in its kitchen the university's "Laboratorium für Physiologische Chemie" was installed in 1818. There, in 1868/69, Friedrich Miescher detected, and isolated from pus cells, his nuclein (=nucleic acid). (Copyright by K.-W. Mundry, 1950.)

References

Anderer, F. A., Uhlig, H., Weber, E., and G. Schramm. (1960). Primary structure of the protein of tobacco mosaic virus. *Nature* **186**:922-925.

Anderer, F. A., Wittmann-Liebold, B., and H. G. Wittmann. (1965). Weitere Untersuchungen zur Aminosäuresequenz des Proteins im Tabakmosaikvirus. *Zeitschrift für Naturforschung* **20b**:1203-1213.

Friedrich-Freska, H., Melchers, G., and G. Schramm. (1946). Biologischer, chemischer und serologischer Vergleich zweier Parallelmutanten phytopathogener Viren mit ihren Ausgangsformen. *Biologisches Zentralblatt* **65**:187-222.

Gierer, A. (1957). Structure and biological function of ribonucleic acid from tobacco mosaic virus. *Nature* **179**:1297-1299.

Gierer, A. and G. Schramm. (1956). Infectivity of ribonucleic acid from tobacco mosaic virus. *Nature* **177**:702-703.

Gierer, A. and K.-W. Mundry. (1958). Production of mutants of tobacco mosaic virus by chemical alteration of its ribonucleic acid *in vitro*. *Nature* **182**:1457-1458.

Melchers, G. (1942). Über einige Mutationen des Tabakmosaikvirus und eine "Parallelmutation" des Tomatenmosaikvirus. *Naturwissenschaften* **30**:48.

Melchers, G., Schramm, G., Trurnit, H., and H. Friedrich-Freska. (1940). Die biologische, chemische und elektronenmikroskopische Untersuchung eines Mosaikvirus aus Tomaten. *Biologisches Zentralblatt* **60**:524-556.

Mundry, K. W. (1959). The effect of nitrous acid on tobacco mosaic virus: mutation, not selection. *Virology* **9**:722-726.

Mundry, K. W. (1965). Interactions between Plant Viruses and Host Cells. *In*: M. Locke (ed.): "Reproduction: Molecular, Subcellular and Cellular", pp. 65-94. Academic Press, New York.

Mundry, K. W., Watkins, P. A. C., Ashfield, T., Plaskitt, K. A., Eisele-Walter, S., and T. M. A. Wilson. (1991). Complete uncoating of the 5' leader sequence of tobacco mosaic virus RNA occurs rapidly and is required to initiate cotranslational virus disassembly *in vitro*. *J. Gen. Virol.* **72**:769-777.

Schramm, G. (1943). Über die Spaltung des Tabakmosaikvirus in niedermolekulare Proteine und die Rückbildung hochmolekularen Proteins aus den Spaltstücken. *Naturwissenschaften* **31**:94-96.

Schramm, G. (1947). Über die Spaltung des Tabakmosaikvirus und die Wiedervereinigung der Spaltstücke zu höhermolekularen Proteinen. II. Versuche zur Wiedervereinigung der Spaltstücke. *Zeitschrift für Naturforschung* **2b**:249-257.

Schramm, G. and H. Müller. (1940). Zur Chemie des Tabakmosaikvirus. Über die Einwirkung von Keten und Phenylisocyanat auf das Virusprotein. *Hoppe-Seyler's Zeitschrift für physiologische Chemie* **266**:43-55.

Schramm, G. and H. Müller. (1942). Über die Bedeutung der Aminogruppen für die Vererbungsfähigkeit des Tabakmosaikvirus. *Hoppe-Seyler's Zeitschrift für physiologische Chemie* **274**:267-275.

Schramm, G., Schumacher, G., and W. Zillig. (1955). An infectious nucleoprotein from tobacco mosaic virus. *Nature* **175**:549-550.

Schuster, H., Schramm, G., and W. Zillig. (1956). Die Struktur der Ribonucleinsäure aus Tabakmosaikvirus. *Zeitschrift für Naturforschung* **11b**:339-345.

Schuster, H. and G. Schramm. (1958). Bestimmung der biologisch wirksamen Einheit in der Ribosenukleinsäure des TMV auf chemischem Wege. *Zeitschrift für Naturforschung* **13b**:697-704.

Wittmann, H. G. (1959). Vergleich der Proteine des Normalstammes und einer Nitritmutante des Tabakmosaikvirus. *Zeitschrift für Vererbungslehre* **90**:463-475.

Wittmann, H. G. (1960). Comparison of the tryptic peptides of chemically induced and spontaneous mutants of tobacco mosaic virus. *Virology* **12**:609-612.

Wittmann, H. G. and B. Wittmann-Liebold. (1963). Tobacco mosaic virus mutants and the genetic coding problem. *Cold Spring Harbor Symposia on Quantitative Biology* **XXVIII**:589-595.

Wittmann, H. G. and B. Wittmann-Liebold. (1966). Protein chemical studies of two RNA viruses and their mutants. *Cold Spring Harbor Symposia on Quantitative Biology* **XXXI**:163-172.

Biographical Sketch

K.-W. Mundry studied botany at the university in Göttingen from 1946-49. In 1950, he moved to Tübingen to work on plant viruses with Georg Melchers at the Max Planck-Institut für Biologie. In 1954, he received his Ph.D. at the university in Tübingen with a thesis on the *in vitro* and *in vivo* mutability of TMV. In the fall of the same year, he became head of the virology department of the "Institut für Landwirtschaftliche Technologie und Zuckerindustrie an der Technischen Hochschule Braunschweig" to work with sugar beet viruses (beet yellows and beet mosaic). In 1957, he accepted an assistant position with G. Melchers at the Max-Planck-Institut für Biologie in Tübingen. From 1959-1961, he spent two years at the Biology Division of the California Institute of Technology in Pasadena, California (U.S.A.) with Robert Sinsheimer. Returning to the Max-Planck-Institut für Biologie, he became a staff member in 1968. In 1972, he was appointed Professor of Botany at the university in Stuttgart and, in 1993, he became Professor Emeritus.

(*Reprinted from Nature, Vol.* 177, *pp.* 702-703, *April* 14, 1956)

Infectivity of Ribonucleic Acid from Tobacco Mosaic Virus

In their experiments with bacteriophages, Hershey and Chase[1] have shown that only the nucleic acid component plays a part in the intracellular multiplication. There are also indications that in simple viruses containing ribonucleic acid the nucleic acid plays a dominant part in the infection. Thus, experiments with tobacco mosaic virus have shown that the protein can be changed chemically without affecting the activity and the genetic properties[2]; recently, it was even found[3] that part of the protein can be removed from tobacco mosaic virus without destroying the activity.

We have now obtained evidence that after complete removal of the protein, the ribonucleic acid itself is still infectious.

The protein was extracted from tobacco mosaic virus with phenol by a procedure elaborated by Schuster, Schramm and Zillig (to be published). After a short treatment at low temperatures, a preparation of ribonucleic acid is obtained which has a high molecular weight during the first few hours but depolymerizes in the course of time. Its physical properties will be described elsewhere.

A solution of 10 per cent tobacco mosaic virus in 0·02 *M* phosphate buffer of *p*H 7·3 is shaken for 8 min. at 5° C. with an equal amount of water-saturated phenol. The aqueous phase which contains the ribonucleic acid is separated by centrifugation, and the process of extraction with phenol is repeated at least twice for 2 min. The phenol is then extracted by ether from the aqueous phase. The whole procedure is carried out at 5° C. and takes about 50 min.; it is followed immediately by testing the infectivity.

For that purpose, five to ten plants of *Nicotiana glutinosa* with five leaves each were inoculated with a diluted solution of the ribonucleic acid, and an equal number of plants with a standard solution of tobacco mosaic virus. The number of local lesions produced by ribonucleic acid and tobacco mosaic virus are compared in Table 1. It is found that 10 μgm. of ribonucleic acid produces about the same number of lesions as does 0·2 μgm. tobacco mosaic virus. The infectivity of the ribonucleic acid preparation is thus about 2 per cent of that of the native virus.

Table 1. COMPARISON OF THE INFECTIVITY OF RIBONUCLEIC ACID AND TOBACCO MOSAIC VIRUS IN 0·1 *M* PHOSPHATE BUFFER

*p*H	Ribonucleic acid μgm./ml.	lesions	Tobacco mosaic virus μgm./ml.	lesions
6·1	10	153	{ 0·09	95
			{ 0·8	445
7·3	10	815	0·27	1,048
7·3	1	524	0·05	795
7·5	10	998	0·27	685

The following experiments, the results of which are collected in Table 2, have been carried out to show that the infection is due to the nucleic acid rather than to contamination of the ribonucleic acid with native virus.

(*a*) In the ribonucleic acid solution protein was not detectable by chemical methods (Schuster, Schramm and Zillig, unpublished work); thus the amount must be less than 0·4 per cent of the ribonucleic acid content. By serological methods (complement fixation) it was shown that the ribonucleic acid contains less than 0·02 per cent of native tobacco mosaic virus protein.

(*b*) Treatment of both the ribonucleic acid and tobacco mosaic virus with normal rabbit serum (concentration 3×10^{-3}, applied for 10 min. at 4° C.) somewhat reduces the infectivity. There is no significant further reduction if the ribonucleic acid is treated with the same amount of tobacco mosaic virus antiserum, whereas with antiserum the infectivity of the virus itself is almost completely destroyed.

(*c*) Incubation of the stock solutions of ribonucleic acid (0·3 per cent) and tobacco mosaic virus (0·06 per cent) with 2 μgm. per ml. of ribonuclease at 4° C. for 10 min. reduces the activity of the ribonucleic acid to 0, whereas that of the virus remains almost unaffected.

(*d*) The sedimentation constant of the ribonucleic acid is 12–18 *S*, compared with 180 *S* for tobacco mosaic virus. We have centrifuged the stock solution of ribonucleic acid for 30 min. at 50,000 rev. per min. and found the supernatant liquid to be only a little less active than the original solution. If the solution of the virus is treated in the same manner, the activity of the supernatant liquid is very low.

(*e*) The ribonucleic acid is known to be unstable; and, as would be expected, its infectivity is much reduced after 48 hr. at 20° C., whereas that of the virus is much less affected.

These experiments show that protein, if present at all, is only there in very small amounts and does not resemble closely the native protein of tobacco

Table 2. COMPARISON OF RIBONUCLEIC ACID (10 μgm./ml.) AND TOBACCO MOSAIC VIRUS (0·27 μgm./ml.) IN 0·1 *M* PHOSPHATE BUFFER OF *p*H 7·3

(Infectivity expressed as lesions per 30 leaves)

	Ribonucleic acid	Tobacco mosaic virus
Normal	488	629
With normal serum	180	117
With antiserum	145	0
With ribonuclease	0	473
After ultracentrifugation	367	31
After 48 hr. at 20° C.	2	130

mosaic virus. We are thus led to conclude that the infectivity is due to the nucleic acid itself.

The infectivity of the ribonucleic acid preparation is about 0·1 per cent of that of the same amount of ribonucleic acid contained in native tobacco mosaic virus. Whether this relatively low value is due to a large inactive fraction of the ribonucleic acid preparation, or to low efficiency of the mechanism of infection, has still to be determined.

Studies on the combination of the ribonucleic acid with proteins are being carried out and may elucidate the connexion between our findings and the reactivation experiments of Fraenkel-Conrat and Williams[4], of Lippincott and Commoner[5], and of Hart[6].

We are much indebted to Prof. H. Friedrich-Freksa for helpful discussions, to Mr. R. Engler and Dr. H. Schuster for their co-operation, and to Miss A. Kleih for assistance.

A detailed account of this work will be published in the *Zeitschrift für Naturforschung*.

A. GIERER
G. SCHRAMM

Max-Planck-Institut für Virusforschung,
Tübingen.
Feb. 10.

[1] Hershey, A. D., and Chase, M., *J. Gen. Physiol.*, **36**, 39 (1952).

[2] Schramm, G., and Müller, H., *Hoppe Seylers Z. physiol. Chem.*, **266**, 43 (1940); **274**, 267 (1942). Miller, G. L., and Stanley, W. M., *J. Biol. Chem.*, **141**, 905 (1941); **146**, 331 (1942). Harris, J. I., and Knight, C. A., *J. Biol. Chem.*, **214**, 215 (1955).

[3] Schramm, G., Schumacher, G., and Zillig, W., *Nature*, **175**, 549 (1955).

[4] Fraenkel-Conrat, H., and Williams, R. C., *Proc. U.S. Nat. Acad. Sci.*, **41**, 690 (1955).

[5] Lippincott, J. A., and Commoner, B., *Biochim. Biophys. Acta*, **19**, 198 (1956).

[6] Hart, R. G., *Nature*, **177**, 130 (1956).

Printed in Great Britain by Fisher, Knight & Co., Ltd., St. Albans.

(*Reprinted from Nature, Vol.* 182, *pp.* 1457–1458, *Nov.* 22, 1958)

Production of Mutants of Tobacco Mosaic Virus by Chemical Alteration of its Ribonucleic Acid *in vitro*

WHILE many agents are known to act as mutagens on the living cell, there have been no convincing reports that the genetic character of isolated particles, such as viruses or transforming principles, have been changed *in vitro*. Recent results with tobacco mosaic virus help to overcome this difficulty. The infective unit of the virus is its ribonucleic acid component, which can be isolated in active form[1]. It is made up of 6,000 nucleotides that probably form a single chain[2].

Recently, Schuster and Schramm[3] have used nitrous acid to convert adenine, guanine and cytosine into hypoxanthine, xanthine and uracil, respectively, while the ribonucleic acid strand remains intact. It was shown that the alteration of any one of about 3,000 nucleotides is lethal. Possibly, changes of other bases are mutagenic.

On the basis of this work we studied a simple symptom of mutation, the production of necrotic lesions on a tobacco variety (Java) on which the untreated strain produces chlorotic lesions only.

Table 1. MUTAGENIC EFFECT OF NITROUS ACID ON RIBONUCLEIC ACID. 2 vol. of 0·19 per cent ribonucleic acid extracted from tobacco mosaic virus *vulgare* by phenol (ref. 1) were mixed with 1 vol. 4 *M* sodium nitrite and 1 vol. 1 *M* acetate *p*H 4·8 at 22° C. After time t, the ribonucleic acid is diluted to 19 μgm./ml. with *M*/15 phosphate, and assayed. Control dilutions of untreated ribonucleic acid contain an equivalent amount of nitrite. Time τ (equation 1) is 18·5 min. Total infections are given as numbers of lesions (per leaf) on Xanthi, and as the sum of chlorotic and necrotic lesions on Java; mutant infections as numbers of necrotic lesions per leaf on Java

t (min.)	t/τ	Total infections		Mutant infections	Mutant infections (per cent)
		Xanthi	Java	Java	Java
1	0·05	72·8	183	1·4	0·8
4	0·22	72·4	130	2·5	1·9
8	0·43	89·8	188	4·5	2·4
16	0·86	59·3	97	5·4	5·6
32	1·73	36·1	63	6·6	10·5
64	3·46	10·4	21	2·1	9·8
96	5·2	2·7	3·5	0·5	15·5
Untreated controls, μgm./ml. ribonucleic acid					
19		155	138	0·3	0·2
1·9		31·2	42	0·1	0·3
0·19		3·1	10	0	—
× 0·019		0·3	0·8	0	—

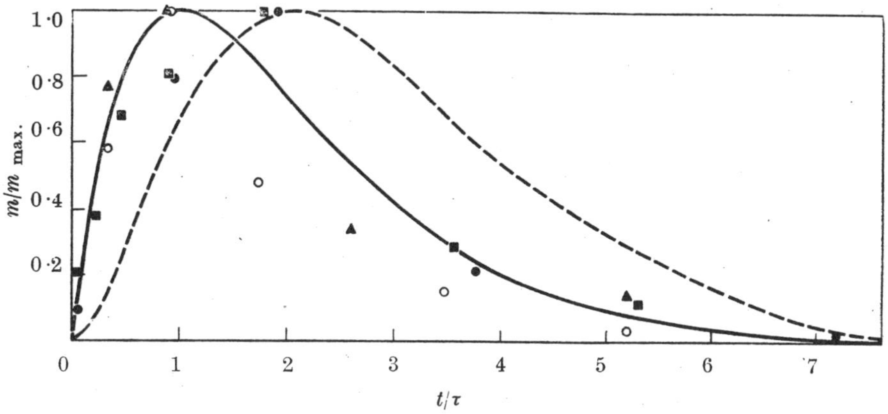

Fig. 1. Dependence of the concentration of mutants (number of necrotic lesions on Java tobacco, relative to maximum value) on the time of incubation with nitrous acid, related to τ. ———, Single-hit curve (equation 3); – – –, double-hit curve; ▲, tobacco mosaic virus *vulgare* (assayed at 0·7 μgm./ml.); ●, ribonucleic acid from tobacco mosaic virus *vulgare* (assayed at 39 μgm./ml.); ■, ribonucleic acid from tobacco mosaic virus *vulgare* (assayed at 19 μgm./ml.); ○, ribonucleic acid from light-green strain *B*11 (assayed at 9·5 μgm./ml.)

Virus or isolated ribonucleic acid was treated with nitrous acid. After times t, samples were diluted to a constant concentration for assay on Xanthi and Java tobacco plants. The conditions and results are given in Table 1. The total concentration n of infective particles was measured by the number of lesions on Xanthi, and of chlorotic and necrotic lesions on Java tobacco, as compared with standard dilution curves. n decreases exponentially with t:

$$n = n_0 \exp(-t/\tau) \qquad (1)$$

where τ is the average time for one lethal conversion of a base per ribonucleic acid molecule.

The necrotic lesions on Java tobacco are due to mutated particles, and their number is expected to be nearly proportional to their concentration m. If the alteration of a single base is mutagenic, and if the average number of such mutations per ribonucleic acid molecule in the time τ is p, the concentration of such mutants is:

$$m = n_0 p \frac{t}{\tau} \exp(-t/\tau) = np\frac{t}{\tau} \qquad (2)$$

or in terms of the maximum value of m:

$$\frac{m}{m_{\text{max.}}} = e\,\frac{t}{\tau} \exp(-t/\tau) \qquad (3)$$

Fig. 1 represents measurements under different conditions with tobacco mosaic virus and with ribonucleic acid derived from two strains. For each case, τ was determined experimentally, using equation (1). The numbers of necrotic lesions on Java tobacco, divided by the maximum value reached in time t, were plotted against t/τ. The experimental results were in agreement with the theoretical relation (3).

The results lead to the following conclusions.

(1) Ribonucleic acid and tobacco mosaic virus treated with nitrous acid give rise to a much larger number of necrotic lesions on Java than untreated material assayed at the same concentration; it is twenty times the spontaneous level when total infectivity has dropped to only half the original value (Table 1). This appearance of mutants at low rates of inactivation excludes selection, and cannot be due to pre-existing mutants, or to mutants arising in the plant.

(2) The mutagenic effect on tobacco mosaic virus and on its isolated ribonucleic acid is equal when the fraction of bases altered in the ribonucleic acid, measured by t/τ, is the same. Also, tobacco mosaic virus treated with nitrous acid has the same proportion of mutants as ribonucleic acid isolated after the treatment. Thus, the mutants are produced by chemical reactions with ribonucleic acid rather than by aggregation phenomena, or by changes in the protein.

(3) The production of mutants is at first linear with t and follows a single-hit curve, irrespective of varying conditions (Fig. 1). Therefore, alterations of single nucleotides are mutagenic.

At $t = \tau$, where one out of about 3,000 nucleotides is altered[3], 6 per cent of the primary infections on Java plants infected with *vulgare* are found to be necrotic (Table 1). If the plating efficiency is assumed to remain unchanged upon mutation, the alteration of any one out of 180 of the total 6,000 nucleotides in the ribonucleic acid strand would be mutagenic, leading to necrotic lesions on Java; this assumption may be subject to later correction.

(4) A variety of other mutants is also produced by nitrous acid. In order to detect them, we isolated individual lesions produced by ribonucleic acid treated with nitrous acid on Xanthi tobacco ($t/\tau = 5{\cdot}2$, Table 1) and assayed them on Java and Samsun plants. Out of 60 lesions, 20 produced no infection, 33 led to a variety of altered symptoms and only 7 gave apparently unchanged symptoms on Samsun tobacco. The altered symptoms proved genetically stable in a transfer experiment. From a control dilution of untreated ribonucleic acid, 64 out of 65 lesions produced infection and no altered symptoms were detected.

We are led to conclude that the chemical alteration of individual bases of the ribonucleic acid molecule *in vitro* is mutagenic. A variety of different mutants may be produced in this way. It remains undecided which of the three types of chemical change caused by the nitrous acid (see above) are mutagenic, and how the chemically altered ribonucleic acid is related to its progeny. In the case of a conversion of cytosine into uracil, it is conceivable that the altered ribonucleic acid undergoes identical reproduction in the host cell.

We are much indebted to Prof. G. Melchers for generous support and advice, and to Prof. G. Schramm and Dr. H. Schuster for helpful discussions. A detailed account of this work is being published in the *Zeitschrift für Vererbungslehre*.

ALFRED GIERER

Max-Planck-Institut für Virusforschung,
Tübingen.

KARL-WOLFGANG MUNDRY

Max-Planck-Institut für Biologie,
Abt. Melchers,
Tübingen.

Aug. 26.

[1] Gierer, A., and Schramm, G., *Nature*, **177**, 702 (1956); *Z. Naturforsch.*, **11b**, 138 (1956).

[2] Gierer, A., *Nature*, **179**, 1297 (1957).

[3] Schuster, H., and Schramm, G., *Z. Naturforsch.* (in the press).

Printed in Great Britain by Fisher, Knight & Co., Ltd., St. Albans.

Reprinted from VIROLOGY, Volume 12, No. 4, December 1960

Comparison of the Tryptic Peptides of Chemically Induced and Spontaneous Mutants of Tobacco Mosaic Virus

The genetic information of tobacco mosaic virus (TMV) resides in the sequence of the bases within its ribonucleic acid (RNA). The infecting TMV-RNA induces the cell to produce new virus particles consisting of RNA and a specific protein. There must therefore exist a structural relation between the RNA and the corresponding protein. One way to study this relation is to change the sequence of the RNA bases and to study the effect on the corresponding protein.

Treatment of TMV-RNA with nitrous acid converts certain bases into others, e.g., uracil to cytosin (*1*); a change of one nucleotide is sufficient to produce a mutation (*2*). By variation of the experimental conditions it is possible to vary the statistically defined average number of nucleotide changes within the RNA. Investigation of the protein structure of these chemically induced mutants promises to give information on the coding problem between nucleic acids and the corresponding proteins. The results of a comparison of the tryptic peptides of 26 chemically induced and spontaneous TMV mutants will be described here. Part of this work has been presented before the National Academy of Sciences, Washington, D. C. (*3*). Comparisons of the amino acid composition (*4*) and of the composition of the tryptic peptides (*5*) of the normal TMV strain and a mutant induced by nitrous acid treatment have been published.

To solve the coding problem, it is essential to have a method capable of detecting with certainty a change in only one out of the 157 amino acids making up the TMV protein. Therefore the purified TMV was split into RNA and protein and (besides determining the amino acid composition of the virus protein in an automatic amino acid analyzer) the protein was digested with trypsin; the resulting 12 tryptic peptides were isolated and purified by column and paper chromatography and paper electrophoresis. Their amino acid composition was analyzed in amino acid analyzers according to Spackman *et al.* (*6*). The methods were essentially the same as those described previously in detail (*5, 7*).

Figure 1 gives a scheme showing derivation of the mutants whose tryptic peptides have been studied. Those mutants designated by "Ni" were produced by nitrous acid treatment; all others were spontaneous mutants. The isolation of the spontaneous mutants A 7 to E 83 was done in such a way as to reduce to a minimum the probability of getting more than one mutation step (*8*). This principle was not stressed as much in isolating the spontaneous mutants designated by Latin names [*flavum*

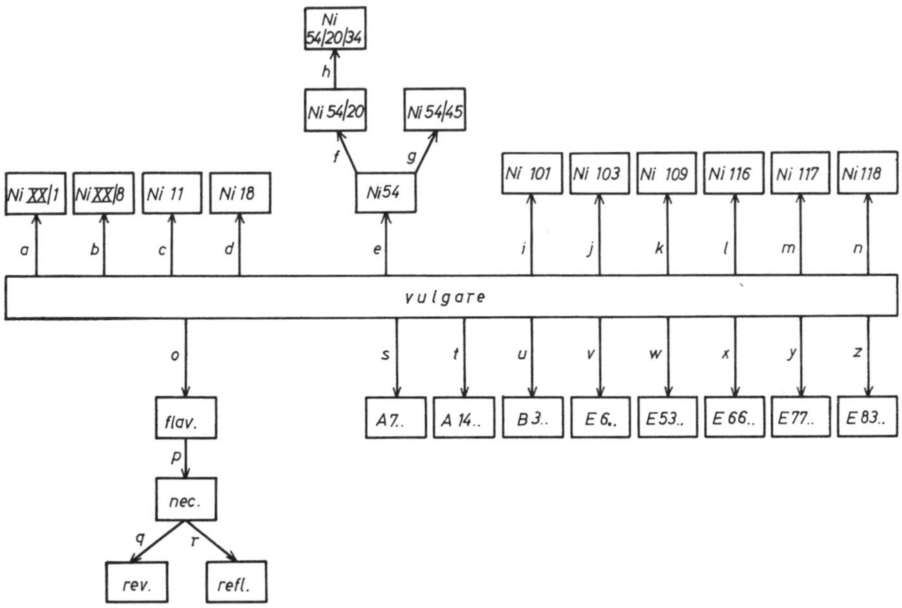

FIG. 1. Scheme of the TMV mutants studied. The differences in the amino acid composition of the tryptic peptides are given in the text.

(flav.), *necans*, (nec.), *revirescens* (rev.), and *reflavescens* (refl.)], some of which were isolated many years ago (*9*). The amino acid composition of the latter mutants has been analyzed before, without splitting the protein into tryptic peptides (*10*).

When t is the time of treatment and τ the time after which the infectivity decreases to $1/e$ of the value before treatment [see (*2*)], then for both Ni XX/1 and Ni XX/8, $\frac{t}{\tau} = 1.1$; for Ni 11, Ni 18, and Ni 54, $\frac{t}{\tau} = 5.2$ [see (*11*)]; for Ni 54/45 and Ni 54/20, $\frac{t}{\tau} = 2.6$; for Ni 54/20/34, $\frac{t}{\tau} = 6.5$; and for Ni 101 to Ni 118, $\frac{t}{\tau} = 3.0$. The 6 mutants Ni 101 to Ni 118 and the 2 Ni XX/ . . . were produced by treatment of the virus nucleoprotein with nitrous acid whereas all other Ni mutants were isolated after nitrous acid treatment of the TMV-RNA. The mutants *flavum*, Ni XX/8, Ni 54/20/34, and E 83 are yellow strains which cannot be differentiated by symptoms on "Samsun tobacco".

The analysis of the tryptic peptides of the 26 chemically induced and spontaneous mutants listed, gives the following results [cf. Fig. 1; peptides numbered as in (*7*)]:

Steps a, c, d, e, f, g, s, u, v, w, y: no change
Steps b, h, o, z: Asp → Ala in peptide XII[1]
Steps i, j: Asp → Gly in XII
Step k: Glu → Gly in VII
Step l: Asp → Gly in IX
Step m: Ser → Phe in IV; Glu → Val in XII
Step n: Pro → Leu in XII
Step p: Ala + Phe → Val + Leu in XII; Ser → Phe in IV
Step q: Val + Leu → Asp + Phe in XII
Step r: Leu → Phe in XII
Step t: Ileu → Thr in XI
Step x: Asp → Lys in IV

The results confirm the earlier finding (*5*) that there exist mutants which have the same amino acid composition of their tryptic peptides although they show very striking differences in symptoms; this is true for both spontaneous and chemically induced mutants. This can be explained by assuming that the region within the TMV-RNA which codes for the 157 amino acids of the virus protein is smaller than the 6500 nucleotides making up the TMV-RNA. Only when nucleotides within this region are changed does an alteration of amino acids in the viral protein occur, whereas changes of nucleotides outside this region result

[1] Peptide No. XII (with 41 amino acids) contains 4 Asp [see (*7*)]. ASP means aspartic acid or asparagine. Glu means glutamic acid or glutamine.

in alterations that lead to different symptoms without changing the viral protein. By investigation of more nitrous acid mutants than have been studied hitherto, the number of nucleotides within the "gene" for the viral protein could be determined and the question answered how many nucleotides code for one amino acid. The elucidation of the total sequence of the 157 amino acids of the TMV protein (*12*) makes it possible to pinpoint the alteration of a certain amino acid at a certain position within the protein chain. This opens up the possibility of seeing whether the points of amino acid replacement are distributed randomly over the protein chain or whether there exist regions with a high frequency of replacement, and also of answering the question whether a certain amino acid can be substituted only by a limited number of others; this information would give additional help in resolving the code.

ACKNOWLEDGMENT

The technical assistance of Miss Ingrid Hindennach, Miss Brigitte Ostertag, and Miss Julita Huf is gratefully acknowledged.

REFERENCES

1. Schuster, H., and Schramm, G., *Z. Naturforsch.* **13b,** 697–704 (1958).
2. Gierer, A., and Mundry, K. W., *Nature* **182,** 1457–1458 (1958).
3. Wittmann, H. G., *Symposium on Genetic Determination of Protein Structure, Ann. Meeting Natl. Acad. Sci. Washington April, 1960.*
4. Tsugita, A., and Fraenkel-Conrat, H., *Proc. Natl. Acad. Sci. U. S.*, **46,** 636–642 (1960).
5. Wittmann, H. G., *Z. Vererbungslehre* **90,** 463–475 (1959).
6. Spackman, D. H., Stein, W. H., and Moore, S., *Anal. Chem.* **30,** 1130–1206 (1958).
7. Wittmann, H. G., and Braunitzer, G., *Virology* **9,** 726–728 (1959).
8. Melchers, G., unpublished data.
9. Melchers, G., *Naturwissenschaften* **36,** 48–49 (1942).
10. Aach, H. G., *Z. Naturforsch.* **13b,** 425–433 (1958).
11. Mundry, K. W., and Gierer, A., *Z. Vererbungslehre* **89,** 614–630 (1958).
12. Anderer, A., Uhlig, H., Weber, E., and Schramm, G., *Nature* **186,** 922–925 (1960).

Max-Planck-Institut für Biologie
Abt. Melchers
Tübingen, Germany

H. G. Wittmann

Received October 21, 1960

Comments on Takebe and Otsuki

Milton Zaitlin
Department of Plant Pathology, Cornell University, Ithaca, NY

This paper has had a truly monumental and lasting impact on plant virology as it opened the door to detailed studies on replication and gene function of many plant viruses. It is safe to say that there is hardly a paper published today studying either replication or gene function of plant viruses which does not use protoplasts. Up to the time of this remarkable discovery, plant virus replication studies were constrained because of the lack of a 'single cycle' analysis of virus replication; i. e., studies involving the complete cycle of replication from infection to virus maturation in synchronously-infected cells. Before protoplasts, to study virus replication, leaves were inoculated mechanically, initiating at most a few thousand foci of infection. The virus would then spread radially from the initial foci, infecting other cells in sequence. Thus, at any given time the investigator was examining replication in a mixed population of cells, some of which had completed virus replication, others of which were in various stages of the replication process, and a few of which were newly-infected. The overwhelming numbers of uninfected cells also affected the studies.

The 1968 Takebe and Otsuki paper changed that situation dramatically. Their work allowed for studies in which large numbers of cells were infected simultaneously. This paper and its progenitor (Takebe *et al.*, 1968) are really remarkable in that they not only discovered how to prepare isolated leaf cells, how to remove the cell walls to yield viable protoplasts, how to culture them, but also how to infect them. Ask any plant virologist today how difficult and erratic protoplast isolation and infection can be at times, even utilizing a well-established method. Takebe and Otsuki found a good source of non-toxic enzymes, and worked out a truly unique method of infecting and maintaining the protoplasts to allow replication to proceed.

In a paper submitted in December of 1967, Takebe *et al.* (1968) described the conditions to get functional tobacco mesophyll cells and, although the title does not mention it, they also got functional protoplasts from these cells. This paper also shows that the cells and protoplasts could support virus replication; however, the cells were isolated from infected leaves. The paper reproduced here went one step further and demonstrated that the protoplasts could be infected with TMV virions, and in a paper published shortly thereafter, were infectible by TMV RNA (Aoki and Takebe, 1969).

I have some personal experience with this area of experimentation, and had an early contact with Takebe and Otsuki. In 1959, I published a paper on the isolation of tobacco leaf cells capable of supporting virus replication (Zaitlin, 1959). I isolated separated leaf cells by treatment of the tissues with pectinase. These cells were then incubated in ^{14}C-tagged amino acids, and radiolabeled TMV was isolated as evidence for TMV replication. In retrospect, this is a pretty poor paper as the cells were barely alive. I never appreciated, in contrast to Takebe and Otsuki, that healthy cells were plasmolyzed cells; thus I had utilized isotonic sucrose to prepare and maintain my cells. I note that I never gave the labeling data in my paper, and the future publication promised in this manuscript never appeared.

In August 1967 I attended a symposium in Tokyo, "Biochemical Regulation in Diseased Plants or Injury". (My symposium paper did not utilize my separated cells for replication studies; rather, we labeled leaf slices.) Another participant at that meeting, T. O. Diener, knowing my interest in leaf cell isolation, told me that I should visit Chiba, near Tokyo, to see the work going on there. I did indeed make the visit. Dr. Otsuki (who is now the Head of the Department of Tea Production at the National Research Institute of Vegetables, Ornamental Plants and Tea, in Kanaya, Japan), recently pinpointed the day of my visit for me as August 30, 1967, having found my signature in the Record of Visitors book. He recalled that in anticipation of my visit his boss, the late Dr. Itaru Takebe, told him to prepare a batch of protoplasts for the "world famous scientist" (his words, not mine). Dr. Otsuki remembered that when I viewed them in the microscope I said "Pretty good". Later Dr. Takebe told him that he could not understand if "Pretty good" really meant "Wonderful". I should have said "Wonderful", for it truly was.

Takebe and his coworkers not only developed the protoplast system for TMV, but in succeeding years they applied the technique to other viruses, and they utilized the technique effectively to study TMV replication. I will not discuss their achievements; they are chronicled in a very fine review from Takebe (1975), describing protoplast studies in his laboratory and in other laboratories. Briefly, however, Takebe's group looked at the one-step growth curve of TMV in protoplasts from both systemic and hypersensitive (NN genotype) hosts and found that they were the same. They developed a technique for detecting the virus by fluorescent antibody staining, and used this procedure to follow the fate of two tobamoviruses inoculated into the same cell simultaneously and in sequence. Later they showed that the protoplasts could be infected by electroporation. Further, they studied the effect of virus replication on the RNA and protein synthesis of the host cell, and the fine structure of the infected protoplasts.

One very significant paper with implications far beyond plant virology, came from a collaboration with G. Melchers in Tübingen (where Takebe was a visiting scientist at the time). They demonstrated that tobacco mesophyll protoplasts could be regenerated into mature, fertile plants. This was an outcome from Takebe's earlier finding (Usui and Takebe, 1969) that the protoplasts could regenerate cell walls and could divide and differentiate in culture. Regeneration from single cells has become very important in plant genetics, plant molecular biology and in producing transgenic plants – another significant milestone for Itaru Takebe and his colleagues.

References

Aoki, S. and I. Takebe. (1969). Infection of tobacco mesophyll protoplasts by tobacco mosaic virus ribonucleic acid. *Virology* **39**:439-448.

Takebe, I. (1975). The use of protoplasts in plant virology. *Annu. Rev. Phytopathol.* **13**:105-125.

Takebe, I., Otsuki, O., and S. Aoki. (1968). Isolation of tobacco mesophyll cells in intact and active state. *Plant Cell Physiol.* **9**:115-124.

Takebe, I., Labib, G., and G. Melchers. (1971). Regeneration of whole plants from isolated mesophyll protoplasts of tobacco. *Naturwissenschaften* **58**:318-320.

Usui, H. and I. Takebe. (1969). Division and growth of single mesophyll cells isolated enzymatically from tobacco leaves. *Development, Growth and Differentiation* **11**:143-151.

Zaitlin, M. (1959). Isolation of tobacco leaf cells capable of supporting virus multiplication. *Nature* **184**:1002-1003.

Biographical Sketch

Milton Zaitlin graduated from the University of California, Berkeley with a degree in plant pathology in 1949. He worked as a technician at the California Institute of Technology for one year where he met Sam Wildman, who was moving to a position in the Botany Department at UCLA. He entered graduate school in 1950, and received his Ph.D. under Wildman in 1954 in botany, with a minor in biochemistry. Irving Rappaport and Albert Siegel, postdoctoral fellows in the Wildman laboratory, were also significant in his training. He spent about 4 years working at the Commonwealth Scientific and Industrial Research Organisation in Canberra, Australia, with Keith Boardman, Don Spencer and Paul Whitfeld. He returned to the United States to a position in the Department of Horticulture at the University of Missouri with the late Dan Millikan. In 1960 he worked as a postdoctoral fellow with Albert Siegel at the University of Arizona, Tucson, and rose in the ranks to professor. In 1973 he accepted a position as Professor in the Department of Plant Pathology at Cornell University. He also served as both Associate Director and Director of the Cornell Biotechnology Program. He received Emeritus status in 1997.

Reprinted from the PROCEEDINGS OF THE NATIONAL ACADEMY OF SCIENCES
Vol. 64, No. 3, pp. 843–848. November, 1969.

*INFECTION OF TOBACCO MESOPHYLL PROTOPLASTS BY TOBACCO MOSAIC VIRUS**

BY ITARU TAKEBE AND YOSHIAKI OTSUKI

INSTITUTE FOR PLANT VIRUS RESEARCH, AOBACHO, CHIBA, JAPAN

Communicated by Hitoshi Kihara, August 21, 1969

Abstract.—It was proved that the protoplasts prepared from mesophyll of *Nicotiana tabacum* are infected by tobacco mosaic virus. The infection occurred when purified tobacco mosaic virus particles were added to a protoplast suspension in the presence of poly-L-ornithine. The virus multiplied in these protoplasts to a level of 10^6 virus particles per infected protoplast during 24 hours of incubation. The efficiency of infection was remarkably high, exceeding that by mechanical inoculation of tobacco leaves.

Protoplasts can be prepared from plant cells by mechanical or enzymatic removal of cell walls.[1, 2] These isolated naked cells should be far more suitable than conventional tissue materials for experiments of quantitative nature, since they can be used as a homogeneous suspension in liquid medium.[3] The isolation of protoplasts from mesophyll was accomplished for the first time in this laboratory by a newly developed enzymatic technique.[3, 4] Mesophyll protoplasts were prepared on a large scale by cellulase treatment of isolated mesophyll cells which had been obtained by digesting leaves with pectinase. The effectiveness of this technique prompted us to explore the potentialities of mesophyll protoplasts as a new experimental material for studying various activities of plant cells.

Recently, RNA from tobacco mosaic virus (TMV) inoculated into tobacco mesophyll protoplasts was shown to give rise to a synchronous multiplication of this virus in these protoplasts.[5] This finding indicated that the mesophyll protoplasts should be invaluable for studying plant virus multiplication at a cellular level. However, it is obvious that studies of the entire process of infection, including penetration and uncoating of infecting virus, are possible only when infection is caused by complete virus particles. In the following, we present evidence to show that complete particles of TMV infect tobacco mesophyll protoplasts and multiply within them.

Materials and Methods.—Protoplasts of palisade parenchyma cells were prepared from mesophyll of healthy tobacco plants (*Nicotiana tabacum* var. Bright Yellow) according to the procedures previously reported[3] with slight modifications.[5] Figure 1 shows microscopical appearance of these protoplasts. The protoplasts were washed with and resuspended in 0.8 *M* mannitol at a concentration of 1 to 4 $\times 10^6$/ml, as measured by a Coulter electronic counter.[5] For infection, purified TMV (OM, a common strain[6]) was dissolved to a concentration of 2 μg/ml in 0.02 *M* potassium citrate buffer, pH 5.0, containing mannitol to 0.8 *M* and poly-L-ornithine (mol. wt. $\sim$130,000, Pilot Chemicals, Inc., Watertown, Mass.) to 2 μg/ml. After standing at 25°C for 10 min, the TMV solution was added to an equal volume of the protoplast suspension, and the mixture was kept at 25°C with occasional gentle swirling. After 60 min of infection, the protoplasts were separated from the inoculum virus by low speed centrifugation[5] and were washed several times with sterile 0.8 *M* mannitol containing $CaCl_2$ to 0.1 mM. The washed protoplasts were transferred aseptically into an incubation medium[5] to yield a concentration of 1 to 4 $\times$ 10^5/ml, and incubated in 10 ml portions as described in a previous paper.[3] After an appropriate period of incubation, the protoplasts were harvested by centrifugation, washed, and homogenized as described previously.[5] The extract thus prepared was made up to 10 ml with 0.1 *M* phosphate buffer, pH 7.0, and assayed for infectivity on a half leaf of *Nicotiana tabacum* var. *Xanthi* nc. Each assay was calibrated with a standard TMV solution inoculated on the other half leaf.[3]

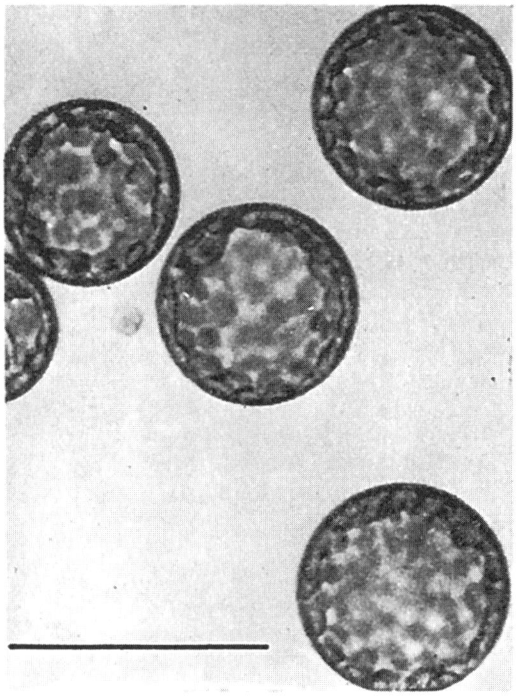

FIG. 1.—Protoplasts isolated from tobacco mesophyll and suspended in 0.8 *M* mannitol. Cytoplasm around central vacuole is packed with chloroplasts, which obscure nucleus and other structures. The scale represents 50 μ.

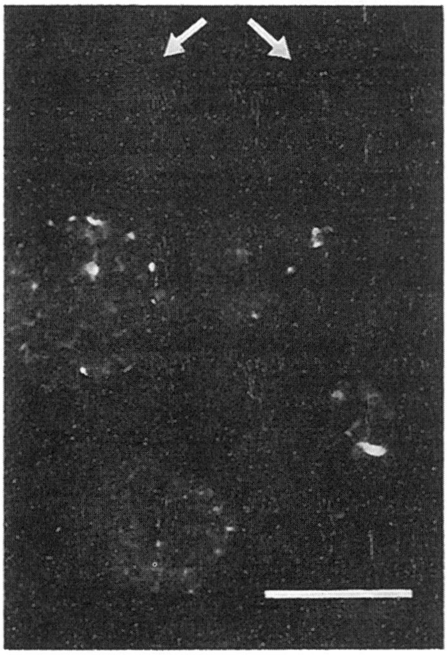

FIG. 2.—Fluorescence micrograph of protoplasts infected by TMV. Protoplasts were incubated for 27 hr after infection and stained with TMV-antibody labeled with fluorescein isothiocyanate. Four protoplasts in this figure contain stained material, while two in the upper part (arrows) do not. The scale represents 50 μ.

Results.—Development of infectivity: During infection under the conditions described above, only 1 per cent of the added virus became associated with protoplasts and was found in their extract as measured by infectivity assay. Correspondingly, the added virus was recovered, without detectable decrease in infectivity, in the infection medium from which the protoplasts were removed. It was thus found that approximately 1 per cent of the inoculum TMV particles were adsorbed to protoplasts under these conditions. Average number of virus particles adsorbed per protoplast was roughly estimated to be 10^2.

When these protoplasts were incubated for 27 hours, the infectivity extractable from the protoplasts increased strikingly (Table 1). The magnitude of increase varied somewhat from one experiment to another, but an increase of over 1000-fold was frequently obtained. As shown in Table 1, the development of infectivity was prevented by adding either 2-thiouracil or cycloheximide to the incubation medium. In contrast, actinomycin D showed apparently no effect

TABLE 1. *Influence of inhibitors on rise of infectivity in protoplasts infected by TMV.**

Inhibitor added	Infectivity of Extract of Protoplasts‡ 0 hr	27 hr
	1.5 ± 1.7	
None†		426 ± 174
Actinomycin D (10 μg/ml)†		558 ± 312
2-Thiouracil (10 μg/ml)†		52 ± 26
None		636 ± 198
Chloramphenicol (100 μg/ml)		636 ± 240
Cycloheximide (10 μg/ml)		3.5 ± 3.2

* Infected protoplasts were suspended in 70 ml of incubation medium at a concentration of 3.9×10^5/ml. Ten ml were removed as 0 hr sample and the rest was incubated in 10 ml portions for 27 hr with and without addition of inhibitor.

† Incubation was performed in the dark.

‡ Number of lesions produced by extract of protoplasts on a half leaf of *Nicotiana tabacum* var. *Xanthi* nc. Mean of nine determinations with standard deviation.

at the concentration which inhibited RNA synthesis in non-infected protoplasts by 85 per cent.[7] A high concentration of chloramphenicol also did not interfere with the development of infectivity (Table 1). These results show that the development of infectivity in the protoplasts requires DNA-independent RNA synthesis as well as cytoplasmic protein synthesis. The same pattern of response to these inhibitors is known for TMV multiplication in tobacco leaves,[8-10] in isolated tobacco leaf cells,[3] and in tobacco protoplasts infected by TMV-RNA.[5] The results in Table 1 thus indicate that TMV multiplied in the protoplasts as the result of infection by the added virus particles.

Fluorescent antibody staining of protoplasts: The protoplasts infected with TMV particles and subsequently incubated for 27 hours were stained with fluorescent antibody against TMV by the method recently described.[11] As illustrated in Figure 2, a number of protoplasts contained tiny masses of material which were stained by the fluorescent antibody. None of the protoplasts contained such material when they were sampled immediately after infection. Three types of control staining experiments[11] confirmed that this staining is specific for TMV antigen. It was thus demonstrated that the protoplasts infected under the present conditions accumulated TMV antigen within them during subsequent incubation.

Percentage of the protoplasts in which infection was established could be estimated by scoring, under the fluorescent microscope, those protoplasts which contained the stained material. The percentage was 13.8, 25.1, 25.3, 30.4 and 31.2 in five separate infection experiments.

Demonstration of progeny particles: Production of complete progeny TMV particles in these protoplasts was ascertained by electron microscopic examination of their extract. Numerous particles of the same morphology as TMV were found in the extract of protoplasts which were incubated for 27 hours, after infection (Fig. 3). Extracts of protoplasts prepared immediately after infection rarely contained similar particles, which should represent the infecting virus.

Time course of TMV multiplication: The multiplication of virus in the protoplasts infected by TMV particles followed a time course illustrated in Figure 4. During the initial period of incubation, there was a drop of infectivity in protoplasts. This was followed by a very rapid rise of infectivity between 8 and 22

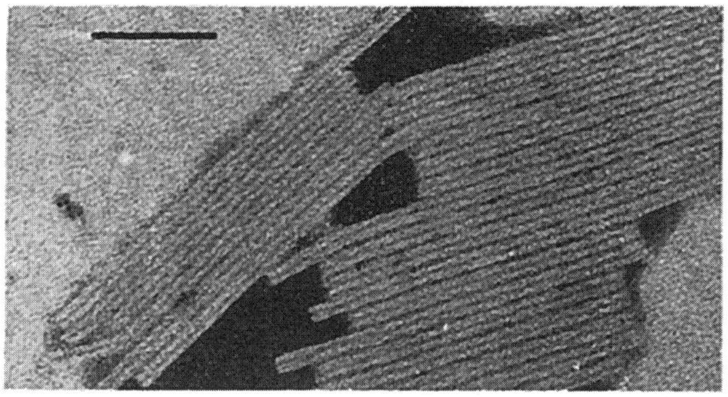

FIG. 3.—Electron micrograph of particles found in the extract of protoplasts infected by TMV and incubated for 27 hr. Negatively stained with phosphotungstic acid.[5] The scale represents 10 mμ.

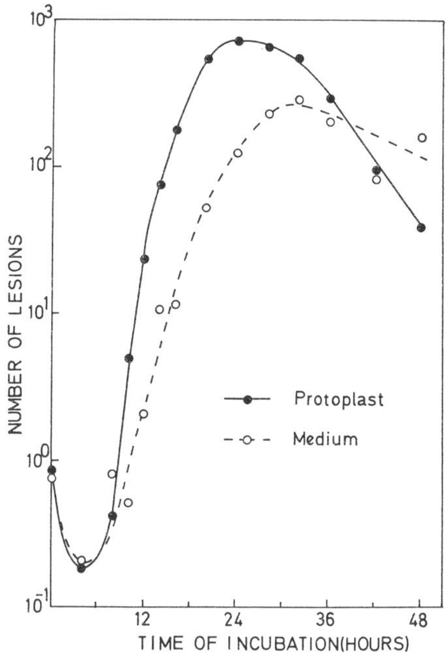

FIG. 4.—Time course of TMV multiplication in protoplasts. Protoplasts infected by TMV were suspended in 150 ml of incubation medium at a concentration of 2.0×10^5/ml and incubated in 10-ml portions. At indicated times one tube was harvested, extract of protoplasts was prepared and assayed for infectivity. To determine infectivity in medium, the incubation medium from which protoplasts were removed by centrifugation was dialyzed first against water and then against 0.1 *M* phosphate buffer, pH 7.0, and assayed for infectivity. Each point in the figure represents mean of eight to ten determinations.

hours of incubation. The infectivity in protoplasts reached a maximum at 24 hours and then decreased. The infectivity in medium, which probably reflects disintegration of some protoplasts, increased roughly in parallel to that in protoplasts during early periods of incubation, but reached a maximum somewhat later.

Factors necessary for infection: When protoplasts were treated under the condition of infection without added TMV particles, no infectivity was found in the protoplasts throughout subsequent incubation (Table 2). This indicates that the inoculum TMV is responsible for the development of infectivity in protoplasts. The amount of infectivity developing in protoplasts during the incubation depended upon the amount of TMV added as inoculum. However, higher concentrations of inoculum TMV than specified above did not result in the development of larger amounts of infectivity.

Poly-L-ornithine was necessary for infection of protoplasts by TMV particles. As shown in Table 2, the adsorption of inoculum virus occurred even in the absence of this polycation. However, the adsorbed virus apparently failed to establish infection, since no increase in infectivity was obtained by subsequent incubation of the protoplasts (Table 2). The effect of poly-L-ornithine depended upon its concentration, but it caused damage to the protoplasts at higher concentrations than used here.

TABLE 2. *Necessity of inoculum TMV and poly-L-ornithine for infection.**

Component omitted	Infectivity of Extract of Protoplasts† 0 hr	38 hr
None	1.2 ± 1.6	1315 ± 186
TMV	0	0
Poly-L-ornithine	1.9 ± 1.8	1.5 ± 1.6

* Protoplasts were treated under the condition of infection with full components, without addition of TMV, or without addition of poly-L-ornithine. The protoplasts were then incubated at a concentration of 1.6 × 10^5/ml for 38 hr.

† Number of lesions produced by extract of protoplasts on a half leaf of *Nicotiana tabacum* var. *Xanthi* nc. Mean of eight determinations with standard deviation.

Discussion.—The experiments described above indicate that TMV particles added with poly-L-ornithine established infection and multiplied in the tobacco mesophyll protoplasts. The time course of development of infectivity in these protoplasts was similar to that observed in the protoplasts infected by TMV-RNA,[5] suggesting that a synchronous multiplication of TMV is taking place also in the present system. The only difference was that, in the present system, the infectivity due to infecting virus was detectable in the protoplasts at the beginning of incubation and that it decreased prior to the development of infectivity due to progeny virus (Fig. 4). Whether this "eclipse" reflects uncoating of infecting virus particles is at present not clear. The nature of the decline of infectivity in protoplasts which occurred in the later phase of incubation also remains to be investigated.

Average yield of progeny virus per infected protoplast could be estimated from the maximal infectivity in protoplasts and the number of infected protoplasts which was determined by fluorescent antibody staining.[5] The yield in the protoplasts infected by TMV particles ranged 0.5–1.1 × 10^6 particles per protoplast, being somewhat higher than that obtained by infection with viral RNA.[5]

The most remarkable aspect of infection in the present system is its high efficiency. In a typical experiment, 3 × 10^5 protoplasts were infected with 2 μg of purified TMV as inoculum. This amount of virus should contain 3 × 10^{10} particles, assuming 40 million for the "molecular weight" of TMV.[12] Infection in this system has, therefore, an efficiency of one infection per 10^5 inoculum particles, which is ten thousand times higher than that for the infection of protoplasts by TMV-RNA.[5] The efficiency is higher even than that by mechanical inoculation of tobacco leaves, since the same amount of TMV produces only 10^3 necrotic lesions on a half leaf of *Nicotiana tabacum* var. *Xanthi* nc. By virtue of this high efficiency, as much as 30 per cent of whole protoplast population was infected in this system.

The infection of protoplasts by TMV particles requires the presence of poly-L-ornithine at the time of infection (Table 2). This polycation is known to strongly enhance pinocytic uptake of proteins by cultured mammalian cells.[13] On the other hand, protoplasts of tomato fruit cells are reported to take up particles of TMV by a pinocytic process.[14] It seems, therefore, that the penetration of inoculum particles into our system is also attributable to the pinocytic activity of plant protoplasts.

Cocking and Pojnar recently showed by an electron microscopic study that TMV is taken up by and multiplies in the protoplasts prepared from tomato fruits.[15] Their system, however, requires the presence of a very high concentration of inoculum virus for a long period, indicating that the efficiency of infection is quite low. Moreover, the rate of virus multiplication in their system is apparently much slower than that in ours.

As discussed previously,[5] the mesophyll protoplasts have definite advantages over conventional tissue materials as the experimental system for studying plant virus multiplication. The present study disclosed another advantage of this host cell system, namely that it can be inoculated with a high efficiency which was not attainable with systems thus far available. The mesophyll protoplasts should thus provide a powerful tool for elucidating the processes underlying the infection and the multiplication of plant viruses.

We are indebted to Dr. Y. Saito of this Institute for taking the electron micrograph.

* This work was supported in part by a research fund from the Ministry of Education.

[1] Cocking, E. C., in *Viewpoints in Biology*, eds. J. D. Carthy and C. L. Duddington (London: Butterworths, 1965), vol. 4, p. 170.

[2] Ruesink, A. W., and K. V. Thimann, *Science*, **154**, 280 (1966).

[3] Takebe, I., Y. Otsuki, and S. Aoki, *Plant & Cell Physiol.*, **9**, 115 (1968).

[4] Otsuki, Y., and I. Takebe, *Plant & Cell Physiol.*, in press.

[5] Aoki, S., and I. Takebe, *Virology*, in press.

[6] Nozu, Y., and Y. Okada, *J. Mol. Biol.*, **35**, 643 (1968).

[7] Sakai, F., and I. Takebe, unpublished results.

[8] Francki, R. I. B., and R. E. F. Matthews, *Virology*, **17**, 22 (1962).

[9] Sänger, H. L., and C. A. Knight, *Biochem. Biophys. Res. Commun.*, **13**, 455 (1963).

[10] Zaitlin, M., D. Spencer, and P. R. Whitfeld, in *Biochemical Regulation in Diseased Plants or Injury*, eds. T. Hirai, Z. Hidaka, and I. Uritani (Tokyo: The Phytopathological Society of Japan, 1968), p. 91.

[11] Otsuki, Y., and I. Takebe, *Virology*, **38**, 497 (1969).

[12] Lauffer, M. A. and C. L. Stevens, in *Advances in Virus Research*, eds. K. M. Smith and M. A. Lauffer (New York and London: Academic Press, 1968), vol. 13, p. 1.

[13] Ryser, H. J.-P., *Science*, **159**, 390 (1968).

[14] Cocking, E. C., *Planta*, **68**, 206 (1966).

[15] Cocking, E. C., and E. Pojnar, *J. gen. Virol.*, **4**, 305 (1969).

Identification of the TMV Coat Protein Messenger RNA

Andrew O. Jackson
Department of Plant and Microbial Biology, University of California, Berkeley

My contribution to this anthology relates to the series of events in the Siegel-Zaitlin labs that preceded the classic paper of Hunter *et al.* (1976), in which the messenger RNA activity encoding the TMV coat protein was determined. My role in this discovery began early in 1970 when I initiated a postdoctoral fellowship with Al Siegel and Milt Zaitlin, who at that time were in the Department of Agricultural Biochemistry at the University of Arizona. I had just completed my Ph.D. at the University of Manitoba working on phenylpropanoid and folic acid metabolism of rust fungi where I investigated the possibility that these metabolites might be involved in the hypersensitive response elicited in resistant cultivars. I wanted to broaden my horizons into other aspects of plant–pathogen interactions, so studies of viruses and their infection processes presented a particularly attractive prospect for me. The Siegel-Zaitlin lab consisted of a group of extremely interesting characters who mostly were focused on various aspects of TMV replication. Among these were attempts to identify viral-specific proteins synthesized in plants and analysis of TMV replicase; two areas in hot competition with Fraenkel-Conrat and Bea Singer's lab at Berkeley. Other research involved analysis of the disease phenotypes elicited by TMV coat protein mutants, and the nature and relationships of malformed protein aggregates and abnormal particles formed by the mutant protein, as well as genome complexity experiments being carried out by Al's students with tobacco and cucurbits that resulted in the discovery of host RNAs encapsidated as pseudovirions. Therefore, when I arrived in Tucson, I found a broad spectrum of topics under intense investigation.

My first project involved a study of replicative form and replicative intermediate RNAs found in TMV infected tissue. Upon publication of this work, Richard Francki, who was on sabbatical in Milt's lab, encouraged me to investigate the kinetics of synthesis of these components in separated leaf cells. This was an exciting new area of exploration that was an offshoot of the protoplast infectivity procedures that had just been developed by Takebe's group in Japan. By the time I began to focus on the system, Milt, Al, Richard and Dick Jensen had already conducted a considerable amount of exploratory work on isolated mesophyll cells, and were able to observe CO_2 fixation, amino acid incorporation into proteins, and uridine labeling of RNA. Unfortunately, they were unable to infect protoplasts, but they had observed that isolated cells and protoplasts recovered from TMV-infected leaves actively synthesized viral RNAs and proteins. Therefore, I decided to investigate the synthesis of TMV by comparing incorporation of ^{3}H-uridine into RNAs of uninfected and TMV-infected mesophyll cells.

With the technology available at the time, these experiments were quite laborious, so it may be instructive to review these procedures for history buffs and for modern day postdocs and graduate students who may be spending the time saved on their more efficient experiments to read this commentary. Mesophyll cells were first isolated from leaves by pectinase treatment, pulse-labeled with isotope, and after phenol extraction, aliquots of the RNAs were spotted on filter paper discs to normalize the isotope incorporation. The RNAs were then separated on the recently devised Peacock and Dingman composite acrylamide-agarose gels, which were cast in small (~6 x 100 mm) Plexiglass tubes. Identification of marker RNAs (usually requiring loading about 10 to 15 mg per band) was carried out by scanning in a spectrophotometer equipped with a tube gel scanner. This procedure provided a great advancement in resolution over the previously used sucrose density gradients, but it was quite crude compared to present day photographic visualization of RNA species. For analysis of isotope incorporation, the cylindrical gels were placed on a spatula, frozen on dry ice, placed on parafilm over a glass plate and cut into sections with a gel slicing device holding 101 single edge razor blades spaced one mm apart. To cut sections of the same size, this delicate operation required that the razor blade device be too warm to hold without asbestos gloves, but not so hot that the agarose melted during sectioning!! When performed properly, uniform sections could be obtained by applying firm pressure to the frozen cylindrical gels. In order to remove the resulting 100 sections easily, the gel next was refrozen to just the right temperature by placing the gel slicing device on the block of dry ice for a short period of time. Then the frozen gel slices were removed with a thin weighing spatula and successive duplicate slices were pooled in scintillation vials. It usually took about 30 min for me to complete this whole procedure with a single gel, so a considerable amount of time was spent carrying out this task. However, this provided ample opportunity for planning the next lab experiments and also permitted a great deal of congenial banter with other postdocs and students. Subsequently, one ml aliquots of H_2O_2 were added to each vial, followed by heating overnight before adding scintillation fluid and counting. Finally, the counts printed on long strips of paper from the scintillation counter were normalized and plotted manually on graph paper to visualize uridine incorporation into individual RNA species. Fortunately, these cumbersome procedures were soon superseded by slab gels, visualization of RNAs by ethidium bromide staining, and methods for transfer of radioactivity from gels to filter papers for northern blot analyses.

When I began to plot the counts from the first experiments, I noticed a substantial peak of radioactivity in the TMV infected samples that migrated well in advance of 16 S ribosomal RNA. I initially thought this peak might be an anomaly, but after several repetitions and experiments to demonstrate RNase sensitivity, I became convinced that the band represented a newly synthesized TMV species, so I started calling the component LMC for low molecular weight component. Al initially was more skeptical, but he also became quite excited following repeated observations of the presence of LMC at different stages of the infection cycle, its persistence following actinomycin D treatment of cells, and "pulse-chase" experiments which suggested that LMC accumulates during infection. After a number of confirmatory experiments, we prepared a manuscript describing the synthesis of replicative form, replicative intermediate, TMV RNA and LMC. As I recall, the first rough drafts contained a considerable amount of discussion about the nature of LMC and its possible role in replication. Most of this material was removed subsequently during internal editing to condense the manuscript and to accommodate the reviewer's comments. Therefore, we essentially wound up focusing on the synthesis of the replicative intermediates and TMV RNA and provided only a cursory discussion of LMC without addressing its possible functions. I have always felt disappointed that we were not more assertive at this time in emphasizing LMC, and in expressing our ideas about its potential involvement in replication.

Because I had accepted a second postdoctoral position in Nebraska with Myron Brakke, my contribution to the study of LMC ended at this point. However, several more interesting twists and turns with some missed opportunities occurred afterwards that relate to progress in understanding the function of LMC. After my departure, Al and Milt conducted hybridization experiments to verify that LMC is TMV specific, and they also demonstrated that polyadenylic acid sequences are not present in the RNA. The results also confirmed our previous results showing that only double stranded RNA intermediates corresponding to the genomic RNA accumulate in infected cells. This led to the hypothesis that LMC might be derived directly from the viral genome during replication, rather than being produced from a subgenomic replicative form RNA. Reconstitution experiments also demonstrated that LMC is not encapsidated. From the latter observation, Al and Milt drew the correct conclusion that LMC lacks the 5' of TMV RNA, but this conclusion was based on a common misconception at the time that encapsidation signals were located at the 5' end of the RNA. [See the interesting commentary by Ken Richards for the story surrounding the dissection of the encapsidation signals.] Al and Milt's paper also revived our earlier speculations that LMC might be the coat protein mRNA. This work thus contributed substantially towards the notion that LMC is a subgenomic RNA derivative arising from the 3' end of the TMV genome and that the RNA might have messenger activity.

Then, this very productive collaboration and the central focus on LMC was interrupted because Al accepted a position as Chairman of Biology at Wayne State University, and Milt moved to Cornell shortly afterwards. Al and Milt divided the TMV projects, and the translation experiments planned for LMC went to Detroit where they languished for several years in the hands of a remarkably unproductive postdoctoral fellow. However, Milt's lab flourished and a number of his experiments provided additional hints that LMC might have a role in protein synthesis. For example, Milt and his new postdoc, Roger Beachy, demonstrated that LMC is associated with polyribosomes, and they emphasized again that the RNA could have a role in synthesis of the coat protein. Next, George Bruening (who was on sabbatical at Cornell), Roger Beachy, Rene Scalla and Milt capitalized on the observation by Jack Morris that virions isolated from the cowpea C_C strain of TMV differ from those of the U1 strain by consisting of two nucleoprotein components. The Cornell group initially separated the C_C RNAs and found two major TMV-specific components which they designated L and S, plus a minor component dubbed I_2. From hybridization experiments, they concluded that the I_2 and S RNAs are duplicated in the genomic L RNA. Translation in wheat germ extracts indicated that L RNA directed synthesis of a major product estimated to be 130 kDa and a minor 150 kDa polypeptide, which corresponded to those found in infected plants. Translation of the I_2 RNA resulted in substantial amounts of a new 30 kDa product now known as the TMV movement protein. Gel electrophoresis indicated that the S RNA corresponded in size to LMC, and it served as an mRNA for synthesis of a polypeptide that had several properties characteristic of the coat protein. These included coelectrophoresis of the translation product and authentic coat protein, demonstration that the translation product could reconstitute with TMV RNA, and precipitation of the reconstituted material with antibodies raised against TMV. Bruening *et al.* (1976) also speculated that the I_2 and S RNAs are derived from the 3' end of the

TMV genome, and they postulated for the first time that each of the three viral RNAs are functionally monocistronic. Milt's group also suggested that the C_C TMV S RNA might contain signals specifying encapsidation by the capsid protein that are lacking in the U1 LMC. Each of these predictions were subsequently shown to be correct, so the C_CTMV paper proposed a series of hypotheses that ultimately contributed an enormous amount to our understanding of virus replication and movement.

The extensive series of experiments in the classic paper published in *Nature* by Hunter *et al.* verified most of the previous results with LMC, and finally provided convincing biochemical proof that LMC *per se* is a subgenomic mRNA encoding the TMV coat protein. The Cambridge investigators initially separated the RNA directly from infected tissue by sucrose density gradient centrifugation, and they provided strong evidence that it is not a degradation product of TMV RNA. Further characterization verified the Tucson hybridizations showing that LMC lacks the polyadenylated structure characteristic of eukaryotic mRNAs, and RNA fingerprinting revealed sequence patterns characteristic of those near the 3' end of TMV RNA. Several additional lines of experimentation demonstrated that LMC directs synthesis of the coat protein. First, the major translation product synthesized in the wheat germ system and in *Xenopus* oocytes comigrated with the coat protein. Initial amino acid incorporation experiments indicated that methionine and histidine were lacking in the translation product, as expected from the amino acid profile of the coat protein of the vulgare TMV strain used in these experiments. Additional peptide mapping analyses provided direct chemical evidence that the translation product was indeed TMV coat protein, and reconstitution experiments verified that the product actively encapsidated TMV RNA *in vitro.* The Cambridge authors then went on to provide a model for the replication of TMV in which the larger translation products, now known to be 126 and 183 kDa, were postulated to be viral replicase components that are synthesized prior to synthesis of LMC RNA. Therefore, this classic paper and the paper by Bruening *et al.* (1976) provided the first models to explain involvement of subgenomic mRNAs in temporal and quantitative regulation of 3' encoded gene products.

On reflection, the series of events arising from studies of LMC provide a typical illustration of the stepwise process whereby scientific discovery advances. From our initial observations of LMC using the newly developed isolated cell methodology, a new line of inquiry emerged that eventually culminated in its identification as the mRNA for the TMV coat protein. This understanding immediately generated new concepts about RNA virus gene expression, and contributed more broadly to enunciation of monocistronic rules for translation of eukaryotic mRNAs. Resolution of several additional questions raised by the Cambridge and Cornell groups ultimately had a profound impact on our understanding of the replication and gene expression of TMV, as well as a wide range of other plant and animal viruses. Among the most important of these questions were: How is gene expression regulated temporally and quantitatively?, What mechanisms are used for transcription of the subgenomic mRNAs?, What roles do the I_2 RNA and its translation product have on the infection process?, Why do the 3' genes of plant viruses generally differ from those of bacterial viruses by not being translated directly from genomic RNAs?, How are the replicase subunits translated and how do they function during RNA replication?, What is the function of the cap structure in translation and how is this structure generated?, and Can the mechanisms operating during TMV infection be applied more globally to other groups of plant and animal viruses? Therefore, although our initial identification of LMC represented only a small step in understanding the biology of TMV, the resulting experiments extending this discovery culminated in several important concepts that illuminated numerous aspects of virus replication. A number of new issues were also raised during the course of the LMC investigations, most of which have contributed to our current emphases on infection processes and disease development. I am pleased to have had the opportunity to be involved in the early phases of these discoveries, and I feel fortunate to have been able to initiate my virology training in the nurturing environment provided by the Siegel-Zaitlin labs.

References

Beachy, R. and Zaitlin, M. (1975). Replication of tobacco mosaic virus. VI. Replicative intermediates and TMV-RNA-related RNAs associated with polyribosomes. *Virology* **63**:84-97.

Bruening, G., Beachy, R. N., Scalla, R., and M. Zaitlin. (1976). *In vitro* and *in vivo* translation of the ribonucleic acids of a cowpea strain of tobacco mosaic virus. *Virology* **71**:498-517.

Hunter, T., Hunt, T., Knowland, J., and D. Zimmern. (1976). Messenger RNA for the coat protein of tobacco mosaic virus. *Nature* **260**:759-764.

Jackson, A. O., Mitchell, D. M., and A. Siegel. (1971). Replication of tobacco mosaic virus. I. Isolation and characterization of double-stranded forms of ribonucleic acid. *Virology* **45**:182-191.

Jackson, A. O., Zaitlin, M., Siegel, A., and R. I. B. Francki. (1972). Replication of tobacco mosaic virus. III. Viral RNA metabolism in separated leaf cells. *Virology* **48**:655-665.

Siegel, A., Zaitlin, M., and C. T. Duda. (1973). Replication of tobacco mosaic virus. IV. Further characterization of viral related RNAs. *Virology* **53**:75-83.

Takebe, I. and Y. Otsuki. (1969). Infection of tobacco mesophyll protoplasts by tobacco mosaic virus. *Proc. Natl. Acad. Sci. USA* **64**:843-848.

Biographical Sketch

Andrew Jackson was awarded the Bachelor of Science Degree in 1964 and the Master of Science Degree in botany and plant pathology in 1967 from Oklahoma State University. He obtained the Ph.D. degree in plant pathology and microbiology from the University of Manitoba in 1970 under the direction of D. J. Samborski and Roland Rohringer, where his dissertation involved an analysis of intermediary metabolism of rust uredospores. His career in virology began when he was awarded a postdoctoral fellowship in the labs of Albert Siegel and Milton Zaitlin at the University of Arizona where the initial identification of LMC was carried out. In 1971, he initiated a second postdoctoral fellowship with Myron Brakke at the University of Nebraska, and then was hired as an Assistant Professor at Purdue University in 1973. In 1985, he accepted his current position as a Professor at the University of California-Berkeley.

(Reprinted from Nature, Vol. 260, No. 5554, pp. 759–764, April 29, 1976)

Messenger RNA for the coat protein of tobacco mosaic virus

Tony R. Hunter* & Tim Hunt
Department of Biochemistry, Tennis Court Road, Cambridge CB2 1QW, UK
John Knowland† & David Zimmern
MRC Laboratory of Molecular Biology, Hills Road, Cambridge CB2 2HQ, UK

TMV RNA is not an efficient template for translation of the viral coat protein, in spite of containing nucleotide sequences coding for the protein. Efficient translation requires the prior synthesis within infected cells of a smaller RNA carrying only a portion of the information encoded in the whole genome.

Tobacco mosaic virus (TMV) consists of an infectious single-stranded RNA molecule 6,400 nucleotides long, encapsidated by 2,100 identical coat protein molecules each with a molecular weight of 17,500 (ref. 1). This packaged RNA (virion RNA) is the sense strand for the translation of coat protein, as was first shown by patterns of mutagenesis[2,3] and more recently by direct nucleotide sequence analysis[4]. The coat protein gene must consist of at least 480 residues of virion RNA. Thus less than 10% of the RNA is accounted for by the coat cistron.

As would be expected, the RNA acts as a good message in cell-free translation systems from bacteria[5], plants[6] and animals[7], but analysis of the products of such incubations has repeatedly shown that coat protein is not the major product, and careful studies with the *E. coli* cell-free system eventually proved that ribosomes never passed over the coat protein cistron in the correct phase[8,9]. This might be due to general incompatibility between eukaryotic and prokaryotic mRNA and ribosomes, but studies of the translation of TMV RNA in various eukaryotic protein synthesising systems have given similarly negative results, thus suggesting that translation of coat protein is not straightforward, and raising the problem of how the coding sequences known to be present on the RNA actually are expressed. In the wheat germ cell-free system, for example, the major products are much larger than the coat protein and also very heterogeneous. Nevertheless Efron and Marcus[6] detected labelled coat protein peptides in products which migrated with authentic coat protein on gel filtration, although they represented only a very minor fraction of the total incorporation. In contrast, when Roberts *et al.*[10] used RNA from the TMV mutant Ni 568, they failed to detect significant levels of product with the mobility of coat protein on acrylamide gels. They did, however, find a coat protein peptide among the products of their incubation, at very low yield, and it was suggested[11] that the coat protein is made as part of a large precursor which is processed by proteolysis to yield mature coat, as in the case of picornaviruses. This idea is inconsistent with the data of Efron and Marcus, who failed to find any coat protein peptides in material larger than coat protein.

We have used two other eukaryotic systems to translate TMV RNA, and have found that it directs the synthesis of discrete polypeptides, which simplifies interpretation (ref. 7 and our unpublished results). In *Xenopus laevis* oocytes[7] and in reticulocyte lysates, the products directed by TMV RNA are two large polypeptides with molecular weights of 140,000 and 165,000; the first corresponds to the largest product of the wheat germ system. We have never detected the synthesis of coat protein. The 140,000 polypeptide does not contain coat protein sequences detectable by tryptic peptide mapping[7] and, given the results reported here, it does not seem likely that the coat is contained within the 165,000 molecular weight polypeptide either. We believe that these proteins are authentic translation products of cistrons in TMV RNA other than the coat cistron, for similar sized proteins are made in TMV-infected leaves and protoplasts[12,13]. In synchronously infected protoplasts, these two proteins are readily detectable about 5 h after infection and there is a further lag of about 4 h before coat protein synthesis begins. Late after infection the synthesis of the two large proteins declines, but coat protein synthesis continues at very high levels. It is difficult to interpret these findings, particularly the latter observation, in terms of a precursor–product relationship between the largest proteins and the coat protein.

It thus seems that there is some temporal and quantitative control of gene expression in TMV, as has been found in essentially all RNA viruses (with the exception perhaps of the picornaviruses), and that most of the extra RNA in TMV which does not code for coat protein codes for the proteins of molecular weight 140,000 and 165,000. If this is correct, and cell-free systems and *Xenopus* oocytes make essentially only 'early' TMV proteins, then how is the coat protein made? We extracted TMV-infected tobacco leaves to see if we could find and characterise the mRNA for

* Present address: Salk Institute, PO Box 1809, San Diego, California 92112
† Present address: Department of Anatomy, Medical School, Bristol BS8 1TD, UK

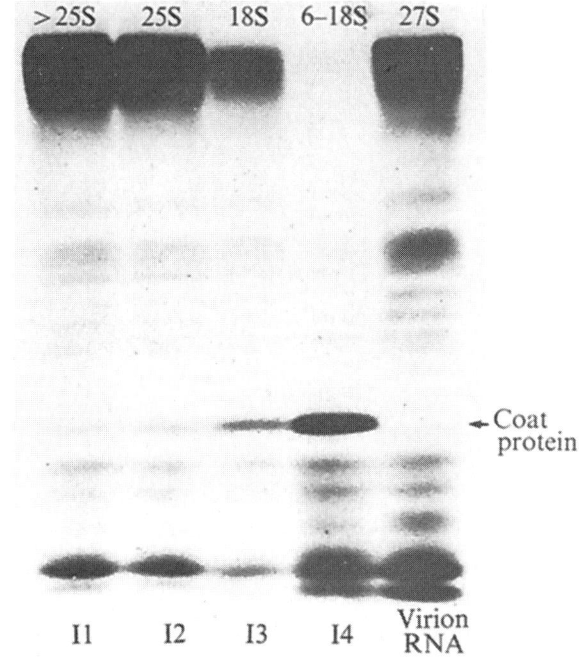

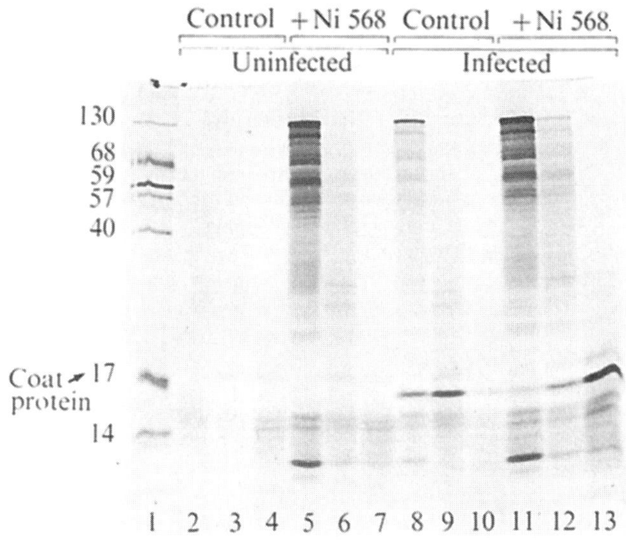

Fig. 1 *a*, Translation of RNA of different size classes from tobacco leaves infected with TMV. Tobacco leaves, infected 7 d previously with TMV vulgare, were washed in water, ground in liquid nitrogen and extracted in an equal mixture of water-saturated phenol and buffer A (10 mM EDTA, 1% SDS, 50 mM Tris–Cl, *p*H 8.8) at 65 °C. The aqueous phase was extracted four more times with phenol, made 0.1 M in NaCl and precipitated with 2 volumes of ethanol. The precipitate was taken up in 1 mM EDTA, 0.5% SDS; warmed for 1 min at 65 °C, and spun at 10,000*g* for 30 min at 4 °C to remove undissolved material. The supernatant was layered over 15–40% (w/v) sucrose gradients in 0.1% SDS, 1 mM EDTA, 20 mM Na acetate, *p*H 5.0, and spun in a Beckman SW 27 rotor for 18 h at 27,000 r.p.m. at 15 °C. The gradients were pumped through a recording spectrophotometer and fractions were pooled as follows: I1 > 25S; I2 ~ 25S; I3 ~ 18S; I4 18S-6S. The RNA fractions were reprecipitated from 0.2 M Na acetate with 3 volumes of ethanol. The precipitates were taken up in water, and stored in liquid nitrogen, except for fraction I4 which was diluted in 1 mM EDTA, 1% SDS, 2 M LiCl and precipitated overnight at 4 °C (ref. 45). The precipitate was recovered by centrifugation at 10,000*g* for 10 min, and taken up in water before ethanol precipitation as above. Each fraction, and RNA extracted from TMV virions, was tested for template activity in a cell-free wheat germ system prepared and incubated as described by Roberts and Paterson[16], except that the ionic conditions in the incubation were: 80 mM KCl, 1.5 mM $MgCl_2$, 0.8 mM spermidine, 20 mM HEPES, *p*H 7.2. The labelled amino acid was *L*-^{3}H-leucine (20 Ci mmol^{-1}, 10 μM). Reactions were stopped by incubation with pancreatic RNase (100 μg ml^{-1}) for 15 min at 30 °C. Four volumes of SDS gel sample buffer were added[17] and the samples were analysed on 20% SDS-acrylamide gels. After fixing and staining, the gels were impregnated with PPO according to Bonner and Laskey[18] and exposed to Kodak RP Royal X-ray film at −70 °C. The fluorogram shows the products directed by fractions I1–I4, and virion RNA. All RNAs were present at 200 μg ml^{-1} in the incubations. A control without added messenger showed no detectable blackening of the film. *b*, Control for degradation of RNA during extraction. RNA was prepared and fractionated from either uninfected leaves (tracks 2–7) or TMV vulgare-infected leaves (tracks 8–13) as described in (*a*), except that 1 mg of purified Ni 568 virus was added per g wet weight of ground leaves to half of each leaf sample just before phenol extraction (tracks 5–7 and 11–13). The RNA was fractionated by size and the three larger size classes were tested for their template activity as described above, except that products were analysed on a 15% SDS–polyacrylamide gel. The markers in track 1 are β galactosidase, bovine serum albumin, pyruvate kinase, glucose-6-phosphatase, creatine kinase, TMV coat protein and lysozyme, all labelled with ^{14}C-iodoacetic acid.

coat protein, since large amounts of coat protein are made in infected cells, and thus some functional messenger of high efficiency clearly exists. This could in theory be the virion RNA complexed with some control protein, but there were indications from earlier work that the actual messenger might be a smaller RNA[14,15]. This could explain the marked discrepancy between the apparently faithful synthesis of the 140,000 and 165,000 molecular weight polypeptides in *in vitro* systems, and the total lack of coat protein synthesis in the same systems when programmed with virion RNA, implying that the coat cistron in virion RNA is repressed.

Preliminary experiments

RNA was extracted from tobacco leaves which had been systematically infected with TMV 7 d previously. They were frozen and pulverised in liquid nitrogen before extraction in hot phenol–sodium dodecyl sulphate (SDS). After several phenol extractions, RNA was precipitated with ethanol, and after resuspension in aqueous buffer, fractionated on sucrose gradients. Fractions from the gradients were pooled according to their sedimentation rate and recovered by ethanol precipitation. The slowest sedimenting fraction was very heavily contaminated with carbohydrate, and was further purified by precipitation with 2M LiCl at 2 °C overnight. Each fraction, as well as corresponding fractions from uninfected plants, was added to cell-free extracts of wheat germ, and incubated essentially according to Roberts and Paterson[16]. The incubations were analysed by SDS–polyacrylamide gel electrophoresis on slabs[17], followed by fluorography[18] to detect ^{3}H-labelled polypeptides. Figure 1*a* shows the result of one such experiment. A strongly radioactive band with the same electrophoretic mobility as coat protein was made only in incubations programmed with RNA from TMV-infected leaves, and not in incubations programmed with virion RNA. Furthermore, maximal synthesis occurred when the smallest RNA fractions were used, suggesting that the active template was an RNA molecule considerably smaller than intact TMV RNA.

An obvious objection to the straightforward interpretation that a small RNA molecule is the natural mRNA for TMV coat protein is that it might have arisen by degradation of virion RNA during extraction of the leaves. We sought to exclude this possibility by the experiment shown in Fig. 1*b*. The mutant TMV strain Ni 568 contains a single methionine residue in its coat protein[2], unlike the wild-type strain vulgare which contains no methionine. We therefore repeated the extraction of RNA from leaves infected with TMV vulgare, but added a large amount of purified TMV strain Ni 568 virus just before the phenol extraction. After extraction and fractionation of the RNA as before, the fractions were tested in the wheat germ system using either ^{35}S-methionine or ^{3}H-leucine as label in parallel incubations. The products were analysed on slab gels as before; the results of the incubations with ^{3}H-leucine as label are shown in Fig. 1*b*. No coat protein-like products were seen in RNA from uninfected plants, whether or not TMV Ni 568 had been added during extraction. In contrast, such a protein was synthesised when RNA from infected plants was used, but it was labelled only with leucine, and not with methionine. This shows that RNA from the added virions is not converted into a form capable of being translated into the putative coat protein by degradation or modification during extraction of the RNA from leaves. The experiment also shows that uninfected leaves do not contain detectable mRNA for this protein. It should be noted that since typical virion RNA-directed products are observed in the track corresponding to the uninfected plants to which Ni 568 RNA had been added, this RNA was extracted

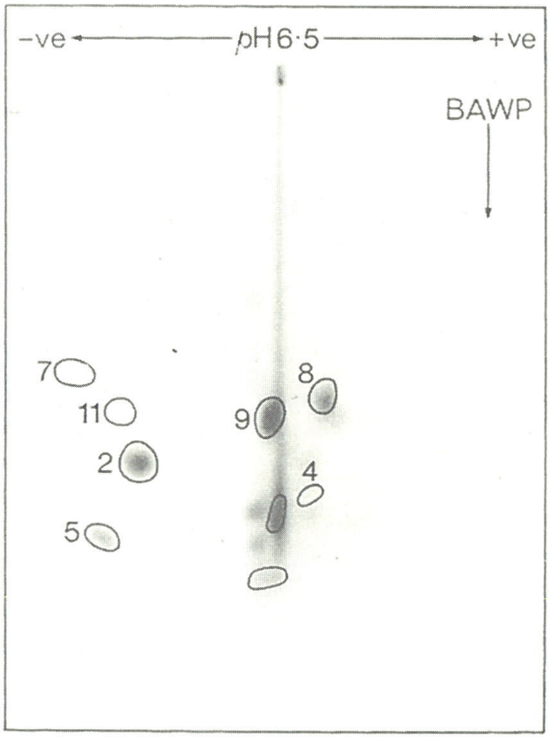

Fig. 2 Fingerprint of putative coat protein synthesised *in vitro*. Fraction I4 RNA from infected leaves was injected into 20 *Xenopus* oocytes, which was incubated in medium containing 20 μCi of L-^{14}C-valine (280 μCi μmol^{-1}) for 24 h. They were homogenised as described previously[7] with 100 μg of authentic TMV vulgare coat protein. The homogenate was centrifuged, and the supernatant fractionated by electrophoresis on an 18% SDS-polyacrylamide gel. The gel was stained with Coomassie blue, dried, and autoradiographed. The stained coat protein band, which was heavily labelled, was cut into small pieces. 2 ml of 10 mM Tris–Cl, *p*H 7.0, 0.1% 2-mercaptoethanol, 2% SDS were added, and the suspension was incubated for 10 min at 95 °C and for 18 h at 37 °C. The bits of paper and gel were filtered off, and the protein was recovered from the filtrate by precipitation with trichloroacetic acid (TCA). It was oxidised and digested with trypsin[7] and the soluble peptides were separated by electrophoresis at *p*H 6.5, 20 V cm^{-2} on a 20-cm^{2} Silica gel G thin layer (250 μm) (Anachem)[46], followed by chromatography in butanol–acetic acid–water–pyridine (30:6:24:20) at right angles to the first dimension. After autoradiography, the carrier TMV coat protein peptides were stained with fluorescamine[47]. The stained peptides (outlined) are numbered from the N-terminus; peptides 4, 7 and 11 do not contain valine.

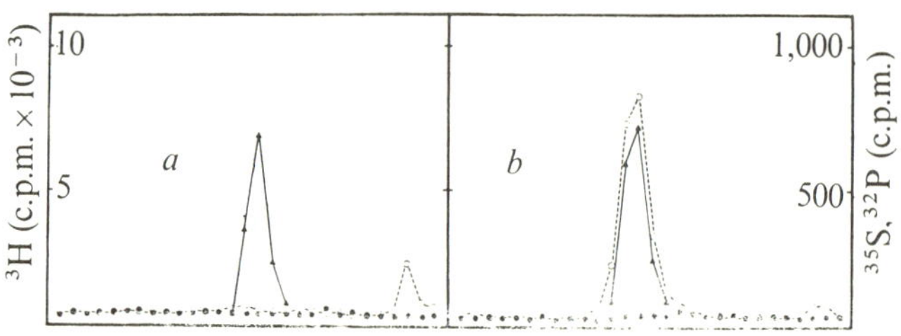

Fig. 3 Assembly of the *in vitro* product. Four cell-free incubations using the wheat germ system programmed with either fraction I4 RNA from TMV-infected leaves or 27S virion RNA, were labelled with ^{35}S-methionine or ^{3}H-leucine. After incubation for 1 h, the ^{3}Hl- and ^{35}S-labelled samples corresponding to each template were mixed together, and 1.8 mg of TMV coat protein was added to both mixtures. The two mixtures were dialysed against 10 mM Tris–Cl, *p*H 8.0, and passed over 0.5-ml columns of DE 52 cellulose equilibrated with the same buffer. After washing with buffer, the coat protein was eluted with 0.2 M KCl, 10 mM Tris–Cl, *p*H 8.0. This material was dialysed against 10 mM Na acetate, *p*H 5.0, to form the helix aggregate[31], which was collected by centrifugation at 100,000*g* for 1.5 h, suspended in Na phosphate buffer, *p*H 7.0, ionic strength 0.1, made 5 mg ml^{-1} in coat protein by a further addition of carrier, and dialysed overnight at 4 °C against the same buffer. The sample was then incubated at 20 °C for 18 h to form disks, at which time TMV RNA (1 mg ml^{-1}) was added, and incubation continued for 10 min at 20 °C. Uncoated RNA was digested with T1 RNase (20 μg ml^{-1}) for 1 h at 20 °C, and the partially reassembled virions were purified by two cycles of preparative CsCl density gradient centrifugation, using added ^{32}P-labelled TMV as marker, in the Beckman SW 50 rotor (CsCl, 1.33 g ml^{-1}, 32 h at 35,000 r.p.m.). The fractions containing marker virus were analysed finally in 8-ml CsCl gradients in the 50 angle head rotor, spun at 40,000 r.p.m. for 24 h. *a*, Result of attempted assembly using the translation product specified by 27S virion RNA; *b*, that using incubations programmed with fraction I4 RNA. Overall recovery of radioactivity from the ^{3}H-labelled incubation programmed with coat mRNA in the peak of CsCl-banded reassembled virus was about 10% while essentially no radioactivity from any other incubation survived to the final analysis. ▲, ^{32}P radioactivity; ○, ^{3}H radioactivity; ●, ^{35}S radioactivity.

efficiently. Furthermore, the amount of products formed suggested that we had added about the same amount of virion RNA as was naturally present in the infected leaves (compare track 5 with track 8 in Fig. 1*b*). We are therefore confident that the RNA which promotes the synthesis of the putative coat protein is not an artefact of degradation on phenol extraction.

Characterisation of the product

The new band programmed by RNA from infected leaves has the same mobility on SDS gels as authentic TMV coat protein, and it lacks methionine. We also found that histidine failed to label the gel band, which would be expected of coat protein, which lacks both methionine and histidine.

When the RNA from TMV-infected leaves was injected into *Xenopus* oocytes, it directed synthesis of a similar band. The yields were higher, and the background lower than in the wheat germ system, and we therefore used oocyte synthesised material to make a tryptic peptide map (Fig. 2). There is excellent correspondence between the *in vitro* synthesised peptides, labelled with ^{14}C-valine, and the fluorescamine-stained carrier peptides, except in the case of coat protein peptides that lack valine. We conclude that this material is chemically indistinguishable from TMV coat protein.

We next asked whether the putative coat protein could be assembled into virus particles *in vitro;* that is whether it resembled coat protein by functional as well as chemical criteria. We therefore partially purified the *in vitro* synthesised coat protein by chromatography on DEAE–cellulose, a procedure which removes basic proteins which may bind nonspecifically to viral RNA during the assembly process. After elution from the column, the protein was concentrated by precipitation at *p*H 5, dissolved in buffer (*p*H 7) and mixed with TMV RNA. After digestion with T1 RNase to remove uncoated RNA tails, the reconstitution mixtures were analysed by CsCl gradient centrifugation with the result shown in Fig. 3. Of four reaction mixtures, using the products labelled with either ^{3}H-leucine or ^{35}S-methionine, programmed by either coat protein mRNA or virion RNA, only the one with coat protein mRNA and ^{3}H-leucine as label gave radioactively labelled nucleoprotein of the same buoyant density as added $^{32}PO_4$-labelled TMV. Furthermore, the labelled protein extracted from this reconstituted nucleoprotein migrated with authentic TMV coat protein on polyacrylamide gels, and no other labelled bands could be detected on the fluorogram. This was in marked contrast to the starting material, which contained many other labelled polypeptides besides coat protein. We conclude from these experiments that a small RNA species present in TMV-infected cells, and not an artefact of extraction, programmes the *in vitro* synthesis of a protein which is chemically and functionally indistinguishable from TMV coat protein.

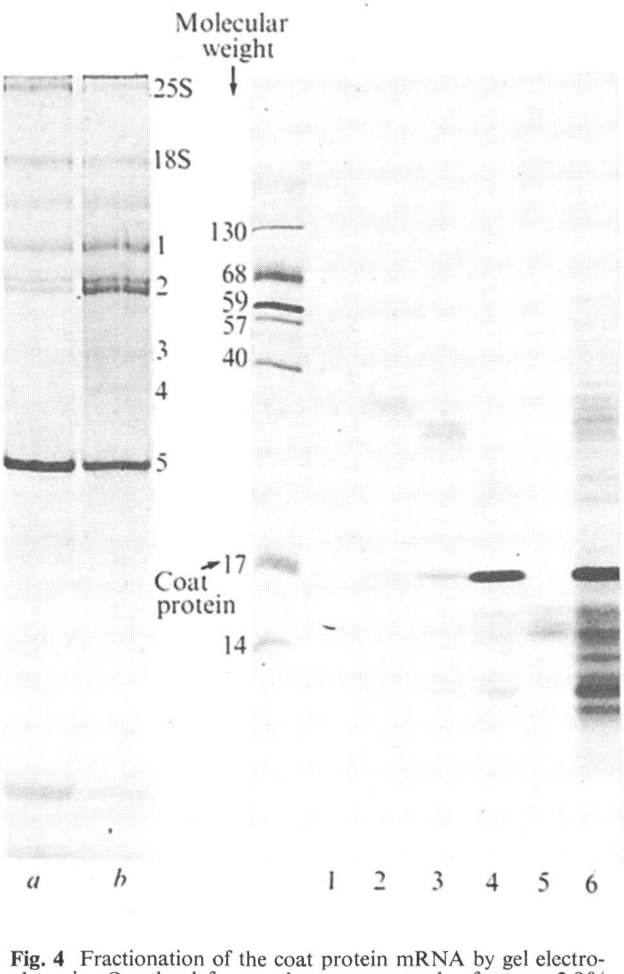

Fig. 4 Fractionation of the coat protein mRNA by gel electrophoresis. On the left are shown two tracks from a 2.8% Agarose/acrylamide gel stained with Stainsall (Eastman Chemicals). *a*, 6–18S RNA from uninfected tobacco leaves; *b*, 6–18S RNA from TMV-infected leaves. The indicated bands from parallel unstained tracks were cut out and the RNA was extracted from them by homogenisation in an equal mixture of phenol and buffer containing 0.1 M NaCl, 10 mM Tris–Cl, *p*H 7.5, 1 mM EDTA, 1% SDS, with 50 μg of carrier yeast tRNA. After extraction for 6 h at 0 °C, the tubes were centrifuged and RNA precipitated from the supernatants with 2.5 volumes of ethanol. The precipitates were washed with ethanol before being taken up in water and tested for template activity in the wheat germ system as described in Fig. 1. The autoradiograph on the right shows the analysis on a 20% acrylamide gel of the products specified by fraction 14 RNA (track 6) and RNA from the indicated gel bands (tracks 1–5). The positions of molecular weight markers are indicated in the left margin of the gel.

Characterisation of the RNA

Up to this point, the fractionation of RNA by sucrose density gradients provided only partially pure coat protein mRNA, but one would expect the coat mRNA activity to reside in a discrete RNA species if it were of genuine biological significance. We therefore took our most active fraction of gradient purified RNA and fractionated it further on an Agarose/acrylamide gel[19]. Figure 4 shows a gel stained with Stainsall[20]; adjacent tracks contain RNA from TMV-infected and uninfected tobacco leaves. We found a reproducible minor stained band in infected leaves which was absent in control leaves. To see if it was coat protein mRNA, a gel with duplicate tracks was run, and one track was stained while the other was cut into bands and the RNA was extracted. We found that RNA from the new gel band was an excellent mRNA for coat protein (Fig. 4). Some of the slower migrating bands also programmed the synthesis of small amounts of coat protein, besides making larger proteins as one would expect. We do not know whether this represents a certain amount of trailing on the gel or of aggregation (since these were not denaturing gels), or whether there may perhaps be larger RNA molecules with coat protein mRNA activity. We estimate that the molecular weight of the major mRNA species is about 2.5×10^{5}, or 750 nucleotides, of which at least 480 would be required to code for coat protein. Coat protein mRNA failed to bind to oligo(dT)–cellulose (data not shown), showing that, like TMV virion RNA, the coat protein message lacks a significant poly(A) region.

From the classical work mentioned in the introduction, we expected the coat protein mRNA to have the same polarity as virion RNA, and since large end products of RNase T1 digestion which come from the coat cistron have already been characterised[4,21], we sought these characteristic markers in coat protein mRNA. Tobacco plants were infected with TMV and labelled by growing them in medium containing $^{32}PO_4$. Coat protein mRNA was extracted as before, and purified on sucrose gradients and Agarose/acrylamide gels. We found the expected marker oligonucleotides in fingerprints of the T1 RNase digests (Fig. 5*b*), thus proving directly that the gel band contained nucleotide sequences from the expected region of virion RNA, so that it is a fragment of the viral (+) strand, and does not come from an unrelated RNA, such as a satellite virus. There was another unexpected bonus from this analysis, in that the fingerprint bore a striking resemblance to that of an RNA extracted from a nucleoprotein fragment of defined length (containing 1,000 nucleotides) which is generated when TMV particles are degraded in mildly alkaline conditions[22,23]. The RNA from this 'alkali stable fragment' has independently been shown to contain the 3′ terminal T1 oligonucleotide of TMV in molar yield[24], and can be charged with histidine by yeast activating enzymes, as is the case for intact TMV RNA[25]. Analysis of the T1 RNase fingerprint of this fragment showed that it also contained the marker oligonucleotides found in the coat protein mRNA (Fig. 5*a*), showing that the part of the coat cistron from which they originate must be within 1,000 nucleotides of the 3′ end of TMV RNA. Furthermore, the purified coat protein mRNA was found to contain a T1 oligonucleotide which gave rise to the pancreatic RNase products (AAU, C_5,G). This is probably identical with the oligonucleotide AAUCCCCCG which occurs about 20 nucleotides from the 3′ end of intact virion RNA[26]. This suggests that the coat protein mRNA probably corresponds to the 3′ terminal 750 nucleotides of virion RNA.

This assignment of the location of the coat protein cistron is at the opposite end of the RNA to that found by Kado

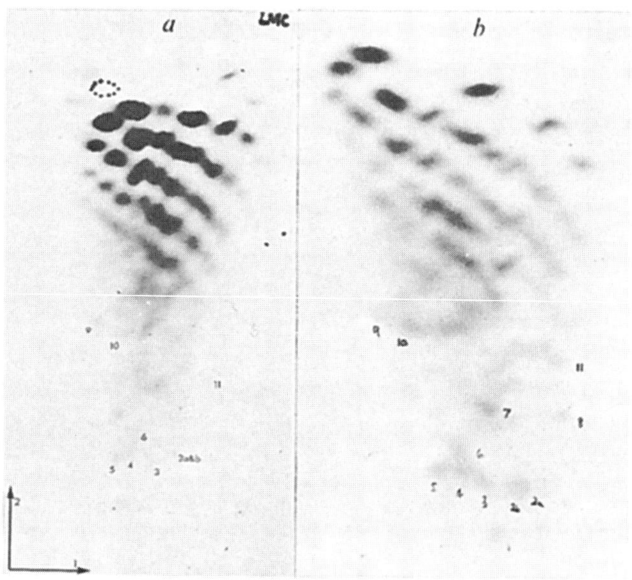

Fig. 5 Comparison of T1 RNase digested coat protein mRNA and 'stable fragment' RNA. ^{32}P-labelled TMV was prepared as described by Zimmern[40] and incubated at 4 °C overnight in 10 mM $NaHCO_3/Na_2CO_3$, *p*H 10.2. Uncoated RNA was digested with micrococcal nuclease (1 µg ml^{-1}) in the same buffer adjusted to *p*H 8.0 with HCl and 1 mM $CaCl_2$. EDTA was added in excess to stop the digestion, and the RNA that had been protected from digestion extracted with phenol–SDS, and purified by sucrose density gradient centrifugation. Fractions containing 'stable fragment RNA' were pooled and precipitated with 2.5 volumes of ethanol. ('Stable fragment' from ^{32}P-labelled TMV was prepared by Michael Wilson.) The precipitate was taken up in 10 mM Tris–Cl, *p*H 7.6, 1 mM EDTA and digested with 3 µl of RNase T1 (1 mg ml^{-1}) in the presence of carrier *E. coli* tRNA (50 µg). The digest was fingerprinted as described by Barrell[44]. Spot 2b is AUAAAUAAUUUAUAGp, which codes for residues 125–129 of coat protein[4]. Spot 5 is AAACCUUCACCACAAGp, which codes for amino acids 53–57[4,21]. Spot 9 is AACCCCACGp, coding for residues 101–103[4]. Spot 4 can also code. Spot 10 here is a mixture, but contains a component giving pancreatic RNase products AAU, Cn, G, chain length 9–10, consistent with AAUCCCCCGp which comes from positions −20 to −28 from the 3′ end of whole TMV RNA (ref. 26). The left hand autoradiograph shows the digest of band 4 extracted from a gel similar to that shown in Fig. 4, except that the RNA was prepared from leaves of plants infected with TMV grown in $^{32}PO_4$. Spots 2a–11 were identified by their pancreatic RNase digestion products as being identical with the equivalent spots in the fingerprint of stable fragment RNA. The more complex appearance of this fingerprint is due to slight pancreatic RNase digestion, which probably occurred during the lengthy extraction process; the mRNA was also slightly contaminated with ribosomal RNA fragments. *a*, Coat protein mRNA; *b*, 'Stable fragment RNA'.

and Knight[27] on the basis of SDS stripping and marker rescue *in vivo*, but agrees with that of Richards *et al.*[28].

Virus replication cycle

On the basis of this evidence, we propose the following account of the replicative cycle of TMV (Fig. 6). After infection of the cell and uncoating of the virion RNA, ribosomes can only translate the cistrons for the 140,000 and 165,000 molecular weight polypeptides, consistent with their early appearance in infected protoplasts. We assume that these functions are required early in infection, and since the subunit molecular weights are close to the apparent native molecular weight of TMV replicase activity in partially purified extracts[29,30], we suggest that both proteins are viral replicases, or their subunits; the function of each being different. For example, one of them may use the (+) strand as a template, the other the (−) strand. Alternatively, one may make coat protein mRNA by partial transcription of the (−) strand, while the other makes full-length transcripts. At least one of them must catalyse the production of complementary copies of the virion RNA, which serve in turn as templates for the synthesis of progeny virion RNA molecules.

The synthesis of coat protein is delayed until its functional mRNA has been produced, because virion RNA does not serve as a template for coat protein synthesis. Two plausible mechanisms for production of the coat protein message are (1) that it is released from virion RNA by specific cleavage, and (2) that it is synthesised by a replicase which initiates transcription at a specific internal site on the complementary strand of TMV RNA. We cannot distinguish between these possibilities at present.

Only when enough coat protein has accumulated to permit the formation of "disk" aggregates[31] can the assembly of virus particles begin[32]. This would delay the encapsidation of virion RNA thereby allowing time for synthesis of early proteins and (−) strands. Once assembly begins the virion RNA will be sequestered and therefore can no longer be a template for either process. But since neither (—) strands nor coat protein mRNA are encapsidated in the case of TMV vulgare, the production of virions can continue for as long as these RNA molecules remain intact, and the replicase(s) active.

This kind of scheme can easily explain how TMV can produce such large quantities of coat protein compared with its other proteins. Being a helical virus, TMV requires many more protein subunits to encapsidate its genome than would a spherical virus with the same genome size[33]. Thus poliovirus, whose RNA is slightly larger than that of TMV, requires only 60 copies[34] of each structural polypeptide compared with about 2,100 in the case of TMV.

Translational control

We do not know why the coat protein cistron is untranslatable in intact TMV RNA whereas it is easily read in the coat protein mRNA; however, at least two precedents offer clues. Initiation of translation of the A cistron of the RNA phages seems to occur only in short growing RNA molecules, since the initiator region of this cistron is involved

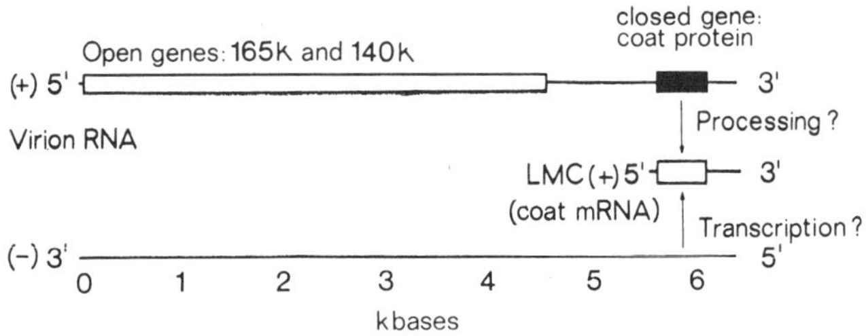

Fig. 6 A tentative map of TMV RNA and its known proteins. The map is drawn to scale. The location of the coat cistron is discussed in the text; the location of the 165,000 and 140,000 molecular weight polypeptides is less certain. Since the sum of their molecular weights exceeds the coding capacity of the virion RNA, they must overlap somehow. We believe they are initiated at different sites since virion RNA can bind two ribosomes simultaneously and independently in the presence of 10^{-4} M sparsomycin, promoting the formation of two methionyl dipeptides: Met–Ala and (probably) Met–Thr (T.R.H. and T.H., unpublished observations). The simplest possibility is that they then terminate at the same site, although this need not necessarily be the case. Their position near the 5′ end of the RNA is suggested by the fact that RNA molecules with small deletions at the 5′ end, constructed by alkali stripping procedures similar to those described in the legend to Fig. 5, fail to synthesise either large polypeptide (T.R.H., M. Wilson, and D.Z., unpublished results). This again is inconsistent with the possibility that the coat protein gene is contained within the large polypeptides (see above). These assignments leave approximately 1,100 residues of RNA unaccounted for, enough to code for up to 40,000 daltons of protein, and this gap has been positioned in the middle of the map. There is evidence from another strain of TMV (the cowpea strain) that a 28,000–30,000 molecular weight protein, whose translation is regulated by a mechanism similar to that described here for TMV coat protein, may be coded in this region[49].

in stable secondary structure in the full-sized RNA molecule[35]. Similar mechanisms may operate in brome mosaic virus (BMV), a multicomponent virus which contains four different RNA molecules. BMV RNA 3 is larger than BMV RNA 4, and just as TMV coat protein mRNA is wholly contained within virion RNA, so BMV RNA 4 is contained within BMV RNA 3 (ref. 36); furthermore, BMV RNA 4 codes efficiently for BMV coat protein, whereas BMV RNA 3 makes another polypeptide but little coat[17]. It seems probable that the coat cistron in BMV RNA 3 is masked by secondary structure, but there is another possibility which may either augment or replace this mechanism. When the sequence of the ribosome binding site of BMV RNA 4 was determined[38], the 5′ end was found to be blocked with m^7G; this unusual end group may be necessary for ribosomes to attach to mRNA[39]. We do not know whether TMV coat protein mRNA has this structure at its 5′ end or not, though intact virion RNA certainly does[40].

Such regulatory mechanisms are not confined to plant and bacterial RNA viruses, however; in both Sindbis and Semliki Forest virus-infected cells small RNAs with the same polarity as the packaged RNA have been found which, unlike their parent molecules, efficiently direct synthesis of the corresponding coat polypeptides *in vitro*[41,42]. Even in the cases of the DNA viruses polyoma and SV40 it seems that larger RNA molecules which code for one protein code for a different protein after cytoplasmic cleavage[43]. Whether such a mechanism regulates the translation of normal cellular messages is unkown. It might account for some cases of 'masked mRNA', for example. We would like to conclude by pointing out that these cases of translational control operate with high efficiency, yet they do not appear to require the action of specific protein synthesis initiation factors.

A preliminary account of these experiments has been published[48].

D. Z. was an MRC Scholar, and T. H. a Beit Fellow. We thank the MRC, Clare College, Cambridge and the Cancer Campaign for research funds, and Jo Butler, Richard Jackson, Bryan Roberts, Ruth Longthorne, Anne Brayley and Christine Harley for assistance. We thank Herman Lehmann for laboratory space.

Received February 3; accepted March 4, 1976.

1 Caspar, D. L. D., *Adv. Protein Chem.*, **18**, 37–121 (1963).
2 Wittman, H. G., and Wittman-Liebold, B., *Cold Spring Harb. Symp. Quant. Biol.*, **31**, 163–172 (1966).
3 Funatsu, G., and Fraenkel-Conrat, H., *Biochemistry*, **3**, 1356–1362 (1964).
4 Guilley, H., Jonard, G., Richards, K. E., and Hirth, L., *Eur. J. Biochem.*, **54**, 135–153 (1975).
5 Tsugita, A., Fraenkel-Conrat, H., Nirenberg, M. W., and Matthaei, J. H., *Proc. natn. Acad. Sci. U.S.A.*, **48**, 846–853 (1962).
6 Efron, D., and Marcus, A., *Virology*, **53**, 343–348 (1973).
7 Knowland, J., *Genetics*, **78**, 383–394 (1974).
8 Aach, H. G., Funatsu, G., Nirenberg, M. W., and Fraenkel-Conrat, H., *Biochemistry*, **3**, 1362–1366 (1964).
9 Schwartz, J. H., *J. molec. Biol.*, **30**, 309–322 (1967).
10 Roberts, B. E., Mathews, M. B., and Bruton, C. J., *J. molec. Biol.*, **80**, 733–742 (1973).
11 Roberts, B. E., Paterson, B. M., and Sperling, R., *Virology*, **59**, 307–313 (1974).
12 Sakai, F., and Takebe, I., *Virology*, **62**, 426–433 (1974).
13 Paterson, R., and Knight, C. A., *Virology*, **64**, 10–22 (1975).
14 Siegel, A., Zaitlin, M., and Duda, C. T., *Virology*, **53**, 75–83 (1973).
15 Beachy, R. N., and Zaitlin, M., *Virology*, **63**, 84–97 (1975).
16 Roberts, B. E., and Paterson, B. M., *Proc. natn. Acad. Sci. U.S.A.*, **70**, 2330–2334 (1974).
17 Anderson, C. W., Baum, P. R., and Gesteland, R. F., *J. Virol.*, **12**, 241–252 (1973).
18 Bonner, W. M., and Laskey, R. A., *Eur. J. Biochem.*, **46**, 83–88 (1974).
19 Peacock, A. C., and Dingman, C. W., *Biochemistry*, **7**, 668–674 (1968).
20 Dahlberg, A. E., Dingman, C. W., and Peacock, A. C., *J. molec. Biol.*, **41**, 139–147 (1969).
21 Garfin, D. E., thesis, Univ. California (1972); Garfin, D. E., Mandels, S., *Virology*, **64**, 388–399 (1975).
22 Schramm, G., Schumacher, G., and Zillig, W., *Z. Naturf.*, **10b**, 481–492 (1955).
23 Perham, R. N., *J. molec. Biol.*, **45**, 439–441 (1969).
24 Perham, R. N., and Wilson, T. M. A., *FEBS Lett.*, **62**, 11–15 (1976),
25 Oberg, B., and Philipson, L., *Biochem. biophys. Res. Commun.*, **48**, 927–932 (1972).
26 Guilley, H., Jonard, G., and Hirth, L., *Proc. natn. Acad. Sci. U.S.A.*, **72**, 864–868 (1975).
27 Kado, C. I., and Knight, C. A., *J. molec. Biol.*, **36**, 15–23 (1968).
28 Richards, K. E., Morel, M. C., Nicolaieff, A., Lebeurier, G., and Hirth, L., *Biochimie*, **57**, 749–755 (1975).
29 Brishammer, S., and Junnti, N., *Virology*, **59**, 245–253 (1974).
30 Zaitlin, M., Duda, C. T., and Petti, M. A., *Virology*, **53**, 300–311 (1973).
31 Durham, A. C. H., Finch, J. T., and Klug, A., *Nature new Biol.*, **229**, 37–42 (1971).
32 Butler, P. J. G., and Klug, A., *Nature new Biol.*, **229**, 47–50 (1971).
33 Caspar, D. L. D., and Klug, A., *Cold Spring Harb. Symp. Quant. Biol.*, **27**, 1–24 (1962).
34 Rueckert, R. R., in *Comparative Virology* (edit. by Maramorosch, K., and Kurstack, E.), 255–306 (Academic, New York, 1971).
35 Fiers, W., *et al.*, *Nature*, **256**, 273–278 (1975).
36 Shih, D. S., Lane, L. C., and Kaesberg, P., *J. molec. Biol.*, **64**, 353–362 (1972).
37 Shih, D. S., and Kaesberg, P., *Proc. natn. Acad. Sci. U.S.A.*, **70**, 1799–1803 (1973).
38 DasGupta, R., Shih, D. S., Saris, C., and Kaesberg, P., *Nature*, **256**, 624–628 (1975).
39 Both, G. W., Banerjee, A. K., and Shatkin, A. J., *Proc. natn. Acad. Sci. U.S.A.*, **72**, 1189–1193 (1975).
40 Zimmern, D., *Nucleic Acid Res.*, **2**, 1189–1201 (1975).
41 Simmons, D. T., and Strauss, J. H., *J. molec. Biol.*, **86**, 397–409 (1974).
42 Clegg, J. C. S., and Kennedy, S. I. T., *Eur. J. Biochem.*, **53**, 175–183 (1975).
43 Prives, C. L., Aviv, H., Gilboa, E., Revel, M., and Winocour, E., *Cold Spring Harb. Symp. Quuat. Biol.*, **39**, (1), 309–316 (1975).
44 Barrell, B. G., in *Procedures in Nucleic Acid Research* (edit. by Cantoni, G. L., and Davies, D. R.), 751–779 (Harper and Row, New York, 1971).
45 Schimke, R. T., *et al.*, in *Methods in Enzymology*, **30** (edit. by Moldave, K., and Grossman, L.), 631–648 (Academic, New York, 1974).
46 Ritschard, W. J., *J. Chromatog.*, **16**, 327–333 (1964).
47 Vandekerkhove, J., and Van Montague, M., *Eur. J. Biochem.*, **44**, 279–288 (1974).
48 Knowland, J., Hunter, T., Hunt, T., and Zimmern, D., *Proc. INSERM-EMBO Colloquium In vitro Transcription and Translation of Viral Genomes, INSERM*, **47**, 211–216 (1975).
49 Bruening, G., Beachy, R. N., Scalla, R., and Zaitlin, M., *Virology*, (in the press).

Printed in Great Britain by Henry Ling Ltd., at the Dorset Press, Dorchester, Dorset

The Tale of the Two-Tailed Monsters

Ken Richards
Institut de Biologie Moléculaire des Plantes, Strasbourg, France

The model of TMV assembly described in the two papers of the Anthology marked for me the climax of an exciting chase with many abrupt twists and blind alleys, leading to that rare event in science: a moment of revelation when a totally unexpected solution emerges. As the reader will discover below, my account can aptly be titled 'The Tale of the Two-Tailed Monsters'.

The story starts for me in 1971 when I began a postdoctoral fellowship with Robley Williams in the Virus Laboratory at Berkeley. The Virus Laboratory was where *in vitro* assembly of TMV was first achieved by Fraenkel-Conrat and Williams and my project there was to learn more about the mechanism of assembly. But upon arrival I discovered that the ground had shifted quite a bit since my initial readings on the subject. In particular, Tony Durham, in Aaron Klug's group at the MRC Laboratory of Molecular Biology in Cambridge, had just produced an in-depth study of the different polymeric forms assumed by the TMV coat protein (CP). His key insight concerned the ~20 S polymerization product which forms at physiological conditions of pH and ionic strength. He convincingly argued that this species was a two-layer disk composed of two rings of 17 CP subunits, with the rings apposed head-to-tail [references to this and other uncited literature may be found in Hirth and Richards (1981)].

The positioning of the 17 CP subunits in each ring of the two-layer disk approximates the arrangement of the 16 2/3 CP subunits in a turn of the viral nucleoprotein helix. Butler and Klug (1971) recognized that the flat surface of the disk provided an ideal platform for simultaneous binding of as many as 51 residues of the RNA during initiation of viral assembly. A coordinated interaction of this sort would account for the observed specificity in assembly of TMV CP for the viral RNA. Binding of RNA to the first layer of the disk was assumed to trigger dislocation of the planar disk structure to a proto-helix and elongation would then occur by successive binding of more CP to the RNA protruding from the business end of the incipient rod.

Butler and Klug's experiments illustrated that virus assembly occurred with great rapidity *in vitro* if the CP was first allowed to form disks, and the role of disks in assembly initiation was quickly accepted by all. The Butler-Klug model, however, further posited that disks were added during the elongation phase of rod growth, which seemed implausible to me on topological grounds. A lively controversy broke out over this issue, pitting Butler and his colleagues against Williams and me and Okado and Ohno, who defended the view that rod elongation involved addition of the CP monomer or small aggregates which exist in equilibrium with the disk. The debate simmers on, but without my participation.

In 1973, I moved from Berkeley to Strasbourg, France, where Léon Hirth led a sizable group of workers interested in TMV assembly. After several months there, I drifted into the orbit of Hubert Guilley and Gérard Jonard, who were trying to characterize the assembly initiation sequence on TMV RNA. Their approach was to partially digest the RNA with T1 RNase, mix the digestion products with limiting amounts of CP under assembly conditions and allow the CP to, hopefully, select out short RNA fragments containing the initiation sequence. They had isolated and laboriously sequenced one such specifically encapsidated RNA fragment (SERF) of 105 nucleotides. The sequence contained a strong third-order repeat of purine residues, particularly G's. A trimodal repeat was expected for a sequence which interacted with the array of nucleotide binding sites (three per subunit) on the disk surface. But the sequence could also be folded into a stable hairpin, which was less expected. Analysis of its coding capacity revealed that the SERF corresponded to the central portion of the CP cistron. This last discovery caused great dismay since the assembly initiation sequence was thought at the time to lie at an extremity of the viral RNA. Further experiments confirmed that the SERF sequence was not involved in initiation of assembly of common strain TMV, although it may initiate assembly for some other strains, such as the cowpea strain.

Not to be discouraged, Jonard proposed that we partially digest the viral RNA with another nuclease, RNase A, before fishing for encapsidated fragments. I must confess that I was skeptical of this idea at the time because I was beginning to think that the 'real' assembly initiation sequence might be relatively unstructured, to facilitate its binding to the disk surface, and such a sequence would not be expected to resist even mild nuclease treatment. As things turned out, however, Jonard's idea was a good one and we were able to isolate and sequence a second SERF. This 139 residue fragment also displayed a trimodal purine repeat and could be folded into a hairpin, but it did not come from the coat protein cistron, the only portion of the TMV RNA sequence for which sequence information could be deduced at the time. Furthermore, the sequence was already encapsidated in partially assembled rods containing as little as 800 nucleotides of the viral RNA, indicating that it was near, if not identical to, the authentic assembly initiation sequence.

While we in Strasbourg were fishing for the assembly initiation sequence, David Zimmern and Butler in Cambridge had taken a different approach. They prepared initiation complexes by allowing about one disk-equivalent of CP to react with the viral RNA. Unencapsidated RNA was then eliminated by nuclease treatment. Analysis of the smallest protected RNA fragments identified a core initiation sequence of about 100 residues, but the small amounts of RNA available precluded determination of the complete sequence. A cautious exchange of information between Cambridge and Strasbourg revealed that the two groups were working on the same sequence and it was decided to publish the papers describing the core initiation sequence 'back-to-back' (Zimmern, 1977; Jonard *et al.*, 1977). As a bonus, analysis of some of the longer protected RNA species (Zimmern and Butler, 1977) established that the direction of rapid CP addition during rod elongation was toward the 5' end of the viral RNA.

Exactly where on the viral RNA was the assembly initiation sequence located? When visualized by electron microscopy, partially assembled TMV rods appeared to have RNA (either a tangled tail or a 'puff', depending upon the specimen preparation procedure) protruding from only one extremity. This had led to the idea, noted above, that assembly initiated at or very near one end of the viral RNA. Data from TMV disassembly experiments confuted this conclusion. In mild alkaline conditions, TMV particles uncoat in a polar fashion starting from the end of the particle containing the RNA 5' terminus. Zimmern and Mike Wilson showed that the assembly initiation sequence was exposed only when the RNA had been alkali-stripped of CP to a point about 1000 residues from the 3' end of the RNA. We now know that the initiation sequence is situated between nucleotides 975 and 845 from the 3' end, within the movement protein cistron.

To summarize, as things then stood TMV assembly was thought to be initiated at a point approximately 1000 nucleotides from the 3' extremity of the viral RNA and rapid rod elongation proceeded from there in the 5' sense. Yet electron microscopy showed unencapsidated RNA at only one end of short, partially assembled rods! The technical breakthrough which resolved the dilemma in Strasbourg was provided by Geneviève Lebeurier and Anatole Nicolaieff. They had been studying the disassembly of TMV particles provoked by dimethylsulfoxide. DMSO disassembled virions from both ends, but evaluation of the length of the two tails of partially uncoated rods by electron microscopy was hindered by their tendency to become entangled. They discovered fortuitously that inclusion in the sample preparation medium of small amounts of DMSO (below the level which provokes particle disassembly) prevented tail-tail interactions. Lebeurier and 'Nico' had also tried to use DMSO to better visualize the unencapsidated RNA in preparations of partially assembled rods, but the results had been disappointing. Most of the rods had only a very short tail or no tail at all and there was much free RNA in the background. [It later became clear that non-optimal conditions for partial assembly and the poor quality of the particular viral RNA preparation were at fault.] Occasionally, particles with two RNA tails protruding from the same extremity of the rod were seen. Such was the force of our preconceptions, however, that such particles (termed 'two-tailed monsters') were dismissed as artefacts which had arisen by binding of a strand of free RNA to the presumably 'sticky' growing end of a one-tailed partially assembled rod.

The lucky break came when Jonard and I prepared what was to be our last sample of partially assembled rods using radioactive viral RNA. To assess its quality, I asked Nico to take a look at it with the electron microscope. Fortunately, he chose to include DMSO in the sample preparation mix. He emerged from the dark room muttering darkly that he had never seen such a poor preparation: nothing but two-tailed monsters! A closer look at the monsters, however, revealed the interesting fact that the shorter tail was always about 700 nucleotides in length, approximately consistent with it being the 3' terminal tail, while the length of the longer tail decreased as the length of the partially assembled rod increased. Suddenly, everything clicked into place and we realized that the only possible explanation was that the longer tail contained the 5' extremity and that, after emerging from the growing end of the nucleoprotein helix, it must loop back through the central channel of the particle to protrude from the opposite end. The results were quickly written up to produce the paper by Lebeurier *et al.* (1977) included in this anthology.

I have no inside knowledge of the events in Cambridge surrounding the experiments described in the Butler *et al.* (1977) paper. I believe, however, that the Cambridge group was already thinking along the proper lines because they had access to the unpublished crystal structure of the disk, in which the two layers of the disk could be seen to gape apart near the central channel so that access to the RNA binding sites of the CP subunits is easiest from the inside. A model in which the initiating RNA is threaded between layers of the disk via the central channel would also account nicely for the hairpin structure of the initiation sequence. Finally, having the long 5' tail of RNA loop back through the central channel would free the growing end of the nucleoprotein helix for addition of two layer protein aggregates such as the disk during elongation, a point that was obviously attractive to the Cambridge group.

The critical experimental evidence again came from electron microscopy. At high resolution, the ends of TMV particles are morphologically distinct, one being slightly concave and the other slightly convex. The convex end was identified as that containing the 3' end of the RNA by visualization of partially alkali-disassembled particles. But when partially assembled rods were examined, the puff of unencapsidated RNA was found to be associated with the convex end of the particle (Butler *et al.*, 1977). Given the known 3' to 5' polarity of rapid rod growth during assembly, a little thought should convince the reader that this finding leads inexorably to the same model for assembly as that deduced independently by the Strasbourg group. Indeed, the two sets of observations are complementary. Although the Cambridge work lacked the compelling visual evidence for the two RNA tails obtained in Strasbourg, it provides direct evidence that the unencapsidated RNA appears at what will eventually become the 3' end of the particle, not at the end where rapid elongation occurs.

TMV almost certainly assembles in an infected cell in the same manner as in the test tube and certain features of the assembly process may have an impact on the infection cycle. For example, formation of disks from smaller CP aggregates is slow and concentration-dependent, and this could provide a means of temporally regulating particle formation relative to translation and replication of the viral RNA. It may also be significant that the portion of the viral RNA containing the CP cistron is slow to be encapsidated. Thus it is entirely possible that TMV assembly may have more surprises in store for us in the future.

References

Butler, P. J. G. and A. Klug. (1971). Assembly of the particle of tobacco mosaic virus from RNA and protein. *Nature New Biol.* **229:**47-50.

Hirth, L. and K. E. Richards. (1981). Tobacco mosaic virus: model for structure and function of a simple virus. *Adv. Virus Res.* **26:**145-199.

Jonard, G., Richards, K. E., Guilley, H., and L. Hirth. (1977). Sequence from the assembly nucleation region of TMV RNA. *Cell* **11:**483-493.

Zimmern, D. (1977). The nucleotide sequence at the origin for assembly on tobacco mosaic virus RNA. *Cell* **11:**463-482.

Zimmern, D. and P. J. G. Butler. (1977). The isolation of tobacco mosaic virus RNA fragments containing the origin for viral assembly. *Cell* **11:**455-462.

Biographical Sketch

Ken Richards received his B. A. from the University of Kansas and continued his graduate training there with Robert Hersh in the Department of Comparative Biochemistry and Physiology. He was awarded a Ph.D. in 1970 and then worked as a postdoctoral fellow with Robley Williams in the Virus Laboratory at the University of California, Berkeley. In 1973, he joined the group of Léon Hirth at the Institut de Biologie Moléculaire et Cellulaire in Strasbourg, France and was recruited as a CNRS scientist in 1976. He is presently a CNRS Research Director at the Institut de Biologie Moléculaire des Plantes, in Strasbourg.

Proc. Natl. Acad. Sci. USA
Vol. 74, No. 1, pp. 149–153, January 1977
Biochemistry

Inside-out model for self-assembly of tobacco mosaic virus

(electron microscopy)

G. LEBEURIER, A. NICOLAIEFF, AND K. E. RICHARDS

Laboratoire de Virologie, Institut de Biologie Moléculaire et Cellulaire du CNRS, Université Louis Pasteur, 15, rue Descartes, 67000 Strasbourg, France

Communicated by Robley C. Williams, October 22, 1976

ABSTRACT Incompletely reconstituted particles of tobacco mosaic virus treated with moderate concentrations of dimethyl sulfoxide prior to electron microscopy have two visible tails of unencapsidated RNA, a long one and a short one. The short tail has a constant length of 720 ± 80 nucleotides in incomplete particles of diverse size and probably corresponds to the 3′OH end of the RNA. The length of the long 5′ tail is inversely related to the length of the unfinished rod. Surprisingly, both tails appear to protrude from the same end of the particle. These observations suggest that the strand of RNA coming out of one end of the particle loops back along the length of the rod so as to appear at the opposite end. The "looped-back" tail almost certainly passes down the empty central channel of the incomplete rod. The findings are compatible with a model of tobacco mosaic virus assembly in which the RNA inserts itself beneath layers of incoming protein from within the central channel.

Tobacco mosaic virus (TMV) is a hollow cylinder of length 300 nm, inner diameter 4 nm, and outer diameter 18 nm. Each particle contains 2200 protein subunits arranged as a right-hand helix with $16\frac{1}{3}$ subunits per turn. The RNA chain (6600 nucleotides) is sandwiched between successive turns of the protein helix.

The self-assembly of TMV *in vitro* from its protein and RNA components is initiated by the binding of a bilayer disk of TMV protein to a site on the RNA molecule. Upon or immediately after binding, the protein subunits of the disk are thought to undergo a coordinated rearrangement to yield a helical nucleoprotein complex (lock-washer) which serves as the substrate for further addition of protein (1).

Previous electron microscopic studies of TMV reconstitution, in which only initiation and a little rod growth were allowed to take place, showed a single "tail" of uncoated RNA protruding from one end of the incomplete particle, leading to the idea that the origin of assembly was at or very near one end of the RNA chain (1–3). Several independent lines of evidence were marshalled in support of this conclusion and all seemed to point to the 5′ end as the starting site (1, 3–5). However, more recent observations indicate that, contrary to the earlier findings, neither the 5′ nor the 3′ extremity of the RNA is encapsidated early in assembly (6, 7). Evidently, the apparent one-tailed appearance of incompletely reconstituted particles deserves reinvestigation.

In earlier publications we have described the use of dimethyl sulfoxide (Me_2SO) to prepare single-stranded RNA as extended filaments for electron microscopy (8, 9) and to uncoat TMV particles (10, 11). In the latter studies single-stranded RNA could be perceived protruding from one or both ends of partially uncoated particles and the lengths of the tails could be measured. Hence we decided to use Me_2SO as an aid in the visualization of the tails of partially reconstituted particles.

In this paper we shall show that incompletely reconstituted particles treated with moderate concentrations of Me_2SO have two tails, one short and one much longer. Interestingly enough, both tails appear to protrude from the same end of the rod. This observation is most plausibly interpreted by the hypothesis that one of the RNA tails must traverse the length of the incomplete rod by passing along the central channel. This leads to a model for TMV assembly in which, during all but the last stages of reconstitution, the virus builds itself from the inside out, that is, the RNA inserts itself into the helix from within the central channel of the rod.

Abbreviations: TMV, tobacco mosaic virus; Me_2SO, dimethyl sulfoxide.

MATERIALS AND METHODS

Partial Reconstitution of TMV. Incompletely reconstituted particles were prepared by reacting TMV RNA with a fourfold weight excess of 25S TMV disk protein in 0.5 IS sodium pyrophosphate, pH 7.25, for 20 min at 24° (12). The particles were collected by sedimentation for 3 hr at 100,000 × *g* in a Beckman 50 rotor. The pellet of incomplete particles was then gently dispersed in 0.01 M potassium phosphate, pH 7.0.

Specimen Preparation. The sample solution to be spread by the Kleinschmidt technique (13) was prepared by adding 10 μl of partially reconstituted particles (1 μg/ml with respect to RNA) to 200 μl of 60% Me_2SO in water. Fifty microliters of cytochrome *c* (500 μg/ml in 90% Me_2SO) was then added and the solution was immediately spread on a distilled water hypophase (11).

A portion of the protein film was picked up on a Siemens platinum grid covered by a carbon-Formvar film and washed a few seconds in water, then in 90% ethanol, and finally in 50 mM uranyl acetate in 90% ethanol. Excess liquid was removed by touching the edge of the grid with a piece of filter paper. Specimens were rotary shadowed with platinum, and micrographs were taken at a nominal magnification of 10,000X with a Siemens 101 electron microscope. The exact magnification was determined with a standard TMV preparation. Micrographs were magnified 20X with a Nikon Profile Projector and the RNA profile was traced. The contour length of the tracing was measured with a curvimeter. Observed tail length was converted to nucleotides by dividing by 0.32 nm per nucleotide (9).

RESULTS

Electron Microscopy of Incompletely Reconstituted Particles. As noted in the introduction, there is now good evidence that, in spite of the previous reports to the contrary, incompletely reconstituted virus particles should have two RNA tails. If the site of initiation for reconstitution is quite near one end of the RNA, one of the two tails could simply be too short to be seen. Alternatively, the two tails could be held together by RNA–RNA interactions so as to appear as one thread of RNA in the electron microscope. In order to investigate this second possibility more fully, we treated incompletely assembled particles with Me_2SO in an effort to dissociate the two tails from one another.

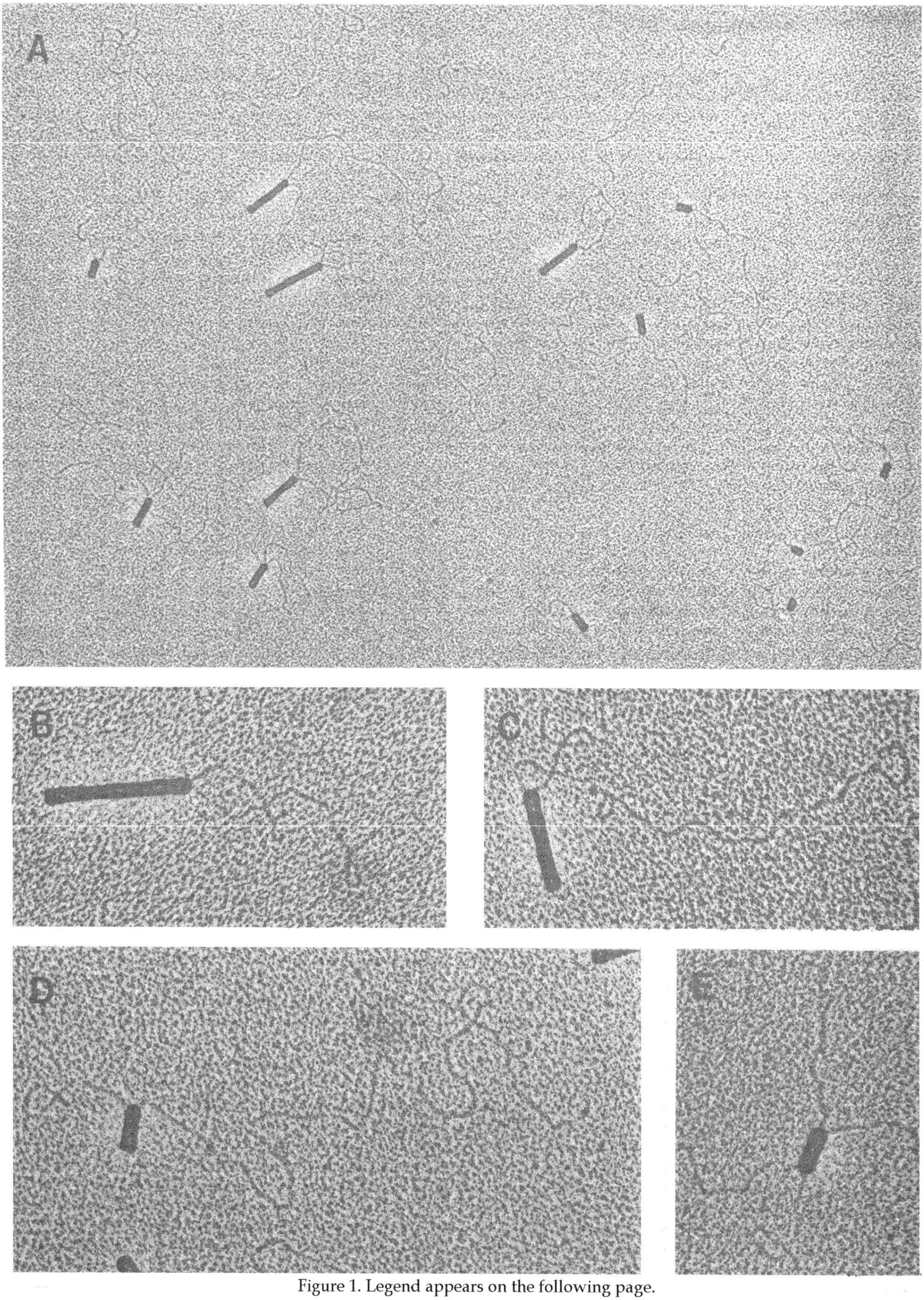

Figure 1. Legend appears on the following page.

In preliminary experiments, TMV RNA was spread in the presence of various concentrations of Me_2SO. It was found that the mean contour length of the RNA spread in 60% Me_2SO (2.12 μm) was the same as that observed when the RNA was spread under more stringent denaturing conditions (90% Me_2SO) (9). At the same time, length measurements of incompletely reconstituted particles spread with and without 60% Me_2SO gave closely similar distributions of rod length. Thus, 60% Me_2SO is sufficient to unravel most of the RNA secondary structure but is below the threshold of Me_2SO concentration at which significant uncoating of the incomplete particles occurs.

Fig. 1A shows a typical field from an electron micrograph of incompletely reconstituted particles treated with 60% Me_2SO before spreading. A gallery of particle images is shown in Fig. 1B–D. It is evident that each incomplete particle has two visible tails, generally of dissimilar length. In every instance the twin tails appear to protrude from the same end of the rod. We have examined many such images and find that essentially all (>95%) of the incomplete particles in a typical preparation have such an appearance. When particles of different size are examined, the length of the longer RNA tail is inversely related to the length of the incomplete rod (Fig. 2A); the average length of the short tail, on the other hand, is independent of particle length (Fig. 2B). The implications of this latter observation will be explored more fully below.

From time to time, incomplete particles were found which appeared to have more than two tails (Fig. 1E). The extra tails may protrude from opposite ends of the rod, as in the example shown in Fig. 1, or all the tails may seem to come out of the same end. We attribute the supplementary tails to fortuitous binding of one or more molecules of free RNA to the end(s) of the incomplete rod. (The ends of such rods may be somewhat "sticky".) Although the many-tailed particles are a distinct minority, their existence raises the question of whether one of the two tails of the two-tailed particles that constitute the major species might arise by similar adventitious binding of free RNA. However, there are several arguments against this possibility; perhaps the most persuasive is that for the two-tailed particles, the sum of the lengths of the short tail and the long tail, when added to the length of the RNA already encapsidated, is usually close to the expected chain length for one molecule of TMV RNA (Fig. 2).

Structural Model for Incompletely Reconstituted Rods. The above observations suggest that in an incompletely assembled rod the strand of RNA coming out of one end of the particle loops back along the length of the rod so as to appear to protrude from the opposite end. The "looped-back" tail could either run back along the outside of the rod or pass down the empty central channel. Against the former possibility is the fact that two-tailed particles produced by uncoating TMV with 74% Me_2SO have the expected appearance, i.e., the two tails are at opposite ends of the rod (11). We take this to mean that in incompletely reconstituted particles, as opposed to partially stripped rods, the second tail is firmly held in the looped-back configuration, as would be the case if it passes back along the central channel of the virus. Although this conclusion seems at first somewhat surprising, we shall show below that it is readily reconciled with other evidence, including recent x-ray crystallographic studies on the structure of the disk (14).

Fig. 2B shows that the mean length of the short tail of incomplete rods corresponds to 720 ± 80 nucleotides. Zimmern and Wilson (personal communication) estimate that the site of

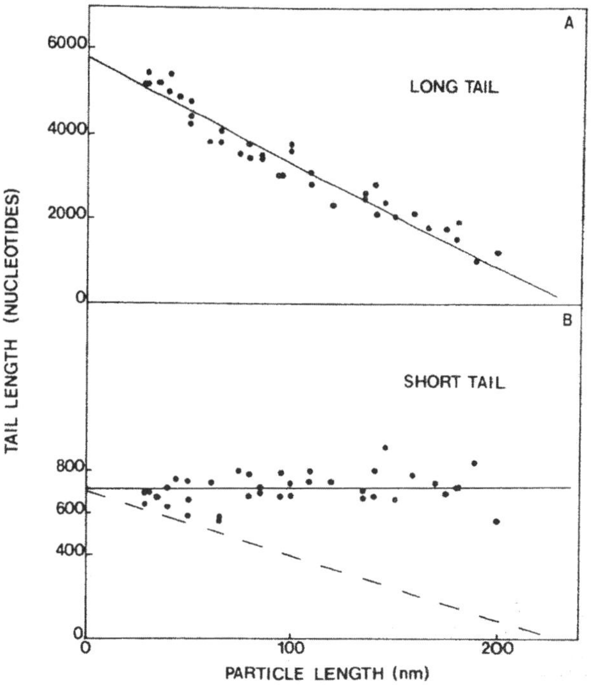

FIG. 2. Length of the long RNA tail (A) and the short RNA tail (B) of incomplete particles as a function of nucleoprotein rod length. Fifty particles of diverse size were selected for measure. In 10 cases the sum of the length of the long and short tails plus the length of the RNA already encapsidated was less than or equal to 5600 nucleotides and presumably represent particles with a broken tail; in two cases the sum was greater than 7600 nucleotides. These particles are not included in the graph. (B) Dashed line is the relation expected between short tail length and particle size if the short rather than the long RNA tail passes down the central channel (see *text*).

initiation of TMV assembly is 900–1300 nucleotides from the 3'OH end of the RNA chain. Hence the short tail almost certainly corresponds to the 3' terminus of the RNA molecule. The fact that the short tail remains constant in length for incomplete particles ranging from 30 to 200 nm in size (Fig. 2B) indicates that, after initiation, addition of protein to the end of the nucleation complex containing the 5' tail is much favored over growth in the 3' direction.

A plausible model for the structure of a short, incompletely reconstituted particle is shown in Fig. 3. In the model, it is the long 5' tail that threads back through the central channel. As the nucleoprotein rod grows, the 5' tail feeds up through the tube to be sandwiched between successive layers of newly added protein. The observed length of the short 3' tail remains unaffected by rod growth (at least during all but the later stages of assembly) and should correspond to the distance between the initiation site for assembly and the 3' end.

The observed constancy in length of the short tail allows us to rule out the alternative model for incomplete particle structure, where it is the 3' tail rather than the 5' tail that passes along the hole in the rod. In such a model the 3' → 5' polarity of elongation would require that rod growth take place at the end of the particle from which the twin tails protrude. As the 5' tail became covered with protein, more and more of the 3' tail would be imprisoned within the central channel of the growing rod. Hence, contrary to observation (Fig. 2), the short as well as the long tail should appear to diminish in length as classes of longer incomplete particles are considered.

FIG. 1 (*on preceding page*). Electron micrograph of incompletely reconstituted TMV particles treated with 60% Me_2SO. (A) typical field, ×68,000; (B–D) selected particles with the two-tailed aspect, ×136,000; (E) rare "four-tailed monster" ×136,000.

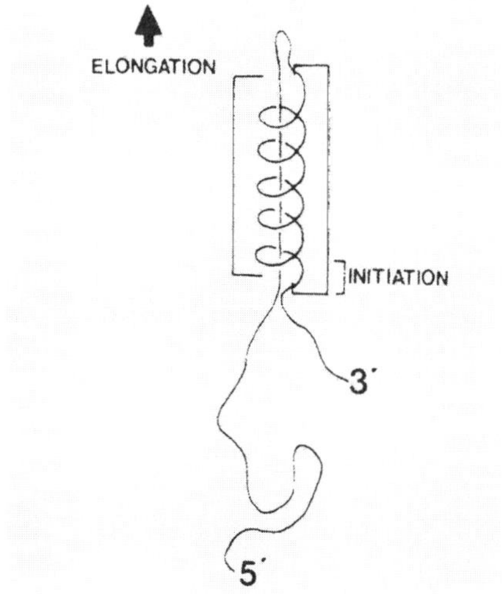

FIG. 3. Schematic diagram of the structure of an incompletely reconstituted particle. The drawing is not to scale.

DISCUSSION

The evidence presented in this paper favors a model of TMV assembly in which the RNA tail containing the 5′ end passes back along the central channel, with protein adding to the growing rod at the end where the RNA turns back on itself.

It is of interest to consider how in the course of assembly the 5′ tail comes to be threaded down the central channel. A recent x-ray crystallographic investigation of the structure of the disk (14) shows that the two layers of the disk are held together by protein–protein contacts at high radius, but that at low radius the layers slew apart. Consequently, access to the RNA binding groove might well be easier from the inside of the disk than from the outside. If, during initiation of reconstitution, the RNA inserts itself between layers of the first disk from the inside, then the region of the RNA involved in initiation would be expected to have a loop structure since it is difficult to imagine how the RNA could enter the central hole of the disk without being folded back upon itself. Hence, it is gratifying that the sequence of the part of the RNA chain involved in nucleation of assembly can be cast into a stable hair-pin loop (Jonard and Richards, personal observations). It is readily seen that an initiation step in which a loop of RNA penetrates between the layers of the first disk from inside will inevitably lead to two-tailed particles of the observed aspect since one of the two arms of the initiating loop must remain in the central channel.

The growing tip of the rod at which addition of protein takes place (Fig. 4) is admirably suited to accept further protein, either in the form of disks (1) or as smaller aggregates (15–17). The part of the RNA chain awaiting encapsidation lies extended down the central channel and so can swivel about its axis in the tube as new protein is added.

In all of the foregoing discussion we have confined ourselves to consideration of rod growth along the long 5′ tail. But eventually the 3′ tail must be encapsidated as well if we are to have a complete particle. This appears to be a "late" event in assembly, since incomplete particles as long as 200 nm have undergone no detectable growth in that direction (Fig. 2B). It may be that the 5′ tail protruding from the central channel hampers addition of protein back toward the 3′ end of the RNA; encapsidation of the 3′-terminal tail could then occur only after the 5′ tail has been pulled up out of the way by reconstitution.

In infected tobacco cells the functional messenger for synthesis of TMV coat protein is a smaller RNA produced from the genome RNA by partial transcription or specific cleavage (18).

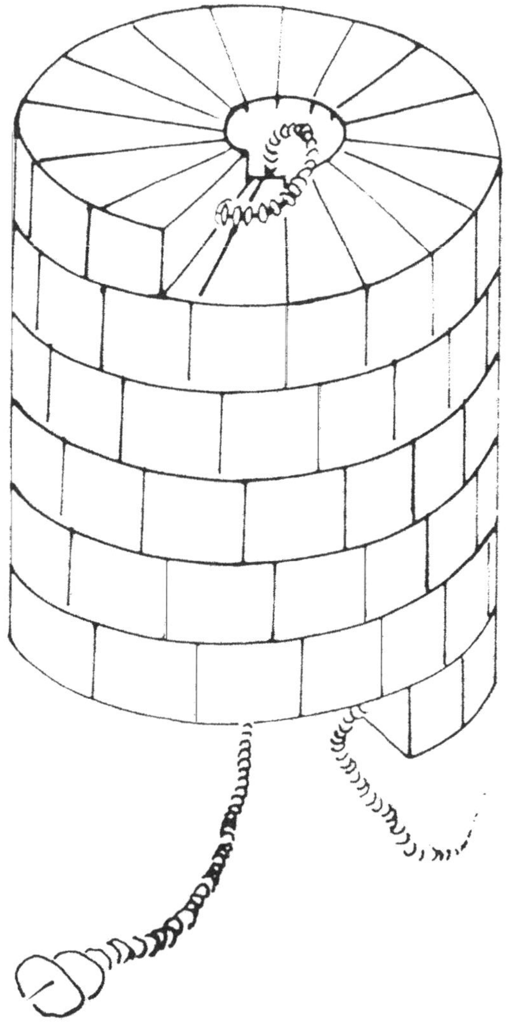

FIG. 4. The growing end of a short, incomplete, reconstituted particle.

The coat protein messenger, known as the low-molecular-weight component (19), corresponds to the 3'-terminal 750 nucleotides of the genome RNA (18). This is, of course, exactly the extent of the short 3' tail of incomplete particles. This circumstance suggests a possible mechanism for the control of protein production. As noted above, the initial burst of synthesis of coat protein in the infected cell must await the production of some low-molecular-weight component. Once some coat protein has been synthesized and disks have formed, reconstitution will begin. As shown in this paper, the portion of the TMV RNA containing the sequence for low-molecular-weight component will be the last part of the RNA to be coated. Hence, production of this component from the uncoated 3' tail of incomplete rods could continue (perhaps by partial replication), leading in turn to more coat protein for reconstitution.

We acknowledge with gratitude the contribution of Prof. L. Hirth, who provided the initial impetus to perform the experiments reported here and subsequently made many useful suggestions. We thank Dr. B. Jacrot for many stimulating discussions, Dr. D. Zimmern for communication of results prior to publication, and Dr. A. C. H. Durham for reading the manuscript.

1. Butler, P. J. G. & Klug, A. (1971) "Assembly of the particle of tobacco mosaic virus from RNA and disk of protein." Nature New Biol. 229, 47-50.
2. Stussi, C., Lebeurier, G. & Hirth, L. (1969) "Partial reconstitution of tobacco mosaic virus." Virology 38, 16-25.

3. Ohno, T., Nozu, Y. & Okada, Y. (1971) "Polar reconstitution of tobacco mosaic virus," *Virology* **44,** 510–516.
4. Thouvenel, J. C., Guilley, H., Stussi, C. & Hirth, L. (1971) "Evidence for polar reconstitution of TMV," *FEBS Lett.* **16,** 204–206.
5. Guilley, H., Stussi, C. & Hirth, L. (1971) "Influence de la phosphodiesterase de rate de porc sur la reconstitution *in vitro* du virus de la mosaïque du tabac," *C. R. Hebd. Seances Acad. Sci. Ser. D* **272,** 1181–1184.
6. Zimmern, D. (1975) "The 5′ end group of tobacco mosaic virus RNA is $m^7G^{5'}ppp^{5'}Gp$," *Nucleic Acids Res.* **2,** 1189–1201.
7. Zimmern, D. (1976) "The region of TMV RNA involved in the nucleation of assembly," *Phil. Trans. R. Soc. London Ser. B,* in press.
8. Oudet, P., Lebeurier, G. & Nicolaieff, A. (1970) "Etude du RNA du virus de la mosaïque du tabac (VMT) par microscopie électronique," *7ème Congrès International de Microscopie électronique, Grenoble* **1,** 597.
9. Nicolaieff, A., Pinck, L., Koenig-Nikes, A. & Hirth, L. (1972) "Electron microscopy of replicative form and single strand RNA of alfalfa mosaic virus," *J. Gen. Virol.* **16,** 47–59.
10. Nicolaieff, A., Lebeurier, G. & Hirth, L. (1974) "Electron microscopy and infectivity studies of the action of dimethylsulfoxide on two strains of tobacco mosaic virus," *J. Gen. Virol.* **22,** 43–53.
11. Nicolaieff, A., Lebeurier, G., Morel, M. C. & Hirth, L. (1975) "The uncoating of native and reconstituted TMV by DMSO: the polarity of stripping," *J. Gen. Virol.* **26,** 295–306.
12. Richards, K. E., Morel, M. C., Nicolaieff, A., Lebeurier, G. & Hirth, L. (1975) "Location of the cistron of the tobacco mosaic virus coat protein," *Biochimie* **57,** 749–755.
13. Kleinschmidt, A. & Zahn, R. K. (1959) "Über desoxyribonucleisaüre-Molekulen in protein Mischfilmen," *Zeitschrift für Naturforschung* **14b,** 770–779.
14. Champness, J. N., Bloomer, A. C., Bricogne, G., Butler, P. J. G. & Klug, A. (1976) "The structure of the protein disk of tobacco mosaic virus to 5 Å resolution," *Nature* **259,** 20–24.
15. Richards, K. E. & Williams, R. C. (1973) "Assembly of tobacco mosaic virus rods *in vitro.* Elongation of partially assembly rods," *Biochemistry* **12,** 4574–4581.
16. Okada, Y. (1975) "Mechanism of assembly of tobacco mosaic virus *in vitro,*" *Adv. Biophys.* **7,** 1–41.
17. Lebeurier, G. & Hirth, L. (1975) "Le virus de la mosaïque du tabac: un modèle de morphogénèse virale *in vitro,*" *Bull. Inst. Pasteur, Paris* 73, 141–165.
18. Hunter, T. R., Hunt, T., Knowland, J. & Zimmern, D. (1976) "Messenger RNA for the coat protein of tobacco mosaic virus," *Nature* **260,** 759–764.
19. Siegel, A., Zaitlin, M. & Duda, C. T. (1973) "Replication of tobacco mosaic virus. IV. Further characterization of virus related RNAs," *Virology* **53,** 75–83.

(Reprinted from Nature, Vol. 265, No. 5591, pp. 217–219, January 20, 1977)

Configuration of tobacco mosaic virus RNA during virus assembly

P. J. G. Butler, J. T. Finch & D. Zimmern

MRC Laboratory of Molecular Biology, Hills Road, Cambridge CB2 2QH, UK

When TMV reassembles, the uncoated RNA is folded back along the growing rod, probably down the central hole. This surprising configuration is essential for rapid elongation—presumably supplying RNA to its site of incorporation while keeping the bulk of the free RNA out of the way.

THE location of the nucleation region for the assembly of tobacco mosaic virus (TMV), from isolated coat protein and RNA, about 1,000 nucleotide residues from the nearer end of the RNA molecule (the 3′-hydroxyl terminus)[1,2] raises questions about the elongation process. Since the RNA is totally enclosed in the mature virus particles, elongation from the original site of nucleation must occur in both directions—towards the 3′- and 5′-hydroxyl termini of the RNA. This presents a paradox in view of the observation[3-7] that partially assembled rods of TMV have RNA tails only at one end, in agreement with the generally accepted view that assembly would be polar. We have shown that the two ends of the TMV rod can be distinguished in the electron microscope because one is convex and the other concave and, taking this together with stripping experiments, we have shown in which direction the RNA runs within the complete nucleoprotein helix[8]. We have now applied this technique to locate the free RNA in the growing nucleoprotein rods.

Preparation of rodlets

Partially assembled rodlets of the desired length range were obtained by allowing an appropriate amount of the protein, supplied as a disk preparation, containing about 80% of the 2-ring disk aggregate[9], to reassemble with the RNA[10] at 20 °C in sodium phosphate buffer, *p*H 7.0, ionic strength 0.1 for 10–15 min[9]. Rods about half the size of complete virus particles were obtained with a protein:RNA ratio of 10:1. Rodlets with free RNA tails at the 5′-hydroxyl end were prepared by partial alkaline stripping as before[8].

Both the rodlet preparations described were contaminated with free protein—either unreacted and possibly damaged in the partially assembled preparation or that which had been stripped off in the partially stripped preparation—and possibly some free RNA. These contaminants were much smaller than the rodlets and were removed before reassembly by centrifugation through sucrose gradients, without allowing the rodlets to pellet, as this alters their reassembly activities[9]. Gradients were prepared in Tris-chloride buffer, *p*H 8.0, ionic strength 0.1, by layering into tubes 2 ml 60% (w/v) sucrose, 2 ml 40%, 3 ml 20% and 2 ml 5% sucrose. Samples (3 ml) were layered on top and the gradients were centrifuged at 40,000 r.p.m. and 5 °C for 1 h in an MSE 6 × 14-ml titanium swingout rotor. Fractions were collected from the bottom of the tubes and the fractions from the turbid layer were pooled and dialysed against sodium phosphate, *p*H 7.0, ionic strength 0.1 containing 1 mM sodium azide.

The turbid suspensions had spectra characteristic of nucleoprotein, while the fractions from the tops of the gradients gave the spectrum of TMV protein, suggesting that little free RNA was present in the original preparations.

Location of RNA tails on growing rods

Although the uncoated RNA is not as clearly visible at the ends of partially assembled rodlets as on partially stripped rodlets, it can nevertheless be seen on many of the particles and the morphologically distinct ends of these particles can be distinguished. From Fig. 1 it seems that the RNA tails are at opposite ends of the two types of rodlet. This was confirmed by inspecting all the particles in

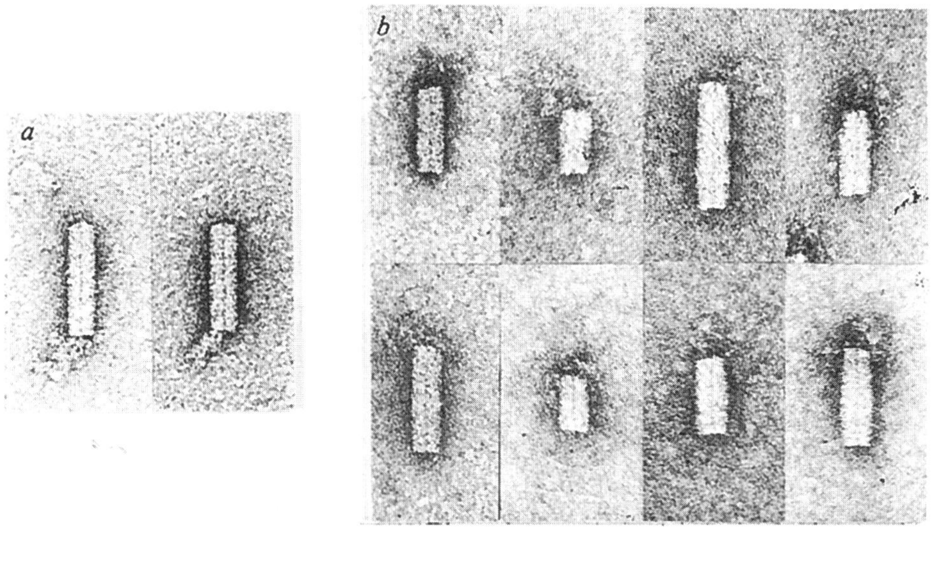

Fig. 1 Electron micrographs of TMV particles: *a*, partially stripped with alkali[8], and *b*, partially reassembled. The images are aligned with the rounded or convex ends at the top and the concave ends at the bottom. The 'puffs' of uncoated RNA can be seen at opposite ends in the two cases (approximately × 200,000). Specimens from the rodlet preparations were prepared by allowing the particles to adsorb on to carbon-coated grids from a drop of solution for about 30 s, followed by rinsing with distilled water and negative staining with 1% (w/v) uranyl acetate. The grids were examined and photographed in a Philips EM 300 electron microscope at a magnification of approximately 25,000.

Table 1 Numbers of particles for which the end with the exposed RNA is morphologically distinguishable

Type of rodlet	End showing RNA Convex	Concave	Both
Partially stripped[8]	1	62	6
Partially assembled	88	4	3

several fields of partially assembled rodlets and assigning the free RNA to the different ends whenever this could be done unambiguously. The counts are shown in Table 1, together with those for rodlets partially stripped by alkali using the same technique[8]. Clearly the RNA tails of the partially assembled rodlets occur at the convex ends of the particles in contrast to the situation found for partially stripped rodlets[8]. (Less than 8% of rodlets, whether single or double tailed, were identified with a tail at the concave end.)

Because TMV RNA is about 6,400 nucleotides long with the nucleation region about 1,000 nucleotides from the 3′-hydroxyl end[1,2] and hence more than 5,000 nucleotides from the 5′ end, the main direction of assembly must be from 3′ to 5′. In partially assembled rodlets the longer tail should therefore contain the 5′ end, which is the end released by alkali stripping[11]. As RNA tails are visible in partially assembled rodlets principally at the ends which contain the 3′ ends of the RNA in the final viruses, the RNA must be folded back on itself in the growing rods because it protrudes only from the end at which least elongation occurs (Fig. 2).

Rates of elongation

We suggest that the looping back of the 5′ end of the RNA in the growing rods, probably down the central hole, is important for the mechanism of elongation of the rods in the dominant direction, from 3′ to 5′. In this case we expected that the rates of elongation of partially assembled rods and of partially stripped rodlets might be different because, although both would be elongating in the 3′ to 5′ direction, the long RNA tail would be tucked down inside the former but protruding from the stripped end of the latter.

This prediction was investigated using rodlets which had been purified and checked in the electron microscope to ensure that most still had RNA tails. Protein for elongation was supplied both as a disk preparation and as A-protein. The results of one such experiment, in which the partially assembled rodlets and two preparations of partially stripped rodlets were purified in parallel, including the final dialysis in the same flask, to eliminate any possible effect of preparative history, are shown in Table 2. Although the large molecular sizes and unknown molecular weights of the preparations make it unsafe to try to calculate the absolute rates of elongation from these rates of increase in turbidity, or even to make a close comparison of the rates between

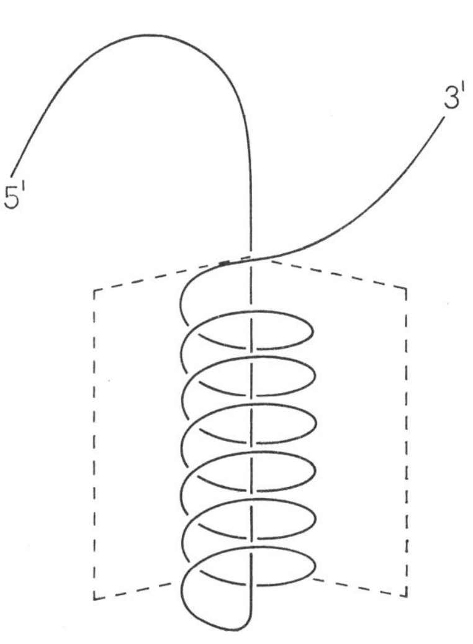

Fig. 2 Schematic picture of the suggested arrangement for the RNA in a growing rod. The dominant direction of assembly (that is, from 3′ to 5′) is downwards—at the end opposite the two RNA tails. The placing of the longer, 5′ tail up the central hole is arbitrary and, while we think this is the most likely configuration, there is no direct proof as to how it reaches the opposite end of the rod.

different preparations, the partially assembled rodlets seemed to elongate more than an order of magnitude faster than did the partially stripped rodlets. This conclusion remains, even allowing for the greater concentration of the former.

A compatible result was obtained when the lengths of partially stripped rodlets were compared by electron microscopy before and after incubation with a disk preparation at 1.0 mg ml^{-1} for 30 min. While partially assembled rodlets would, in these conditions, be expected to elongate by 150–200 nm (refs 9,12), the partially stripped rodlets only increased by about 10 nm. Although a χ^2 analysis on the histograms of lengths of the rodlets showed that even without genuine elongation, such an apparent change could have arisen from sampling with a probability of just under 0.1%, we think that elongation had taken place, as it has a 90% probability of doing. Moreover, rods partially stripped from the same end with SDS[8] will reassemble over several hours to give nuclease resistant infectivity[13], in agreement with the idea that some slow elongation can probably occur even with the RNA tail free at the abnormal end of the growing rod.

Preliminary location of both tails

The observations of the RNA tails described above were made with negatively stained specimens, so that high resolution electron microscopy distinguished the ends of the rodlets, but in these conditions the RNA tails only appear as 'puffs' and it is not

Table 2 Rates of reassembly of partial TMV rodlets

Rodlet preparation	Rate of turbidity increase (D_{310} per min) Disk preparation	A-protein
Partially stripped I	0.00125	0.0006
Partially stripped II	0.00115	0.00065
Partially assembled	0.0316	0.0221

Rodlets were prepared either by partial alkali stripping or by partial reasembly. Elongation in sodium phosphate, *p*H 7.0, ionic strength 0.1 at 20° C, was followed by the increase in turbidity at 310 nm (ref. 9), with either a disk preparation or A-protein added to 1 mg ml^{-1} final concentration. Final rodlet concentrations were 0.20 and 0.195 D_{260} per ml for the partially stripped rodlets I and II and 0.42 D_{260} ml^{-1} for the partially assembled rodlets (corresponding to approximately 0.6, 0.5 and 1.2 mg ml^{-1}, respectively).

possible to distinguish between one or two tails at the same end of a rod, but rather the major tail (the 5′ end) dominates the image. We have occasionally seen partially assembled particles apparently with two RNA tails coming from the same end, by electron microscopy of specimens shadowed after spraying onto carbon coated grids with only bovine serum albumin as a spreading agent. We do not regard these observations as very significant because most of the particles showed only a single tail, and some free RNA was visible in the background. To obtain better spreading and separation of the tails it is probably necessary to add some denaturing agent. Preliminary experiments, however (unpublished results of B. Jacrot and D.Z.), showed that partially assembled rods are very unstable when RNA is denatured for spreading, although partially stripped rodlets could be visualised. This again shows the difference between these types of rodlet, but emphasises the need for extreme caution in interpreting such data because preparation may significantly alter the specimen.

An alternative approach to visualise and distinguish the ends of the RNA is to label them with a bulky molecule. We have done some preliminary experiments using ferritin, activated for reaction with periodate-oxidised RNA[14] by coupling to carbohydrazide with ethyl dimethylaminopropyl-carbodi-imide[15,16], to label both termini. One problem is that the tendency to form ferritin aggregates could cause confusion. Although we have never seen a particle with ferritin molecules at both ends, we have seen several partially assembled rodlets on which two ferritin molecules appear, in the electron microscope, to be attached to the same RNA 'puff' and yet are too far apart to be attached directly to each other. We believe that this indicates that the two RNA termini are indeed at the same end of the assembling rodlet and we are continuing experiments to confirm this.

Implications for assembly

The structure of the protein disk is known to a resolution of 5Å (ref 17) and consists of two rings, each of 17 protein subunits, stacked on top of each other. While these make a firm contact at high radius, they are separated vertically at the RNA-binding radius, like a pair of 'jaws', and the protein has considerable flexibility inside this radius. These features suggested that the RNA might well be intercalated between the two rings of the disk, during both nucleation and elongation, from the central hole of the disk and in this way the two-layer disk could be incorporated into the single helix.

Although there is no firm evidence of whether the doubling back of the longer RNA tail occurs inside or outside the growing nucleoprotein helix, we suggest that it is probably inside with the RNA strand running freely down the central hole (Fig. 2). This configuration in the growing rod would follow smoothly from a mechanism of nucleation by the insertion of a hairpin loop of RNA into the centre of the first disk[18]. Such a geometry will allow access of the relevant region of the RNA to its binding sites on both rings of an incoming protein disk and yet hold the bulk of this RNA strand well out of the way of the region of activity, greatly facilitating the process of rod elongation.

Received September 10; accepted December 1, 1976.

1 Zimmern, D., *Phil. Trans. R. Soc., Lond.*, **B276,** 186–204 (1976).
2 Zimmern, D., and Wilson, T. M. A., *FEBS Lett.*, **71,** 294–298 (1976).
3 Stussi, C., Lebeurier, G., and Hirth, L., *Virology*, **38,** 16–25 (1969).
4 Butler, P. J. G., and Klug. A., *Nature new Biol.*, **229,** 47–50 (1971).
5 Ohno, T., Nozu, Y., and Okada, Y., *Virology*, **44,** 510–516 (1971).
6 Rodionova, N. P., Vesenina, N. E., Kichatova, O. B., and Atabekov, J. G., *Virology*, **46,** 183–191 (1971).
7 Richards, K. E., and Williams, R. C., *Biochemistry*, **12,** 4574–4581 (1973).
8 Wilson, T. M. A., Perham, R. N., Finch, J. T., and Butler, P. J. G., *FEBS Lett.*, **64,** 285–289 (1976).
9 Butler, P. J. G., *J. molec. Biol.*, **82,** 333–341 (1974).
10 Zimmern, D., *Nucleic Acids Res.*, **2,** 1189–1201 (1975).
11 Perham, R. N., and Wilson, T. M. A., *FEBS Lett.*, **62,** 11–15 (1976).
12 Butler, P. J. G., and Finch J. T., *J. molec. Biol.*, **78,** 637–649 (1973).
13 May, D. S., and Knight, C. A., *Virology*, **25,** 502–507 (1965).
14 Steinschneider, A., and Fraenkel-Conrat, H., *Biochemistry*, **5,** 2729–2734 (1966).
15 Carraway, K. L., Spoerl, P., and Koshland, D. E., *J. molec. Biol.*, **42,** 133–137 (1969).
16 Fersht, A. R., and Sperling, J., *J. molec. Biol.*, **74,** 137–149 (1973).
17 Champness, J. N., Bloomer, A. C., Bricogne, G. Butler. P. J. G., and Klug, A., *Nature*, **259,** 20–24 (1976).
18 Butler, P. J. G., *et al.*, *Third John Innes Symp.* 101–110 (1976).

Transport of Tobacco Mosaic Virus From Cell-to-Cell

Masamichi Nishiguchi[1] and Fusao Motoyoshi[2]

[1]National Institute of Agrobiological Resources, 2-1-2 Kan-nondai, Tsukuba, Ibaraki 305-8602, Japan

[2]Research Institute for Bioresources, Okayama University, Chuo 2-20-1, Kurashiki 710-0046, Japan

At the time of these studies, M. Nishiguchi and N. Oshima were working on the characterization of an attenuated strain of tomato mosaic virus (ToMV), $L_{11}A$, which had been isolated to use as a cross-protecting strain for tomato crops against severe ToMV. We found that a temperature shift treatment (at 22°C for 3 days, at 32°C for 2 days and at 22°C for 1 day) of inoculated Xanthi NN tobacco leaves could differentiate $L_{11}A$ from its parent strain, L, because of a morphological difference in the local lesions produced by each of them. After inoculation of Xanthi NN leaves with a mixture of L and $L_{11}A$ followed by the temperature shift treatment, we found two types of local lesions; one was a necrotic local lesion surrounded with a halo of collapsed tissue and the other a necrotic local lesion with no halo. The former type of lesion was confirmed to contain L, since homogenates of the isolated lesions induced severe mosaic symptoms on inoculated plants of Samsun tobacco, while the homogenates of the latter type lesions induced no symptoms on Samsun tobacco, indicating that the lesions contained $L_{11}A$.

In the course of this study, we found a local lesion in an L-inoculated leaf which was morphologically similar to that produced by $L_{11}A$ on Xanthi NN tobacco but contained a virus inducing mosaic symptoms on Samsun tobacco as severe as those induced by L. However, the local lesions produced by this virus on Xanthi nc tobacco at normal greenhouse temperatures were smaller than those produced by L. N. Oshima, who had wide experience in observing various mutants of plant viruses on host plants, suggested that this mutant strain was very unusual in its behavior. We then named it Ls1 because of its deviation from <u>**L**</u> as a <u>**s**</u> mall lesion mutant, and decided to characterize it in more detail. Thus, the finding of Ls1 was rather coincidental, as we were not intending to isolate a mutant with a defect in viral movement function.

It was also lucky that at that time an excellent tomato protoplast system was available, which had been devised by F. Motoyoshi from the tobacco protoplast system established in Dr. Takebe's lab on the same floor in our Institute. We had already known that a ts mutant of tobacco mosaic virus, Ni2519, isolated in Germany and similar to Lsl in lesion morphology, lost its infectivity in protoplasts at a non-permissive temperature. In the first protoplast experiment, however, growth of Ls1 was unexpectedly insensitive to the temperature shift. We then postulated that the temperature sensitivity of Ls1 was related to the ability of virus movement from cell-to-cell; thus, we focused our experiments onto the behavior in intact cells. After several comparative experiments with L and Ls1 in tissues and protoplasts at the permissive and non-permissive temperatures, we finally reached a conclusion: Ls1 replicates in primarily infected cells, but is not able to move from cell-to-cell at the non-permissive temperature. Subsequently, this defect of Ls1 was confirmed by visualization of epidermal tissues stained with ToMV-specific fluorescent antibodies a few days after inoculation (Nishiguchi *et al.*, 1980). Thus, our results revealed that the plant virus itself has a function allowing its cell-to-cell movement. Our finding led us to the belief that a plant virus genome must have a gene or genes necessary for transporting itself from cell-to-cell. This was verified as discussed below.

A minor but significant change in Ls1 was found in the 30 kDa protein translated *in vitro* from a subgenomic RNA fraction (Leonard and Zaitlin, 1982). Subsequently, a single amino acid substitution in the 30 kDa protein (a serine was substituted for a proline) was confirmed in Ls1 from nucleotide sequence data (Ohno *et al.*, 1983). The decisive evidence for movement function eventually came from two groups: transgenic tobacco plants expressing the wild type 30 kDa protein complemented the defect of Ls1 at the non-permissive temperature (Deom *et al.*, 1987), and by using *in vitro* infectious transcripts from Lsl and L, the 30 kDa protein encoded by the ToMV genome was demonstrated to be responsible for virus movement from cell-to-cell but dispensable for viral replication (Meshi *et al.*, 1987). The 30 kDa protein is now referred to as the movement protein (MP).

Since our first publication, a great number of papers dealing with mechanisms of the virus movement and the function of plasmodesmata have been published (for reviews see Carrington *et al.*, 1996; Jorgensen *et al.*, 1998). The MP was observed to be targeted to and localized in plasmodesmata. The ToMV isolate overcoming the ToMV-resistance gene, *Tm-2* or *Tm-2a*, in tomato was found to contain substitutions of amino acids in the MP, suggesting the direct or indirect interaction of these genes with the MP to inhibit the cell-to-cell movement of a wild-type

ToMV. The plasmodesmata in transgenic plants expressing MP permitted a large-sized dextran (9.4 kDa) to pass, which was too large to pass through the plasmodesmata in the non-transgenic plants, suggesting that in the transgenic plants, the size exclusion limit of the plasmodesmata was increased by the MP. The MP could bind to a single-stranded nucleic acid to create an unfolded protein-RNA complex, which is postulated to be narrow enough to pass through the plasmodesmata. The MP can bind directly to cytoskeletal elements, including actin and tubulin. This may be crucial to intracellular movement of the MP from the cytoplasm to the plasmodesmata. The MP studies have also led to the molecular understanding of plasmodesmal functions themselves.

At the very least, our results contributed to partly correcting the classical concept that virus movement from cell-to-cell was a passive process which depended on a host cell function. Our original finding has significantly influenced not only plant virology but also the understanding of physiological aspects and dynamic functions of plasmodesmata.

References

Carrington, J. C., Kasschau, K. D., and S. K. Mahajan. (1996). Cell-to-cell and long distance transport of viruses in plants. *Plant Cell* **8**:1669-1681.

Deom, C. M., Oliver, M., and R. N. Beachy. (1987). The 30-kilodalton gene product of tobacco mosaic virus potentiates virus movement. *Science* **237**:389-394.

Jorgensen, R. A., Atkinson, R. G., Forster, R. L. S. and W. J. Lucas. (1998). An RNA-based information superhighway in plants. *Science* **279**:1486-1487.

Leonard, D. and M. Zaitlin. (1982). A temperature-sensitive strain of tobacco mosaic virus defective in cell-to-cell movement generates an altered viral-coded protein. *Virology* **117**:416-424.

Meshi, T., Watanabe, Y., Saito, T., Sugimoto, A., Maeda, T., and Y. Okada. (1987). Function of the 30 kd protein of tobacco mosaic virus: involvement in cell-to-cell movement and dispensability for replication. *EMBO J.* **6**:2557-2563.

Nishiguchi, M., Motoyoshi, F., and N. Oshima. (1980). Further investigation of a temperature-sensitive strain of tobacco mosaic virus: its behaviour in tomato leaf epidermis. *J. Gen. Virol.* **46**:497-500.

Ohno, T., Takamatsu, N., Meshi, T., Okada, Y., Nishiguchi, M., and Y. Kiho. (1983). Single amino acid substitution in 30K protein of TMV defective in virus transport function. *Virology* **131**:255-258.

Biographical Sketches

Masamichi Nishiguchi attended Kyoto University and obtained a B.S. degree in agrobiology (1972) and an M.S. (1974) in plant pathology. In 1974, he was employed at the Institute for Plant Virus Research to work on plant viruses and their diseases. He earned a Ph.D. in plant pathology under the direction of Prof. M. Yamamoto at Kyoto University in 1982. Following the consolidation of the Institutes in 1984, he worked at the National Institute of Agrobiological Resources (NIAR). In 1987, he was appointed head of the new laboratory at the Kyushu National Agricultural Experiment Station. In 1993, he returned to NIAR. He was awarded an Overseas Research Fellowship from the Science and Technology Agency of Japan to work at Cornell University with Milton Zaitlin in 1984-1985.

Fusao Motoyoshi obtained his B.S. (1960) in agrobiology and his M.S. (1962) and Ph.D. (1966) degrees in genetics under the supervision of Profs. I. Nishiyama and S. Imamura at Kyoto University. He then joined the Institute of Plant Virus Research in 1966 and began working on virus resistance in tomato. He spent two years as a visiting scientist in the Department of Ultrastructural Studies, John Innes Institute, U. K. (1972-74). He moved to the National Institute of Agricultural Sciences to do research on plant biotechnology (1983) and then transferred to NIAR (1984). In 1989, he was offered a professorship at the Research Institute for Bioresources, Okayama University, and was elected its Director in 1998.

J. gen. Virol. (1978), **39**, 53–61
Printed in Great Britain

Behaviour of a Temperature Sensitive Strain of Tobacco Mosaic Virus in Tomato Leaves and Protoplasts

By M. NISHIGUCHI, F. MOTOYOSHI AND N. OSHIMA

Institute for Plant Virus Research, Tsukuba Science City, Yatabe, Ibaraki 300–21, Japan

(*Accepted* 25 *October* 1977)

SUMMARY

A temperature-sensitive (*ts*) strain, Ls1, was isolated from a culture of L, a tomato strain of tobacco mosaic virus (TMV). Ls1 caused smaller necrotic local lesions than L in leaves of hypersensitive tobacco plants. A temperature shift treatment (22 °C for 3 days, 32 °C for 2 days and 22 °C for 1 day) allowed L to produce necrotic lesions surrounded by a collapsed area, whereas Ls1 caused necrotic lesions without a collapsed surrounding area, suggesting that Ls1 did not spread outside the lesions at 32 °C. In tomato leaf discs at 22 °C, infectivity of Ls1 increased in parallel with that of L, but at 32 °C the increase in infectivity of Ls1 was negligible and contrasted with the large increase of infectivity of L. Few mesophyll cells from Ls1-inoculated discs incubated at 32 °C were stained by fluorescent antibody but many of those incubated at 22 °C were stained. In protoplasts at 32 °C, however, Ls1 and L infected and multiplied similarly. When leaves inoculated with Ls1 were kept at 20 to 25 °C for 24 h, and discs were then prepared and cultured for 12 h at 32 or 22 °C, the proportion of mesophyll cells containing virus antigen did not increase at 32 °C but infectivity increased greatly. This suggests that Ls1 multiplied readily at 32 °C in already infected cells. At 22 °C the proportion of infected cells and the infectivity of leaf extracts increased rapidly. The growth curve of Ls1 in leaf discs at 32 °C resembled the so-called 'one-step growth curve' of Ls1 or L in protoplasts, and not the growth curve of L in discs, in which the virus could spread from cell to cell. These results suggest that Ls1 is a *ts* strain that multiplies normally at the non-permissive temperature but has a malfunction in cell-to-cell movement.

INTRODUCTION

It was reported that a severe tomato strain (strain L) of tobacco mosaic virus (TMV) can be differentiated from its attenuated isolate (strain $L_{11}A$) by the type of necrotic local lesion produced in inoculated Xanthi-NN tobacco leaves after incubation in defined conditions that include a specific temperature shift (Nishiguchi & Oshima, 1977). In this case inoculated plants were first kept at 22 °C for 3 days, transferred to 32 °C for 2 days and then maintained at 22 °C for 1 day. This temperature shift treatment differs only slightly from those described by Jockusch (1968) and Peters & Murphy (1975) and can also be used to select temperature-sensitive (*ts*) mutants in cultures of wild type virus. Using this procedure we found an aberrant lesion, which morphologically resembled that of a non-coat protein mutant (Jockusch, 1968), among typical lesions in a leaf inoculated with strain L. Virus from this lesion was designated Ls1. This paper describes the symptoms caused by Ls1 in several host plants and the behaviour of Ls1 in mesophyll protoplasts and leaf discs of tomato.

METHODS

Plants and viruses. Plants used for comparison of symptoms were: *Nicotiana tabacum* cv. Xanthi-nc, cv. Xanthi-NN, cv. Bright Yellow and cv. Samsun, *N. sylvestris, N. glutinosa, Lycopersicon esculentum* cv. Fukuju no. 2 and *Phaseolus vulgaris* cv. Gintebo. They were grown in a greenhouse where the temperature ranged from 22 to 30 °C. Plants were inoculated when 2 months old for Xanthi-nc, Xanthi-NN and *N. glutinosa*, 7 weeks old for Bright Yellow, Samsun and *N. sylvestris*, 3 weeks old for tomato and 2 weeks old for *P. vulgaris*.

Two strains of TMV, L and Ls1, were used. L is the common tomato strain of TMV (Oshima *et al.* 1964, 1971) and Ls1 was isolated from an aberrant lesion among lesions produced on Xanthi-NN by L after the temperature shift treatment. Ls1 was passed through a series of six single lesions in Xanthi-nc and was propagated in Samsun. Viruses were purified as previously described (Moytoyoshi & Oshima, 1975) and the purified virus suspensions (1 mg/ml in 0·01 M-sodium phosphate buffer, pH 7·0) were stored at − 18 °C until use. Sap of infected leaves of Samsun tobacco diluted tenfold with 0·01 M-sodium phosphate buffer, pH 7·0, was used to inoculate the range of plants listed above. Purified virus suspension at appropriate concentrations was used to inoculate leaf discs and protoplasts.

Preparation, inoculation and culture of discs. Fully expanded leaves at the 7 to 8 leaf stage of tomato and the 5 to 6 leaf stage of Samsun tobacco were used. Before inoculation the leaves were washed with tap water and blotted dry with filter paper. The leaves were dusted with 600-mesh Carborundum and inoculated, using a glass spatula, with virus suspension at 500 μg/ml in 0·01 M-sodium phosphate buffer, pH 7·0. The leaves were washed with tap water about 3 min after inoculation and discs, 14 mm in diam., were cut out with the aid of a cork borer. Ten discs in each treatment were floated on incubation medium in a Petri dish, 6 cm in diam. The composition of this medium and the conditions of culture were the same as previously reported (Motoyoshi & Oshima, 1975).

Preparation, inoculation and culture of protoblasts. Preparation, inoculation and culture of protoplasts were as previously reported (Motoyoshi & Oshima, 1975, 1976). The concentrations of virus and of poly-L-ornithine were 1 μg/ml.

Detection of virus antigen in protoplasts. Percentages of protoplasts infected were determined by fluorescent antibody staining as described by Motoyoshi & Oshima (1975). The fluorescent antibody from rabbit antiserum to strain L (Motoyoshi & Oshima, 1975) was used to stain antigens of both L and Ls1.

Virus assay. The concentration of virus in discs and protoplasts was assayed by local lesion tests on Xanthi-nc tobacco. Discs were homogenized with 0·01 M-sodium phosphate buffer, pH 7·0. The homogenate was appropriately diluted with the same buffer and inoculated on ten half-leaves of two Xanthi-nc plants. The standard virus suspension, strain L at 0·1 μg/ml, was inoculated on the opposite half-leaves. The inoculum from protoplasts was prepared by the methods of Motoyoshi & Oshima (1975).

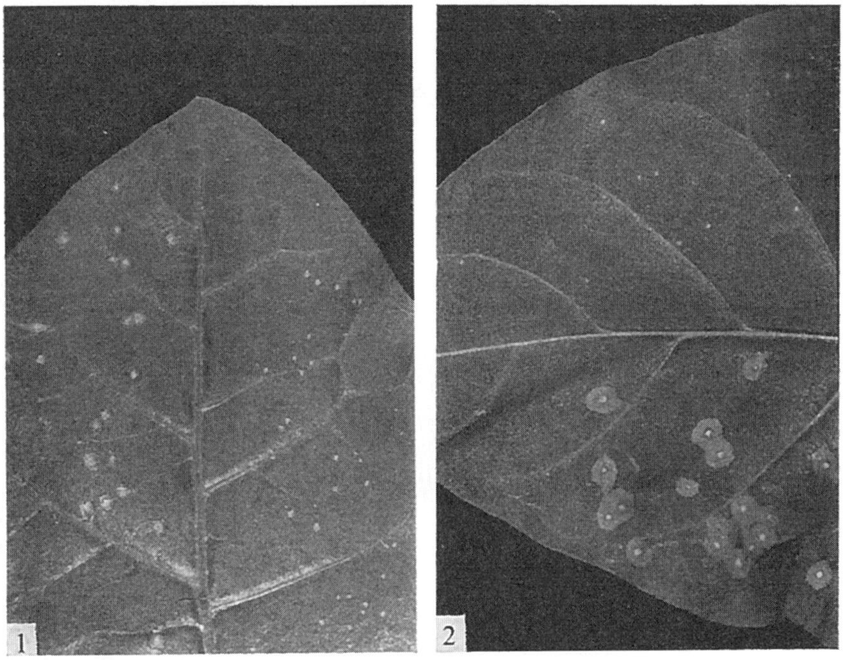

Fig. 1. Lesions produced in an inoculated leaf of Xanthi-nc tobacco plants kept for 5 days in a greenhouse at 22 to 30 °C. The right half was inoculated with Ls1 (0·1 μg/ml) and the left half with L (0·1 μg/ml).

Fig. 2. Lesions produced in a leaf of a Xanthi-nc tobacco plant kept at 22 °C for 3 days, then at 32 °C for 2 days and finally at 22 °C for 1 day. The upper half was inoculated with Ls1 (0·1 μg/ml) and the lower half with L (0·1 μg/ml).

RESULTS

Symptoms on test plants

Both Ls1 and L caused only local necrotic lesions in the inoculated leaves of Xanthi-nc, Xanthi-NN and Bright Yellow tobacco, *N. glutinosa* and *N. sylvestris*. Local lesions of Ls1 were, however, much smaller than those of L (Fig. 1). Ls1 induced yellow mottle in systemically infected Samsun tobacco leaves, whereas L caused a mosaic in them. Both strains produced a systemic mosaic in tomato cv. Fukuju no. 2 and neither induced symptoms in *P. vulgaris*.

When half-leaves of Xanthi-nc tobacco were inoculated with Ls1 and the opposite halves with L and the plants were given the temperature shift treatment described by Nishiguchi & Oshima (1977), L caused lesions having small necrotic centres with a surrounding collapsed area, whereas lesions of Ls1 lacked the collapsed area (Fig. 2). A similar morphological difference between the lesions of the two strains was also observed in Xanthi-NN tobacco, *N. glutinosa* and *N. sylvestris*.

Multiplication of virus in discs and protoplasts

Leaf discs of tomato inoculated with either Ls1 or L were incubated at either 22 or 32 °C and after 3 days the discs were harvested and the infectivities of their homogenates determined. Infectivity was high with either strain at 22 °C or with L at 32 °C, but very low with Ls1 at 32 °C (Table 1). Similar results were obtained using leaf discs of Samsun tobacco. Thus Ls1 is obviously temperature-sensitive in leaf discs.

Table 1. *Growth of strains Ls1 and L in leaf discs of tomato and tobacco**

Plant species	TMV strain	Temperature (°C)	Assay dilution	Total no. of lesions	Relative no. of lesions†
Tomato	Ls1	32	10^{-4}	10	7.2×10
		22	10^{-5}	961	1.6×10^5
		Control‡	10^{-2}	342	4.5×10
Tomato	L	32	10^{-6}	271	2.2×10^5
		22	10^{-5}	1217	1.2×10^5
		Control‡	10^{-2}	419	8.7×10
Tobacco	Ls1	32	10^{-2}	386	8.6×10
		22	10^{-5}	937	8.8×10^4
		Control‡	10^{-2}	667	6.7×10
Tobacco	L	32	10^{-6}	77	1.2×10^5
		22	10^{-5}	1084	1.1×10^5
		Control‡	10^{-2}	316	9.5×10

* Leaf discs (14 mm in diam.) were prepared from leaves inoculated with TMV at 500 μg/ml in 0.01 M-phosphate buffer, pH 7.0, and incubated for 3 days in culture medium under continuous illumination.

† Relative number of lesions defined as:

$$\frac{\text{Total number of lesions} \times \text{dilution factor}}{\text{Total number of lesions with standard suspension of L at 0.1 μg/ml}}.$$

‡ Samples taken before incubation.

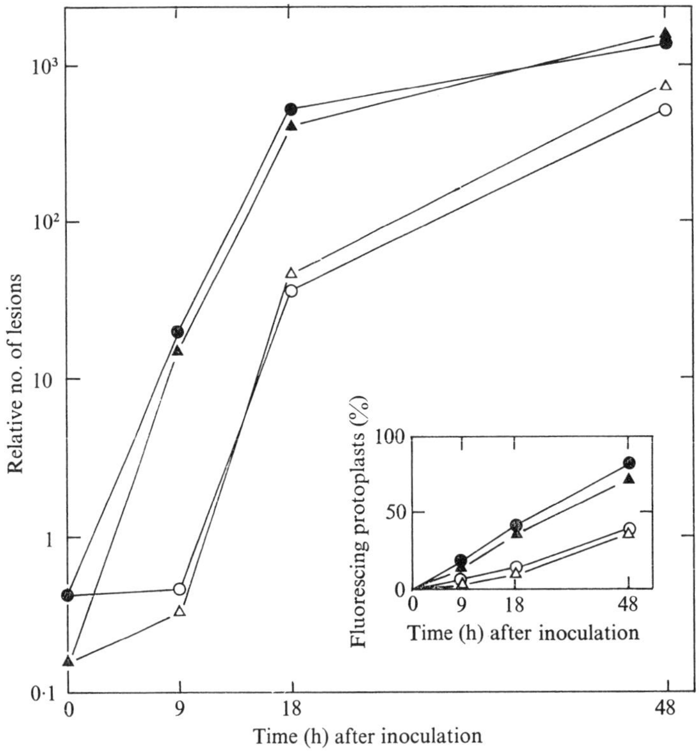

Fig. 3. Growth patterns of Ls1 and L in tomato protoplasts. ●——●, Ls1 at 32 °C; ○——○, Ls1, at 22 °C; ▲——▲, L at 32 °C; △——△, L at 22 °C.

Table 2. *Effect of temperature on virus multiplication and movement from cell to cell in tomato leaf discs**

TMV strain	Temperature for incubation of discs (°C)	% infected mesophyll cells*	Assay dilution	Total no. of lesions	Relative no. of lesions†
Ls1	32	0·8	10^{-3}	1482	737
	22	11·2	10^{-3}	1044	672
	Control‡	1·4	10^{-3}	34	29
L	32	9·4	10^{-4}	1293	7457
	22	11·8	10^{-4}	291	1471
	Control‡	1·6	10^{-3}	26	23

* The design of this experiment and the procedure for estimating % infected mesophyll cells are described in the text.

† Defined as:

$$\frac{\text{Total number of lesions produced by homogenate of discs} \times \text{dilution factor}}{\text{Total number of lesions produced by standard suspension of L at 0·1 } \mu\text{g/ml}}.$$

‡ Samples taken before incubation.

In another experiment, virus-inoculated discs were incubated for 2 days at 22 or 32 °C. Half of the discs were used for infectivity assay and mesophyll protoplasts were isolated from the other half and stained with fluorescent antibody. None of the protoplasts from discs inoculated with Ls1 and incubated at 32 °C were stained, suggesting that no virus accumulated in leaf mesophyll cells. By contrast, 33% of the protoplasts from Ls1-inoculated discs incubated at 22 °C were stained as compared with 65 to 66% of those from discs inoculated with L and kept at either temperature. Infectivity assays gave results very like those in Table 1. These findings suggest that the high temperature greatly limited multiplication of Ls1 in mesophyll cells of leaf discs.

Tomato mesophyll protoplasts were inoculated with Ls1 or L and incubated at 22 or 32 °C for periods up to 2 days. As shown in Fig. 3, the infectivities and percentages of protoplasts infected did not essentially differ between Ls1 and L at either temperature and were higher at 32 °C than at 22 °C. It was apparent that incubation at 32 °C, which limited growth of Ls1 in leaf discs, favoured infection and growth of Ls1 in protoplasts. Virus recovered from homogenates of Ls1-infected protoplasts after incubation at 32 °C for 2 days did not differ in lesion morphology from standard Ls1 in Xanthi-nc tobacco given the temperature shift treatment, suggesting that the progeny virus in protoplasts incubated at 32 °C was Ls1.

From the result of the above experiment, it might be assumed that the temperature-sensitivity of Ls1 involved some very early event in protoplast infection that occurred before inoculated protoplasts reached 32 °C. In one experiment, however, no significant difference in infectivities and percentages of protoplasts infected was observed between Ls1- and L-infected protoplasts which had been inoculated and washed at 32 °C, followed by being cultured at either 32 or 22 °C for 2 days. This result suggests that the temperature-sensitivity of Ls1 is not related to any early event in protoplast infection. Thus it was concluded that the temperature-sensitivity of Ls1 was not expressed in protoplasts.

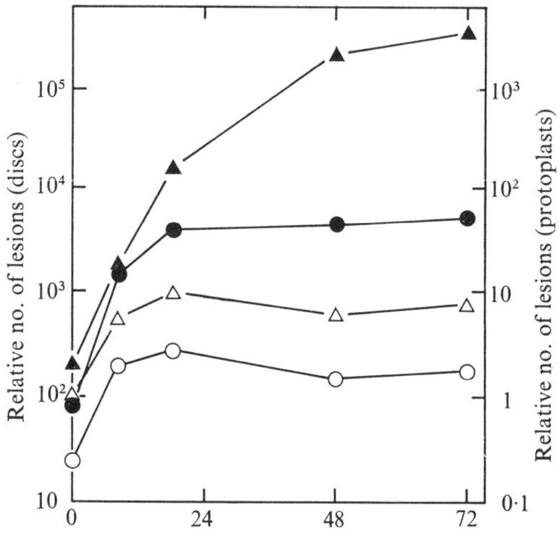

Fig. 4

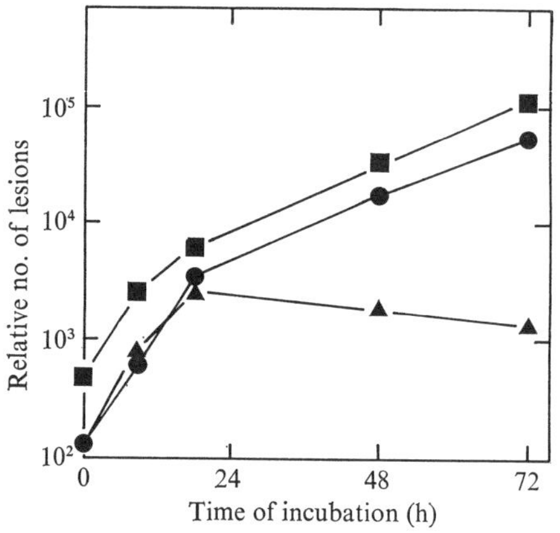

Fig. 5

Fig. 4. Growth patterns of Ls1 and L in leaf discs and protoplasts at 32 °C. Intact leaves on tomato plants were inoculated either with Ls1 or with L and kept at 20 to 25 °C in a greenhouse for 24 h. Discs were then prepared from the leaves, protoplasts were isolated from some of the discs, and the remaining discs and the protoplasts were incubated simultaneously at 32 °C. ●——●, Ls1 in discs; ○——○, Ls1 in protoplasts; ▲——▲, L in discs; △——△, L in protoplasts.

Fig. 5. Growth patterns of Ls1 and L in tomato leaf discs at 22 and 32 °C. Discs were prepared as described in Fig. 4. ●——●, Ls1 at 22 °C; ▲——▲, Ls1 at 32 °C; ■——■, L at 22 °C.

Virus movement from cell to cell

Tomato leaves attached to plants were inoculated with Ls1 or L in the morning and the plants were kept in a greenhouse at 20 to 25 °C under natural light for 24 h to allow the infection to establish. Leaf discs were cut from these infected leaves. The discs infected with Ls1 or L were divided into three batches, each of which consisted of 40 discs. One of the three batches was immediately harvested, and the other two were incubated at 22 and 32 °C, respectively, in the incubation medium and after 12 h they were harvested. Ten of 40 discs in each batch were homogenized immediately after harvest and used for infectivity assay. The rest of the discs were used for isolating mesophyll protoplasts. The protoplasts were cultured at 25 °C for a further 2 days to allow virus to multiply and sufficient antigen to accumulate to be detected by fluorescent antibody staining. Percentages of mesophyll cells which were infected at the time of harvest of the discs were estimated from percentages of fluorescing protoplasts thus obtained. The percentages of mesophyll cells infected with Ls1 and with L were 1·4 and 1·6, respectively, at the start of incubation of discs. During the 12 h incubation period, the percentage of Ls1-infected mesophyll cells increased eightfold at 22 °C but did not increase at 32 °C (Table 2). With L, percentages of infected mesophyll cells increased about six- to sevenfold at both 22 and 32 °C.

Although the percentage of mesophyll cells infected with Ls1 did not increase at 32 °C, infectivity increased by 25 times (Table 2). This increase factor is lower than that of 324 times obtained with L at 32 °C, but as the proportion of mesophyll cells infected with Ls1 was only about one-tenth of those with L, the virus yield of each strain per infected cell was rather similar. The yield per infected cell of Ls1 and L at 22 °C was less than at 32 °C. These results suggest that Ls1 multiplies normally at 32 °C but does not move from cell to cell.

Comparison of growth curves of virus in leaf discs and protoplasts

If Ls1 does not move from cell to cell at 32 °C (Table 2), its pattern of growth in discs at this temperature should be similar to that in protoplasts, because the possibility of secondary infection of protoplasts by progeny virus is negligible (Takebe, 1975).

Tomato leaves attached to plants were inoculated and the plants were kept for 24 h in a greenhouse at 20 to 25 °C under natural light. Discs were then prepared from the leaves. Some of the discs (ten discs in each treatment) were incubated at 32 or 22 °C and protoplasts were isolated from the other discs (30 discs in each treatment) and incubated. At various times, discs and protoplasts were harvested and infectivities were assayed. The pattern of Ls1 growth in discs at 32 °C resembled those of Ls1 and L in protoplasts (Fig. 4). The pattern of Ls1 growth in discs at 22 °C was similar to that of L and clearly different from that of Ls1 at 32 °C (Fig. 5). These findings further support the conclusions already mentioned, with respect to the behaviour of Ls1 at 32 °C.

DISCUSSION

Strain L, a common tomato strain of TMV, produces mosaic on tomato plants, necrotic local lesions on *N. tabacum* cv. Bright Yellow and *N. sylvestris*, and no symptom on *P. vulgaris* cv. Gintebo. The reactions of these test plants to Ls1 are similar to those with L, except that the lesions are smaller. It is possible that Ls1 is a mutant produced from L by a spontaneous mutation.

Jockusch (1968) described two types of *ts* mutant of TMV, one of which had coat malfunction and the other replicative defects. The first type causes local lesions of normal size without producing infectious particles, and the second type, e.g. Ni 2519, causes small local lesions in leaves of Xanthi-nc tobacco at 23 °C, but these stop spreading when the temperature is raised to 32 °C. The behaviour of Ls1 on Xanthi-nc tobacco is similar to that of the second *ts* type described by Jockusch (1968). In spite of this apparent resemblance, Ls1 can replicate at 32 °C as rapidly as a wild type strain in isolated protoplasts, and in mesophyll cells of leaf discs of tomato if the virus has already entered these cells, whereas Ni 2519 produced coat protein but not intact virus particles at the non-permissive temperature (Bosch & Jockusch, 1972; Oishi *et al.* 1975). As regards its replication, Ls1 is not temperature sensitive.

One of the *ts* mutants, II-27, selected by Peters & Murphy (1975) also resembles Ls1 in its behaviour in Xanthi-nc tobacco. II-27 caused small lesions at 23 °C and, when exposed to the temperature shift procedure, formed necrotic lesions without a halo. From this lesion morphology Peters & Murphy assumed that the mutant had a defect in its ability to move from cell to cell. Our results show more definitely that Ls1 lacks the ability to move from cell to cell at the high temperature.

In parenchymatous tissue virus particles probably move from cell to cell through plasmodesmata (Gibbs, 1976; Weintraub *et al.* 1976), but it is also likely that non-encapsidated TMV-RNA can move through them (Siegel *et al.* 1962). Ls1 may therefore have a *ts* ability to move through plasmodesmata, either in the form of particles or RNA. Two of the possible causes of the inability of Ls1 to move from cell to cell are the following: (1) the host plants may be induced to produce material that blocks virus movement through plasmodesmata. Callose production is known to be induced in tissue that responds hypersensitively to virus infection, with the result that virus movement through plasmodesmata is blocked (Allison & Shalla, 1974). No tests, however, have yet been made on whether callose production is induced by Ls1 in leaf discs of tomato and Samsun tobacco at the high temperature. (2) The blocking of virus movement from cell to cell may be caused by a functional defect in a virus-coded protein. There is no direct evidence for such a protein but it is not improbable that genetic information for a protein controlling cell to cell movement is carried by the virus genome. From this point of view it is intriguing that, besides coat protein and the possible replicase, TMV produces at least one kind of protein with no known function (Zaitlin & Hariharasubramanian, 1972; Sakai & Takebe, 1974).

Some *ts* mutants of TMV and cowpea chlorotic mottle virus (CCMV) have been studied using protoplast systems (Oishi *et al.* 1975; Dawson *et al* 1975). Oishi *et al.* (1975) reported that tobacco protoplasts could be infected with Ni 118 or Ni 2519, two *ts* mutants isolated by Jockusch (1968). Each mutant produced progeny virus at 24 but not at 33 °C. When protoplasts were inoculated with Ni 2519 and incubated at 33 °C, large inclusion bodies unusual in shape were produced and these could be stained with fluorescent antibody (Oishi *et al.* 1975) although no infectivity was detected in protoplast extracts (I. Takebe, personal communication). A *ts* mutant of cowpea chlorotic mottle virus (CCMV) could infect tobacco protoplasts and multiply at 25 °C, but at 35 °C no encapsidation occurred, possibly because of a functional defect in its coat protein; the mutant RNA however could be encapsidated with the coat of wild type CCMV at 35 °C when the protoplasts were infected simultaneously with *ts* and wild type virus (Dawson *et al.* 1975). In contrast to these *ts* mutants, the high temperature did not cause any detectable effect differentiating Ls1 from L in the protoplast system. The present study thus suggests that protoplast systems are especially useful for classifying and characterizing different types of *ts* mutant of plant viruses.

We are grateful to Dr I. Takebe and Dr T. Shimomura for helpful advice.

REFERENCES

ALLISON, A. V. & SHALLA, T. A. (1974). The ultrastructure of local lesions induced by potato virus X: a sequence of cytological events in the course of infection. *Phytopathology* **64**, 784–793.

BOSCH, F. X. & JOCKUSCH, H. (1972). Temperature-sensitive mutants of TMV: behaviour of a non-coat protein mutant in isolated tobacco cells. *Molecular and General Genetics* **116**, 95–98.

DAWSON, J. R. O., MOTOYOSHI, F., WATTS, J. W. & BANCROFT, J. B. (1975). Production of RNA and coat protein of a wild-type isolate and a temperature-sensitive mutant of cowpea chlorotic mottle virus in cowpea leaves and tobacco protoplasts. *Journal of General Virology* **29**, 99–107.

GIBBS, A. (1976). Viruses and plasmodesmata. In *Intercellular Communication in Plants: Studies on Plasmodesmata*, pp. 149–164. Edited by B. E. S. Gunning and A. W. Robards. Berlin, Heidelberg and New York: Springer-Verlag.

JOCKUSCH, H. (1968). Two mutants of tobacco mosaic virus temperature-sensitive in two different functions. *Virology* **35**, 94–101.

MOTOYOSHI, F. & OSHIMA, N. (1975). Infection with tobacco mosaic virus of leaf mesophyll protoplasts from susceptible and resistant lines of tomato. *Journal of General Virology* **29**, 81–91.

MOTOYOSHI, F. & OSHIMA, N. (1976). The use of tris-HCl buffer for inoculation of tomato protoplasts with tobacco mosaic virus. *Journal of General Virology* **32**, 311–314.

NISHIGUCHI, M. & OSHIMA, N. (1977). Differentiation of a tomato strain of tobacco mosaic virus from its attenuated strain by the lesion type. *Annals of the Phytopathological Society of Japan* **43**, 55–58.

OISHI, M., OTSUKI, Y., TAKEBE, I., HONDA, Y., KAJITA, S. & MATSUI, C. (1975). Inclusion bodies formed by temperature-sensitive mutants of TMV. *Annals of the Phytopathological Society of Japan* **41**, 287–288.

OSHIMA, N., GOTO, T. & SATO, R. (1964). A strain of tobacco mosaic virus (TMV-L) isolated from tomato. *Research Bulletin No. 83 of the Hokkaido National Agricultural Experimental Station*, pp. 87–99.

OSHIMA, N., OHASHI, Y. & UMEKAWA, M. (1971). Studies of some tobacco mosaic virus pathogenic to crucifer plants. I. Physical and chemical studies. *Annals of the Phytopathological Society of Japan* **37**, 319–325.

PETERS, D. L. & MURPHY, T. M. (1975). Selection of temperature-sensitive mutants of tobacco mosaic virus by lesion morphology. *Virology* **65**, 595–600.

SAKAI, F. & TAKEBE, I. (1974). Protein synthesis in tobacco mesophyll protoplasts induced by tobacco mosaic virus infection. *Virology* **62**, 426–433.

SIEGEL, A., ZAITLIN, M. & SEHGAL, O. P. (1962). The isolation of defective tobacco mosaic virus strains. *Proceedings of the National Academy of Sciences of the United States of America* **48**, 1845–1851.

TAKEBE, I. (1975). The use of protoplasts in plant virology. *Annual Review of Phytopathology* **13**, 105–125.

WEINTRAUB, M., RAGETLI, W. J. & LEUNG, E. (1976). Elongated virus particles in plasmodesmata. *Journal of Ultrastructure Research* **56**, 351–364.

ZAITLIN, M. & HARIHARASUBRAMANIAN, V. (1972). A gel electrophoretic analysis of proteins from plants infected with tobacco mosaic and potato spindle tuber viruses. *Virology* **47**, 296–305.

(*Received* 16 *August* 1977)

Cotranslational Disassembly of Tobacco Mosaic Virus

T. Michael A. Wilson
Scottish Crop Research Institute, Invergowrie, Dundee, U. K.

The first three years of my 25-year association with tobacco mosaic virus (1973-1998) were focused, directly or indirectly, on the task of defining and mapping virus particle assembly and disassembly processes *in vitro*. At that time, in Cambridge, the groups with whom I "trained in research" at the Biochemistry Department (Richard Perham and Richard Jackson) and at the MRC Laboratory of Molecular Biology (Aaron Klug and Jo Butler) were actively unravelling much of the higher order structural biology of the virus, and were developing *in vitro* translation systems to reveal the details of eukaryotic translational control mechanisms, at times using the polycistronic RNA genome of TMV as a convenient model. However, as none of us were plant pathologists, we gave relatively little thought to questions such as how the virus structure related to, or could aid, its function as a pathogen *in vivo*.

Nevertheless, it was deeply perplexing how such an allegedly simple, robust and apparently inert capsid structure, which could protect a delicate but infectious 6.4 kb single-stranded RNA molecule for decades *ex planta* could also, seemingly within hours of inoculation onto host leaves, disassemble to initiate TMV RNA translation and replication. Of course, more recent and elegant studies by John Shaw and Xiaojie Wu have shown that TMV disassembly and the replication cycle commence far more rapidly *in vivo* (Wu *et al.*, 1994; Wu and Shaw, 1996) – revelations which would have made the structure-function paradox of TMV even more daunting 15-20 years ago!

The possibility that TMV particle uncoating *in vivo* could be an active process, dependent on viable cell metabolism or a component of the subcellular machinery, and not driven "passively" simply by some physico-chemical property of the intracellular environment, the cell wall or a wound site, was alien to those of us who had considered the virus predominantly from an abstract, structural perspective. Years of thinking in terms of H-bonds, hydrophobic patches, electrostatic interactions, coat protein-coat protein and coat protein-RNA contacts, and the possible effects of ionic strength, pH and laboratory reagents on virus structure and its relative (in)stability did not prepare the imagination for cotranslational disassembly. After all, TMV was the paradigm for biological self-assembly. Why, if all the necessary information to build the structure was present in the RNA and coat protein molecules alone, could the virus not also pre-program its own dissociation under the relatively mild physiological conditions presumed to exist within the transiently damaged plant cell? Moreover, how could any host cell specificity exist in these early virus-cell interactions (e.g., receptors), and how could one possibly study such events *in vivo*?

Probably the only clue which encouraged speculation on a specific mechanism by which TMV particle disassembly might begin *in vivo*, was the fact that many artificial agents appeared to uncoat the common (U1) strain of the virus predominantly, if not exclusively, from the 5' end of the RNA. Therefore, if this end of the helical nucleocapsid was intrinsically less stable, the messenger-sense RNA genome would be exposed in a polar fashion which made biological "sense", as it would be appropriate for subsequent (if not necessarily, concomitant) genome translation. Moreover, the first protein(s) to be synthesised would be components of the proposed TMV replicase, an essential early viral function in the newly infected cell. But the deeper questions which then emerged were, how might the TMV particle be destabilised, activated or partially uncoated *in vivo* and to what extent did this happen prior to translation of the RNA?

"... but what on Earth possessed you to add intact virus particles to a cell-free translation cocktail?" Variations on this question were asked repeatedly at seminars and haunted me for several years after the *Virology* publication. Even with hindsight, I find it difficult to recall precisely, or to justify why these experiments were done. I can only conclude that the very first was a simple, high-risk, low additional effort, "Friday afternoon" experiment using up some surplus translation cocktail (a Scottish trait perhaps!) and some TMV stock.

The history of my Ph.D. training and later research experiences, and an increasing involvement with virus-plant interactions, now approaching true pathology, had doubtlessly raised my curiosity once again to see whether a stable, old preparation of TMV could be dissociated in an active rabbit reticulocyte system. Any exposed RNA would be revealed by acting as a template for protein synthesis. The first part of the 'Results' section of the 1984 *Virology* paper describes what happened and some of the naive experiments and alternative explanations I offered at the time for the tantalizing, but misleading preliminary result. During my Ph.D. on TMV disassembly *in vitro*, a 3'-coterminal nested family of partially stripped virus (PSV) particles were routinely size-fractionated in sucrose density gradients buffered at pH 8.0-8.2. This pH prevented particles from forming clumps, produced sharp separations and did not appear to cause further extensive, or even detectable, uncoating of the RNA. So it was, that, prior to repeating the cell-free particle translation experiment, I elected to "clean-up" the old TMV stock by dialysis, or dilution and sedimentation, or even micrococcal nuclease treatment and pH 8 sucrose density gradient centrifugation. In all the latter cases, the pretreated TMV particles behaved as efficient templates to produce the 126 kDa replicase polypeptide, not only in large amounts, but also remarkably cleanly compared to the smear of N-coterminal "early quitter" peptides frequently observed in translations of naked TMV RNA, particularly with wheat germ extracts.

This work began in late 1983, shortly after I had moved to the John Innes Institute, Norwich, where I greatly enjoyed a stimulating and competitive atmosphere, with many excellent, productive and very bright molecular plant virologists. With hindsight, it amuses me to recall how skeptical I was about these TMV results at the time. Despite an incredible and unnecessary number of replications, modifications, embellishments and additions to the fundamental experiment, to test the obvious explanation and assuage my own incredulity, I would repeatedly challenge my new colleagues (and even the Director, then Harold Woolhouse) to explain "what I had done wrong, or what control was missing?" Some of the more rational and helpful experiments, such as time-courses, found their way into the final paper. Eventually, with skilled help from Kitty Plaskitt of the electron microscopy group at JII, translation inhibitor studies revealed the now-infamous "striposome" complexes shown in the *Virology* paper. To my amusement, this descriptive name, worthy of the tabloid press, stuck.

Nevertheless, such was my lingering self-doubt that the paper was first prepared as a Short Communication for *Virology* and submitted with some trepidation. I feared that such a bizarre conclusion, i.e., that eukaryotic ribosomes could actively (cotranslationally) uncoated the large 5' proximal open reading frame on TMV RNA, would elicit scorn and rejection. Much to my surprise, I promptly received a letter from George Bruening, then the Editor, asking if I would present the paper as a regular article, in view of the need to fully discuss its potential significance for plant virology.

Since the first report of the phenomenon of cotranslational disassembly involved rabbit reticulocyte extracts, this effectively ruled it out as a potential host-specific event, as did subsequent papers using wheat germ extracts, an *Escherichia coli* S30, *Xenopus* oocytes or pea leaf epidermal cells. Nevertheless - thank you, George!

Also at this time, a name I had known from his excellent and heroic work unravelling early TMV infection events in leaves and protoplasts, John Shaw, had been visiting the JII. Word of the "striposomes" reached John, and thus began a most rewarding and productive collaboration and friendship.

At first, we embarked on the quest for evidence of "striposomes" *in vivo* using inoculated tobacco leaf epidermal cell extracts. I had to be taught how to peel fast and collect the epidermal strips in a vat of liquid nitrogen. Furthermore, as if to steady my frozen tweezer hand and reduce stress, the protocol also required us to inoculate $^{32}P/^{3}H$ double-labelled TMV with excitingly high specific activities! The results of these extensive and, thanks to John, extremely carefully controlled experiments were reported in 1986 (Shaw *et al.*, 1986).

Since this early work, the Kentucky group have invested huge efforts, patience and skill, and made dramatic progress using greatly improved techniques, culminating in what I would call "time-lapse PCR". They have revealed the rapid timing (now in minutes) and molecular detail, not only of 5' to 3' co-translational virus disassembly and replicase synthesis, but also the subsequent 3' to 5' co-replicational TMV RNA uncoating process (Wilson, 1988), during which the nascent minus-strand is copied from the parental (i.e. inoculum) virus genome template (Wu *et al.*, 1994; Wu and Shaw, 1996). The sensitivity of these methods and the significance of the conclusions for early infection events by other animal or plant viruses are clear.

In part, my lingering discomfort with the cotranslational disassembly hypothesis arose from the eager willingness of many colleagues to accept it as an obvious (with hindsight) and convenient explanation for an essential event in the virus infection cycle which had, until then, been an intractible and paradoxical conundrum. Indeed, besides the Kentucky group, it was Ron Fraser and Dick Verduin who, independently and reassuringly, first verified the TMV result. Verduin's group actually measured the number of 80 S ribosomes in a series of arrested striposome complexes, as well as the diminishing length of residual TMV nucleocapsid (Verduin, 1992). From the inverse linear relationship, and the known parameters of TMV, they calculated that each ribosome was associated with 140 nucleotides of exposed TMV RNA.

Later, details of the extent of 5' sequence which is easily and rapidly uncoated when TMV particles are diluted, pH 8-treated and/or centrifuged, suggested that 40 S ribosome subunit scanning alone is probably not involved in nucleocapsid disassembly at the concave end of the TMV rod, and confirmed that AUG-mediated 80 S formation and active, ATP/GTP-dependent 126 kDa elongation and RNA translocation are crucial (Mundry *et al.*, 1991).

One of the key questions remaining, of course, is what happens during the systemic spread of

TMV (or any other virus) infection, when intact particles must move through the phloem and companion cells to arrive at undamaged leaf parenchyma/mesophyll cells which they must enter and again become uncoated.

Shortly after the *Virology* paper above, cotranslational disassembly, or at least particle swelling and cotranslational release of the ssRNA genome, was reported for icosahedral nucleocapsids of southern bean mosaic virus by Michael Brisco, Roger Hull and myself. Further elegant work was done by Dick Verduin and his group on cowpea chlorotic mottle virus (Roenhorst *et al.*, 1989). However, recent detailed structural studies on isometric plant virus capsids and mutants rendered unable to swell have complicated this earlier simple story.

The serendipitous nature of the first experiment which began the cotranslational disassembly saga is a testament to the axiom that, while many reasons can be proposed against doing a particular experiment, it is actually far more exciting, potentially productive, and always thought-provoking simply to do the experiment and worry about the meaning of the result later.

TMV has played a key role in many fundamental technical and conceptual advances in the biological sciences, including virology, over the past 100 years. Most had a more rational experimental foundation and justification than the striposome story, but few sought to span the gap between classical structural studies and dynamic functional approaches!

References

Mundry, K. W., Watkins, P. A. C., Ashfield, T., Plaskitt, K. A., Eisele-Walter, S. and T. M. A. Wilson. (1991). Complete uncoating of the 5' leader sequence of tobacco mosaic virus RNA occurs rapidly and is required to initiate cotranslational virus disassembly *in vitro*. *J. Gen. Virol.* **72**:769-777.

Roenhorst, J. W., Verduin, B. J. M., and R. W. Goldbach. (1989). Virus-ribosome complexes from cell-free translation systems supplemented with cowpea chlorotic mottle virus particles. *Virology* **168**:138-146.

Shaw, J. G., Plaskitt, K. A. and T. M. A. Wilson. (1986). Evidence that tobacco mosaic virus particles disassemble cotranslationally *in vivo*. *Virology* **148**:326-336.

Verduin, B. J. M. (1992). Early interactions between viruses and plants. *Sem. Virol.* **3**:423-431.

Wilson, T. M. A. (1988). Structural interactions between plant RNA viruses and cells. *Oxford Survey of Plant Molecular and Cellular Biology* **5**:89-144.

Wu, X. and J. G. Shaw. (1996). Bidirectional uncoating of the genomic RNA of a helical virus. *Proc. Natl. Acad. Sci. USA* **93**:2981-2984.

Wu, X., Xu, Z., and J. G. Shaw. (1994). Uncoating of tobacco mosaic virus RNA in protoplasts. *Virology* **200**:256-262.

Biographical Sketch

Michael Wilson attended Edinburgh University from 1969-1973, where he obtained a first class honor's degree in biochemistry. He then moved to Cambridge University (St. John's College) where he worked with Dr. Richard N. Perham on a project entitled 'Protein-protein and protein-RNA interactions in tobacco mosaic virus' for three years to obtain his Ph.D. degree. His Ph.D. examiners were Sir Aaron Klug and Dr. Robin Offord. After a brief MRC Postdoctoral Research Fellowship working on bacterial exotoxin production, he went to Liverpool University in 1978 to a faculty position in the Biochemistry Department. Then, from 1983-1989 he was a member of the John Innes Institute. He joined the AgBiotech Center at Rutgers University as Professor in 1989. Since 1992 he has been at the Scottish Crop Research Institute, first as Head of Virology (1992-1995), and more recently (1995-present) as Deputy Director.

VIROLOGY **137**, 255-265 (1984)

Cotranslational Disassembly of Tobacco Mosaic Virus *in Vitro*

T. M. A. WILSON

Department of Virus Research, John Innes Institute, Colney Lane, Norwich NR4 7UH, England

Received January 24, 1984; accepted May 30, 1984

Heterogeneous nucleoprotein preparations of tobacco mosaic virus (TMV), stored at pH 7.0, were treated briefly at pH 8 and then incubated in mRNA-dependent rabbit reticulocyte lysate. The treatment at pH 8.0, under conditions which did not cause detectable polar disassembly as judged by exposure to micrococcal nuclease, caused TMV to stimulate *in vitro* translation up to 100-fold over background. High-molecular-weight products, characteristic of TMV, appeared, albeit at a significantly slower rate than when naked TMV RNA was used as template. Ribonucleoprotein material from cycloheximide-treated incubations was examined in the electron microscope to reveal a subpopulation of rodlets (approx 5%) with clusters of ribosomes at their concave 5′ ends. It is proposed that pH 8 treatment removes the last turn of protective coat protein subunits from the rods and "exposes" at least a 49-nucleotide segment of the efficient 5′-leader sequence of TMV RNA for interaction with 40 S ribosomal subunits. Ribosome translocation during protein synthesis presumably dislodges further coat protein subunits progressively while continuing protection of the incoming TMV RNA.

INTRODUCTION

Little is known about the site or mechanism of uncoating of tobacco mosaic virus (TMV)[1] *in vivo* (for a review, see Shaw, 1984). Uncoating of TMV *in vitro* by alkali (Perham and Wilson, 1976), SDS (Wilson *et al.*, 1976), urea (Blowers and Wilson, 1981), and DMSO (Wilson *et al.*, 1981) occurs predominantly, if not exclusively, from the 5′ end of the rod. To free the 5′-proximal, 126K cistron for early translation, it is axiomatic that the helically symmetrical 300-nm rods must disassemble for at least 160 nm, to residue 3419 from the 5′ end (Goelet *et al.*, 1982). In this context, it is interesting to recall an early report (Helms and Zaitlin, 1970) in which TMV particles were treated briefly at pH 10 to remove (on average) between 1.7 and 3.3% of their coat protein. Treated rods had a significantly higher infectivity on *Phaseolus vulgaris* var. Pinto, than untreated virus. The authors believed they were overcoming the failure of most virus particles to initiate uncoating early during infection *in vivo*.

In this paper, I report the results of several independent experiments which indicate that if a very short 5′ sequence of TMV RNA is "exposed" by mild alkaline treatment, 80 S ribosomes can disassemble the helical nucleoprotein particles of TMV in a progressive manner, during translocation. Cotranslational disassembly results in a slower rate of polypeptide elongation and a smearing of the previously sharp peak of "intact" (300-nm) virus particles on sucrose density gradients. Unique and characteristic ribonucleoprotein complexes are visible by negative staining in the electron microscope. By analogy with descriptive terms such as polyribosomes and "informosomes" (Spirin, 1969) I have found it convenient to refer to these complex particles, as "striposomes" or "divestisomes."

The association of plant ribosomes with

[1] Abbreviations used: TMV, tobacco mosaic virus; SDS, sodium dodecyl sulphate; SDS-PAGE, electrophoresis in polyacrylamide gel permeated with SDS; TCA, trichloroacetic acid; NMV, narcissus mosaic virus; PMV, papaya mosaic virus; DMSO, dimethyl sulphoxide; EGTA, ethyleneglycol-bis-(β-aminoethyl ether)*N,N′*-tetra acetic acid; PPO, 2,5-diphenyloxazole; K, kilodalton; (kD; molecular weight $\times 10^{-3}$).

0042-6822/84 $3.00

partially uncoated TMV RNA had been reported previously (Kiho, 1970, 1972; Hashimoto and Okamoto, 1975), but no causal relationship was suggested between polyribosome formation and virus uncoating. The two processes were clearly thought to be distinct (Machida and Kiho, 1970). Much of the early data are extremely contradictory and based on erroneous assumptions (see Shaw, 1984; and Discussion below).

The implications of this study for the process of viral disassembly *in vivo*, during plant cell infection, are discussed.

MATERIALS AND METHODS

Virus purification. TMV *vulgare* (U1) strain, Cambridge isolate, was grown in *Nicotiana tabacum* cv. Samsun and harvested from 10- to 12-week-old plants by the method of Leberman (1966). Partially purified virus had been stored at 4° for at least 12 months in sodium phosphate buffer, pH 7.0, I = 0.1.

Treatment of virus for translation. In some control experiments, TMV, stored at pH 7.0, was sedimented by centrifugation at 35,000 rpm for 120 min in a Beckman Type 40 rotor at 4° and resuspended at approx 40 $mg \cdot ml^{-1}$ (final) in sterile distilled water. Treatment with mildly alkaline buffers (10–100 m*M* Tris-HCl, pH 8.0–8.2 or 10–100 m*M* sodium phosphate, pH 8.0) was originally by dialysis of 0.5-ml samples of virus (at 40 $mg \cdot ml^{-1}$) against 1–2 l of buffer at 4° for 48–60 hr. Later experiments indicated that similar translation effects could be seen with virus which had simply been diluted 20-fold with 10–100 m*M* (final) Tris-HCl, pH 8.1, at 4° and immediately sedimented and resuspended as described above. In all cases, pellets of treated virus were resuspended in sterile distilled water to a final concentration of 40–50 $mg \cdot ml^{-1}$ (estimated by A_{254}). For some experiments, partially pure virus preparations were treated either *before* or *after* the pH 8-dialysis step with micrococcal nuclease at up to 600 $U \cdot ml^{-1}$ (final concentration) in 1–2 m*M* $CaCl_2$ (final) at room temperature for 30 min. Reactions were terminated with 5–10 m*M* EGTA (K^+ salt), pH 7.5, before proceeding as described above.

In vitro translation and product analysis. The mRNA-dependent rabbit reticulocyte lysate was prepared and used according to Pelham and Jackson (1976). Standard incubations (25 μl) contained either TMV RNA (80–100 $\mu g \cdot ml^{-1}$, final), or partially purified virus kept at pH 7.0 or treated at pH 8.0 as described (8–10 $mg \cdot ml^{-1}$, final concentration). Sucrose density-gradient purified "intact" TMV rods (300 nm) were also used successfully as templates in some translation experiments. Sucrose gradients (15–30%, w/v) were buffered with 0.1 *M* Tris-HCl, pH 8.0, at 4° and used according to Lomonossoff and Butler (1979). The mRNA-dependent reticulocyte lysate itself contributed between 30 and 50 m*M* K^+ and 1 m*M* Mg^{2+} which was supplemented during translation by addition of 100 m*M* K^+ and 0.5 m*M* Mg^{2+}. The lysate also contained 50 $\mu g \cdot ml^{-1}$ calf liver tRNA (Boehringer, Mannheim) and a final concentration of 250 $\mu Ci \cdot ml^{-1}$ L-[^{35}S]methionine (Amersham International, plc). All incubations were at 30° for the times stated in the text. Where appropriate, 3 vol of sample buffer containing 2% (w/v) SDS (Laemmli, 1970) were added before boiling for 3 min. Proteins were resolved by discontinuous SDS–PAGE (Laemmli, 1970) on 15-cm-long 15% (w/v) polyacrylamide slab gels run at 60 V for 16 hr. [^{35}S]Met-labeled polypeptides were visualized by fluorography (Bonner and Laskey, 1974) or autoradiography on Fuji RX X-ray film at −80°.

Electron microscopy of "striposomes." TMV rods, dialyzed at pH 8.2, collected by centrifugation, and resuspended in sterile water as described above were incubated for 15 min in a scaled-up volume (125 μl) of rabbit reticulocyte lysate under the conditions described.

At t = 15 min, 0.1 vol of 0.1 *M* cycloheximide was added to arrest translation. Samples (60 μl) were diluted 50-fold in buffer A (20 m*M* Tris-HCl, 5 m*M* $MgCl_2$, 1 m*M* cycloheximide, pH 7.5), and the particles collected by ultracentrifugation in 3-ml polycarbonate tubes in an MSE Prepspin 50 at 90,000 *g* for 3 hr at 4°. In control experiments, equivalent amounts of pH 8-treated TMV were added to 60-μl samples of a template-free incubation at $t = 15$ min, after addition of cycloheximide. Sedimentation produced small, translucent pellets, each of which was resuspended gently in 50 μl of buffer A at 4° for several hours. Alternatively, 60-μl samples of lysate incubation were diluted fivefold in buffer A and sealed in a microdialysis cell through which was pumped 100 vol of buffer A in preparation for electron microscopy. In both cases, further dilution in buffer A was often necessary for electron microscopy. Nucleoprotein material was negatively stained with uranyl acetate and viewed in a Siemens Elmiskop.

Sucrose density gradient analysis of "striposomes." Rabbit reticulocyte lysate incubations each scaled up to 62.5 μl were programmed with 0.5 mg pH 8 treated, and centrifuged TMV. Virus was added either at time zero, or *after* stopping a template-free incubation with cycloheximide. All experiments were stopped at 15 min with 10 m*M* (final) cycloheximide, diluted to 250 μl with buffer A (as above), and loaded onto 10-ml linear sucrose density gradients (15–30%; RNA-free sucrose, Serva, Heidelberg) at 4°, containing buffer B (50 m*M* Tris-HCl, 5 m*M* $MgCl_2$, pH 7.5). Gradients were centrifuged in a Beckman SW41 rotor at 35,000 rpm for 80 min at 4° and fractionated through an LKB Uvicord S scanning continuously at 254 nm. Forty 0.25-ml fractions from each gradient were solubilized with 1 ml "Soluene" (Packard Instrument Co. Inc., Ill.) and counted for [^{35}S]methionine in a Triton X-100/toluene/PPO-based scintillation fluid.

RESULTS

In Vitro Translation of Tobacco Mosaic Virus

Heterogeneous nucleoprotein preparations of TMV (Leberman, 1966), which had been stored in pH 7.0 buffer at 4° for 12 months, were found to stimulate incorporation of [^{35}S]methionine into TCA-precipitable polypeptides in the mRNA-dependent rabbit reticulocyte cell-free translation system (Pelham and Jackson, 1976), approximately two- to fivefold over mRNA-free control incubations. This surprising phenomenon could be attributed to bacterial contamination, partial stripping of rods during storage, or copurification of stored mRNP particles or polyribosomes from the original host plants during virus isolation procedures (Leberman, 1966) which involve polyethylene glycol precipitation and differential centrifugation. Certainly the labeled products (Fig. 1, lane 3), while of high molecular weight (20K–100K), appeared unlike those expected from TMV RNA (Fig. 1, lane 2). Stored virus preparations (pH 7) were centrifuged at 105,000 *g*, resuspended in sterile distilled water (as described above), and added to the lysate. This treatment did not appear to alter either the yield or nature of the polypeptides made *in vitro* (Fig. 1, lane 4).

Partially purified TMV preparations which had been dialyzed at 4° against 0.1 *M* Tris-HCl, pH 8.0–8.3, or sodium phosphate, pH 8.0 (see Materials and Methods) stimulated *in vitro* protein synthesis to a far greater extent (in the range from 10 to 100 times that of template-free controls). A similar effect was observed for virus preparations stored in pH 7.0 buffers which were simply diluted about 20-fold (to approx 2 $mg \cdot ml^{-1}$) with 10–100 m*M* Tris-HCl, pH 8.1, at 4°, then immediately sedimented, resuspended in sterile water, and used as templates. In these experiments, the major [^{35}S]Met-labeled polypeptide products (Fig. 1, lanes 5, 6) comigrated with those produced by translation of purified TMV RNA (Fig. 1, lane 2). This result was entirely unexpected. The template activity was found to be associated with the sharp, single peak of 300-nm rods purified by 15–30% (w/v) sucrose density gradient centrifugation as previously described (Perham and Wilson, 1978; Lomonossoff and Butler, 1979). The possibility that naked or partly exposed TMV RNA had survived storage for 12 months at 4° in a crude plant extract seemed remote. Furthermore, virus preparations

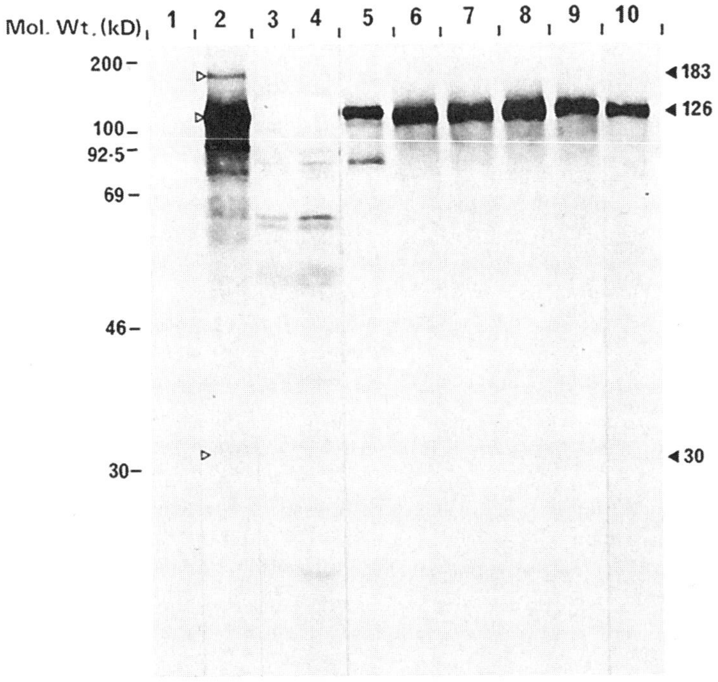

FIG. 1. Fluorograph of electrophoretically resolved polypeptides synthesized in rabbit reticulocyte extracts in response to various exogenous templates. The translation system was incubated for 90 min with [^{35}S]methionine and no added message (lane 1) or various potential templates: (lane 2) 80 μg/ml TMV RNA; (lane 3) partially purified TMV that had been stored for 12 months at 4° in sodium phosphate buffer, pH 7; (lane 4) the same preparation of TMV, sedimented and resuspended in water immediately prior to translation; (lane 5) purified TMV dialyzed for 60 hr against pH 8.2 Tris-HCl buffer or (lane 6) pH 8.0 sodium phosphate buffer. TMV treated as for lanes 5 and 6 also was treated with micrococcal nuclease before (lanes 7 and 8, respectively) or after (lanes 9, 10) dialysis. Lanes 2 and 5 through 10 received equal amounts of radioactivity, whereas lanes 1, 3, and 4 received equal volumes of translation reaction mixture. Lanes 3 and 4 were exposed for 4 days, the other lanes for 20 hr. Molecular weights ($\times 10^{-3}$) of radioactive marker proteins are indicated on the left; positions of well-known TMV polypeptides are indicated on the right and with open arrowheads on the left.

which were treated with micrococcal nuclease at up to 600 U $\cdot$ ml^{-1} *before* dialysis at pH 8 showed no change in either the observed marked stimulation of translation or the spectrum of products (Fig. 1, lanes 7, 8). Partially uncoated rods in the original TMV preparation are clearly not responsible for this phenomenon.

Extensive disassembly during dialysis at pH 8.0–8.2 could also be ruled out for several reasons. First, the sedimentation profile of TMV after dialysis showed only a single peak (300-nm rods). Second, earlier studies (Perham and Wilson, 1976, 1978; Pelcher and Halasa, 1979), had established that no detectable stripping of TMV or release of TMV RNA occurs during dialysis at 4° in 0.1 *M* buffers below about pH 9.0. Last, and most surprising of all, micrococcal nuclease treatment (600 U $\cdot$ ml^{-1}) of the TMV rods *after* pH 8 dialysis and immediately *before* centrifugation, resuspension, and translation had little effect on either the quantitative or qualitative yield of [^{35}S]Met-labeled polypeptides (Fig. 1, lanes 9, 10). Similarly, in almost all cases tested, micrococcal nuclease treatment of pH 8-dialyzed virus following centrifugation and resuspension in sterile distilled water, did not markedly effect its translational activity (data not shown). It would appear that the translationally active species in these dialyzed TMV preparations has little, if any, exposed RNA.

Time Course of Protein Synthesis

If the cotranslational disassembly model is correct, one might predict that the progressive, energy-dependent removal of individual TMV coat protein subunits in a 5′ to 3′ direction during ribosome translocation would slow the rate of elongation during protein synthesis. Figure 2 shows (a) the time course of incorporation of [^{35}S]methionine into TCA-precipitable material and (b) the corresponding spectrum of products at each time interval for parallel incubations containing either a standard TMV RNA preparation or TMV rods which had been dialyzed for 48 hr against 0.1 *M* Tris-HCl, pH 8.1, at 4°, sedimented, and resuspended in sterile water. The overall rate of incorporation of [^{35}S]methionine and the time of appearance of the highest-molecular-weight product are considerably slower in the incubation which contained the dialyzed TMV rods.

Control time-course experiments, in which TMV RNA was translated in the reticulocyte lysate (at 100 $\mu g \cdot ml^{-1}$, final) in the presence or absence of partially purified TMV (Leberman, 1966) stored at pH 7.0, have confirmed that the slower elongation rate is *not* an artifact resulting from the additional "viscosity" of incubations containing large amounts of exogenous ribonucleoprotein particles (data not shown).

Electron Microscopy of "Striposomes"

Further support for the proposed model of cotranslational disassembly came from

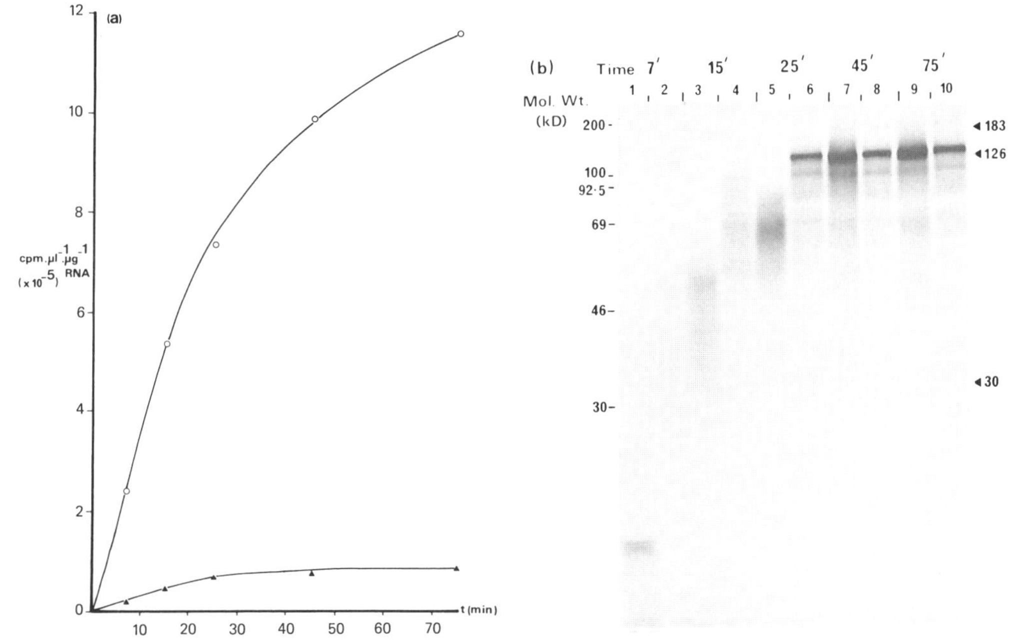

FIG. 2. Time course of polypeptide accumulation in rabbit reticulocyte lysate programmed with TMV RNA or TMV rods. TMV rods were dialyzed against 0.1 *M* Tris-HCl, pH 8.1 for 48 hr, sedimented, and resuspended in water. (a) Samples were removed at times indicated during incubations with 100 µg/ml TMV RNA (○) or 10 mg/ml TMV rods (▲). Incorporation into TCA-precipitated radioactivity is expressed as cpm per µg of TMV RNA or RNA equivalent, corrected to 1 µl of reaction mixture. (b) SDS-PAGE analyses of equal amounts of radioactivity (10^5 cpm) from reaction mixtures to which TMV RNA (even-numbered lanes) or TMV rods (odd-numbered lanes) had been added. Radioactive marker proteins and other zone markings are as in Fig. 1.

electron micrographs of negatively stained material sedimented after cycloheximide treatment of reticulocyte incubations (see Materials and Methods). Selected areas and individual particles from a few micrographs are shown in Fig. 3. Control experiments in which pH 8-dialyzed TMV rods were added to template-free incubations after translation had been stopped with 10 m*M* (final) cycloheximide revealed no particles of the type shown in Fig. 3. Estimates of the frequency of occurrence of these "striposome" complexes suggested that they constituted approximately 5% of the rod population. This is probably an underestimate, as many of these structures will have been lost by shearing the TMV RNA tails (polyribosomes) during preparative ultracentrifugation and grid preparation. To achieve better estimates,

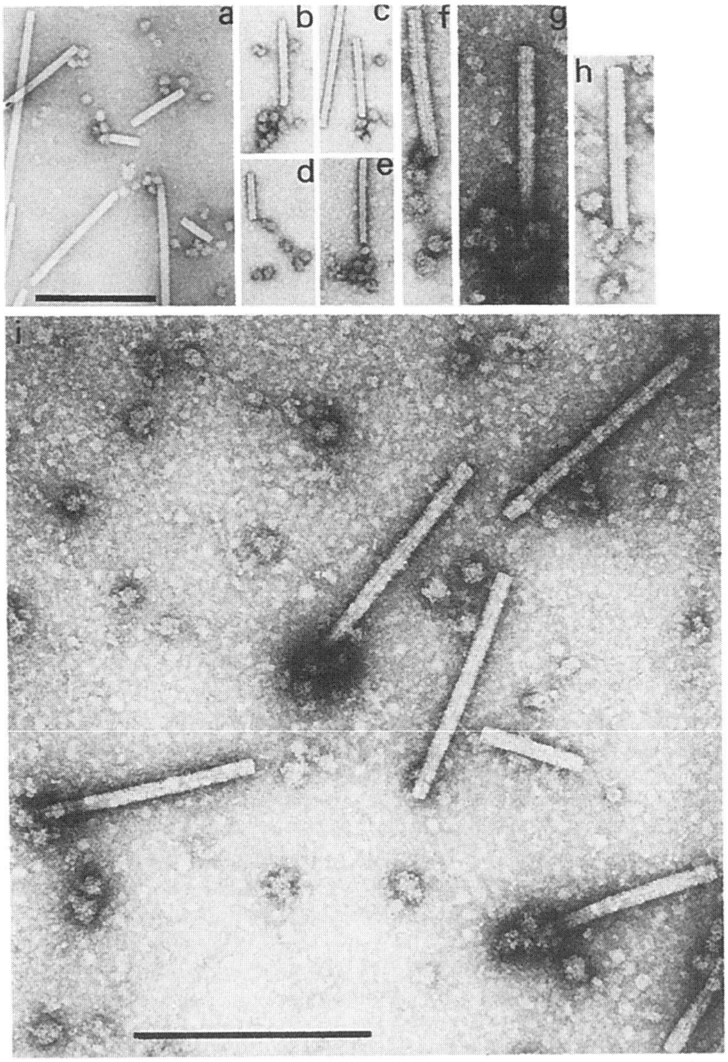

FIG. 3. Electron micrographs of material sedimented from reticulocyte incubations programmed with pH 8.2-treated TMV particles. "Striposome" complexes in panels (a–h) were sedimented and resuspended in buffer A. Panel (i) shows the effect of dialysis against buffer A during preparation of complexes for electron microscopy (see Methods). Magnification factors: ×105,000 (panels a–e); ×210,000 (panels f–i). Bars represent 300 nm (full-length TMV rod).

attempts were made to view negatively stained lysate samples directly or after microdialysis against buffer A (see Materials and Methods). However, resolution was poor, probably due to the high protein content of the lysate which interfered with the staining (Fig. 3i). Rod-length measurements in such complexes (i.e., the residually encapsidated RNA) after 15 min incubation indicated that between one-quarter and one-third of the RNA had become exposed. This corresponds to between 1500–2100 nucleotides, sufficient to code for polypeptides of molecular weight around 50K. However, because of the inevitable problems of rod-shearing mentioned above, some uncertainty must surround such bold correlations. Nevertheless, the estimates agree quite well with the 15-min time-course data shown in Fig. 2b (lane 3).

Sucrose Density Gradient Ultracentrifugation of "Striposomes"

In view of the early work by Kiho (1970, 1972) on the structure of plant polysomes which contained the incoming TMV RNA in TMV-infected leaves, it was of interest to examine in more detail the effect of brief (15 min) incubation on the sedimentation properties of pH 8.2-dialyzed TMV rods added to the reticulocyte lysate. The result is shown in Fig. 4a. Material was seen to pellet in this gradient whereas no pellets were visible in the gradient experiments shown in Figs. 4b or c. Kiho (1970) claimed that membrane-bound polysome structures, sedimenting more rapidly than TMV and still containing some ^{35}S-labeled viral coat protein, were derived from between 2 and 6% of the original population of rods. Template-free, control experi-

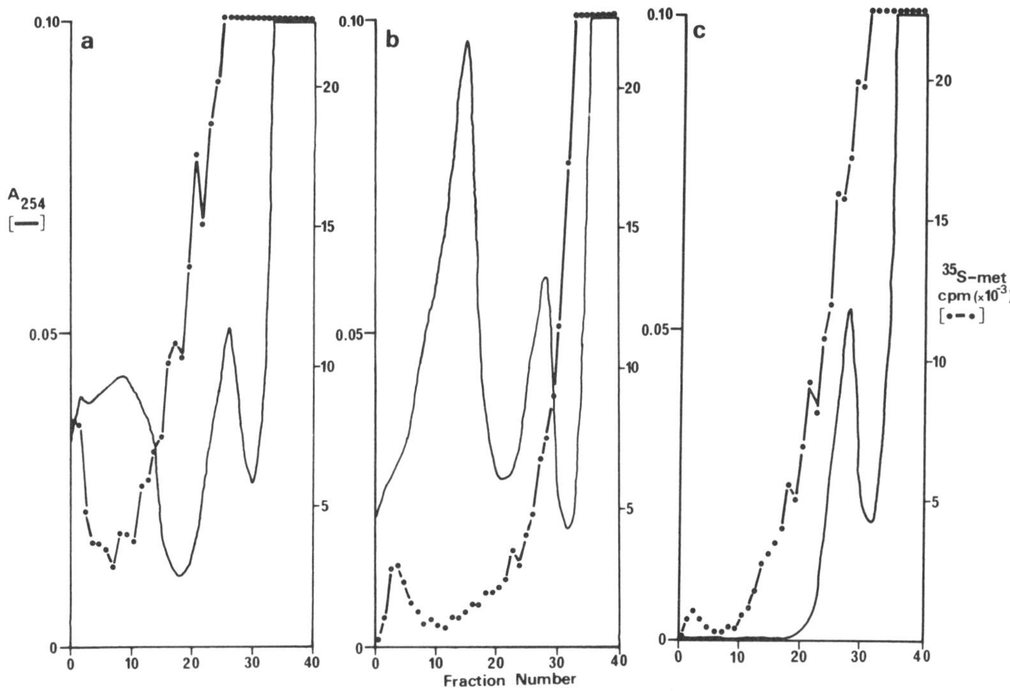

FIG. 4. Sucrose density gradient profiles of rabbit reticulocyte lysate incubations programmed with 0.5 mg pH 8.2-treated TMV rods, either at time zero (a) or *after* stopping a template-free control incubation at 15 min with 10 m*M* (final) cycloheximide (b). A second template-free control incubation, without added rods is shown for comparison (c). Counts per minute in [^{35}S]methionine-labeled nascent polypeptides are plotted (●) for each gradient fraction together with the absorbance profile at 254 nm. Sedimentation is from right to left in each case.

ments (Figs. 4b, c) were incubated for 15 min and in one case, 0.5 mg of pH 8.2-dialyzed TMV rods was added after cycloheximide treatment (Fig. 4b). The results support the conclusion that smearing of the TMV rod peak during incubation is due to formation of structures of the type shown in Fig. 3, rather than a nonspecific association of, or uncoating by, components in the rabbit reticulocyte lysate. The distribution of [^{35}S]methionine in attached, nascent polypeptide chains also confirms that these faster-sedimenting complexes are translationally active.

DISCUSSION

To explain the observations described above, I propose that pH 8.0 treatment causes the 5′-terminal nucleoprotein structure of TMV to become "relaxed" or altered in such a way as to make a portion of the RNA leader sequence available for interaction with 40 S ribosomal subunits to initiate translation. However, the RNA still remains largely protected from nuclease attack. It may be that during self-assembly of TMV *in vivo*, in the major 3′ to 5′ direction, the RNA is completely buried by a final helical turn of coat protein subunits. Maintenance of the helical configuration in this final turn of coat protein subunits might not be favored at pH 8. Protein-protein interactions at this extremity would be inherently weaker as a consequence of each subunit having one normally buried face exposed to the aqueous environment. In addition, these subunits could interact with the RNA only on one surface, thereby reducing the stabilizing contribution from RNA-protein interactions. At the 5′ end of the RNA the unusual "cap" structure (Keith and Fraenkel-Conrat, 1975; Zimmern, 1975), would further perturb the otherwise regular protein-RNA interactions and fit poorly into the repeating structure of the RNA-binding site (Stubbs and Stauffacher, 1981; Holmes, 1982). Above pH 7.8, the high local concentration of negative charges on the triphosphate moiety of the "cap" structure, in concert with the destabilizing features mentioned above, may tip the balance of the assembly-disassembly equilibrium ("breathing") at the 5′ end in favor of a limited degree of disassembly. Sedimentation may further bias the equilibrium by removing the nucleoprotein rods from the pool of released coat protein subunits. However, TMV rods became translatable immediately after pH 8.0 treatment, before centrifugation and resuspension in water (data not shown). An "exposed" portion of the highly efficient leader sequence (Richards *et al.*, 1977), perhaps still lying in the "groove" formed by adjacent RNA binding sites on the new protein surface, might direct the first stage in the formation of a translation initiation complex.

The polypeptide products often appeared less heterogeneous in incubations in which pH 8-treated rods were used as templates (see Fig. 1). This may reflect the protective influence of translating partially encapsidated RNA *in vitro*, or the absence of impeding regions of secondary structure in the RNA template. Interestingly, short rods containing the subgenomic, I_2-RNA (Beachy and Zaitlin, 1977) for the 30K polypeptide, which are a natural feature of crude preparations of TMV (*vulgare* strain), do not seem to participate in the cotranslational disassembly process to the same extent as full-length rods containing the genomic RNA. This observation may be due to the particularly stable interaction between RNA and coat protein in the vicinity of the origin of assembly sequence (Goelet *et al.*, 1982).

Neither narcissus mosaic virus (NMV) nor papaya mosaic virus (PMV), two potexviruses kindly supplied by Dr. M. N. Short, stimulated detectable protein synthesis *in vitro*, despite extensive dialysis (7 and 3 Da, respectively) against 1 m*M* Tris-HCl, pH 8.2. This seems to rule out the possibility that helically packaged viral RNAs are uncoated nonspecifically by components in the rabbit reticulocyte lysate. Initially the NMV and PMV results were disappointing as they appeared to detract from the universality of the proposed mechanism of cotranslational disassembly and its possible significance *in vivo*. However, current thinking on the assembly of potexviruses in general, and

PMV in particular (AbouHaidar and Bancroft, 1978; AbouHaidar and Erickson, 1984), suggests that the 5′ end of the RNA is the site for initiation of monodirectional self-assembly, at least *in vitro*. Thus, although the RNA is "capped," the stabilizing influence of RNA-protein interactions at the 5′ end of the potexvirus rod is likely to be considerably greater than in the case of TMV. In this context, it is also notable that when nucleoprotein particles of the alkali-stable fraction of TMV (Perham, 1969), prepared without a nuclease "tailing" step, were added to the reticulocyte lysate, no stimulation of translation was detected. Many of these particles possess a fully exposed, intact 5′-leader sequence on their short, pendant RNA tails (Wilson and Glover, 1983). So it would appear that whatever structural feature impedes the polar disassembly of this subpopulation of rods at high pH also acts as a block to ribosome translocation *in vitro* (data not shown).

Hashimoto and Okamoto (1975) measured the binding of [^{3}H]Met-tRNA to 80 S ribosomes from wheat germ attached to TMV particles which had been partially disassembled in SDS. They concluded, rather controversially at the time, that the ribosomes were binding to the initiation codon for the internal coat protein cistron. However, in view of the revised polarity of disassembly in SDS (Wilson *et al.*, 1976), it seems probable that they too were measuring ribosome attachment to the 5′-leader sequence of the cistron(s) for the 126K, and the 183K readthrough (Pelham, 1978), polypeptide(s). Initiation complex formation was claimed (Hashimoto and Okamoto, 1975) to be 1.5 times more efficient than with naked TMV RNA (6.4 kb)—perhaps underlining the topological advantages in having a less extensively structured, convoluted, mRNA template for initiation of protein synthesis. No attempt was made in this early report (Hashimoto and Okamoto, 1975) to look at the polypeptide products.

Although it is not possible at this stage to comment on the physiological significance of the disassembly mechanism proposed in this paper, it is undeniably attractive. It would provide continuous protection against plant cell nucleases for the incoming, packaged TMV RNA template during the early stages of plant cell infection. If ribosome translocation were also to provide the energy for the bulk of the uncoating of the TMV rod then it would remove the risk to the virus of having its naked RNA exposed to the cellular *milieu* at times of limited ribosome availability. It may be that membrane phospholipids, local ionic strength, or pH conditions in the infected plant cell fulfill the role of removing the first few "unstable" coat protein subunits at the 5′ end. Several questions immediately arise. Why are newly assembled, progeny virus particles not continuously disassembled *in vivo?* How does disassembly proceed beyond the end of the 183K reading frame? And in strains of TMV which encapsidate "capped," subgenomic mRNA for the coat protein (LMC; Hunter *et al.*, 1976), do these short rodlets also disassemble cotranslationally?

The concept of translation driving the uncoating of TMV rods, and incidentally providing initial copies of the large, nonstructural 126K (and 183K) viral-coded polypeptides which are thought to be involved in viral RNA replication, offers an exciting hypothesis for future experiments.

Recent results from wheat germ cell-free extracts programmed with pH 8-treated TMV rods confirm and extend the significance of many of the results reported in this paper (T. M. A. Wilson, manuscript in preparation).

ACKNOWLEDGMENTS

I am indebted to Mr. P. Jones for virus preparation at the University of Liverpool. Dr. J. W. Davies, Dr. R. Hull, Dr. G. P. Lomonossoff, and Dr. A. J. Maule provided many helpful suggestions and constructive criticism throughout this work. Special thanks go to Miss Kitty Plaskitt for the electron micrographs, and to Muriel Hobbs and Anne Williams for typing the manuscript.

Note added in proof. A recent report [K. Tyc *et al.* (1984) *Eur. J. Biochem.* **140,** 503–511] confirms earlier observations that two 80 S ribosomes bind to the 5′-leader sequence of TMV RNA, one at the first AUG codon [positions 69–71 (Goelet *et al.*, 1982)] and the second at an "entry site" centered on one of the five (or six) AUU codons in the hypervariable 5'-leader sequence (Goelet *et al.*, 1982). Using the 5'-coterminal RNA fragments, Tyc and co-workers (1984) have located the second 80 S ribosome-binding site to the AUU sequence at positions 14-16. Similar results have been reported for other viral leader sequences including brome mosaic virus RNA 3 [P. Ahlquist *et al.* (1979) *Nature (London)* **281,** 277-282]; alfalfa mosaic virus RNA 3 [M. Pinck *et al.* (1981) *Nucl. Acids Res.* **9,** 1087-1099]; Rous sarcoma virus RNA [J. L. Darlix *et al.* (1982) *Nucl. Acids. Res.* **10,** 5183-5196]; and turnip yellow mosaic virus RNA [W. Filipowicz and A. L. Haenni (1979) *Proc. Natl. Acad. Sci. USA* **76,** 3111-3115].

These observations may affect the assumptions made in this paper regarding the length of TMV RNA leader sequence exposed at pH 8.0, and the initial uncoating events originally thought to be driven by 40 S ribosome subunits during "scanning" for the first AUG codon [M. Kozak (1983) *Microbiol. Rev.* **47,** 1-45; M. Kozak (1984) *Nucl. Acids Res.* **12,** 3873-3893].

REFERENCES

ABOUHAIDAR, M., and BANCROFT, J. B. (1978). The initiation of papaya mosaic virus assembly. *Virology* **90,** 54-59.

ABOUHAIDAR, M., and ERICKSON, J. W. (1984). Structure and assembly of papaya mosaic virus. *In* "Molecular Plant Virology" (J. W. Davies, ed.), Vol. I, in press. CRC Press, Boca Raton, Fla.

BEACHY, R. N., and ZAITLIN, M. (1977). Characterization and *in vitro* translation of the RNAs from less-than-full-length, virus-related, nucleoprotein rods present in tobacco mosaic virus preparations. *Virology* **81,** 160-169.

BLOWERS, L. E., and WILSON, T. M. A. (1981). The effect of urea on tobacco mosaic virus: Polarity of disassembly. *J. Gen. Virol.* **61,** 137-141.

BONNER, W. M., and LASKEY, R. A. (1974). A film detection method for tritium-labelled proteins and nucleic acids in polyacrylamide gels. *Eur. J. Biochem.* **46,** 83-88.

GOELET, P., LOMONOSSOFF, G. P., BUTLER, P. J. G., AKAM, M. E., GAIT, M. J., and KARN, J. (1982). Nucleotide sequence of tobacco mosaic virus RNA. *Proc. Natl. Acad. Sci. USA* **79,** 5818-5822.

HASHIMOTO, J., and OKAMOTO, K. (1975). Partially stripped tobacco mosaic virus-stimulated methionyl-tRNA binding to ribosomes in an *in vitro* wheat embryo system. *Virology* **67,** 107-113.

HELMS, K., and ZAITLIN, M. (1970). Enhancement of infectivity of tobacco mosaic virus particles partially uncoated by alkali. *Virology* **41,** 549-557.

HOLMES, K. C. (1982). Flexibility in tobacco mosaic virus. *CIBA Foundation Symp.* **93,** 116-138.

HUNTER, T. R., HUNT, T., KNOWLAND, J., and ZIMMERN, D. (1976). Messenger RNA for the coat protein of tobacco mosaic virus. *Nature (London)* **260,** 759-764.

KEITH, J., and FRAENKEL-CONRAT, H. (1975). Tobacco mosaic virus RNA carries 5'-terminal triphosphorylated guanosine blocked by 5'-linked 7-methylguanosine. *FEBS Lett.* **57,** 31-33.

KIHO, Y. (1970). Polysomes containing infecting viral genome in tobacco leaves infected with tobacco mosaic virus. *Japan. J. Microbiol.* **14,** 291-302.

KIHO, Y. (1972). Association of infecting tobacco mosaic virus ribonucleic acid with polysome. *Japan. J. Microbiol.* **16,** 152-154.

LAEMMLI, U. K. (1970). Cleavage of structural proteins during the assembly of the head of bacteriophage T4. *Nature (London)* **227,** 680-685.

LEBERMAN, R. (1966). The isolation of plant viruses by means of "simple" coacervates. *Virology* **30,** 341-347.

LOMONOSSOFF, G. P., and BUTLER, P. J. G. (1979). Location and encapsidation of the coat protein cistron of tobacco mosaic virus. *Eur. J. Biochem.* **93,** 157-164.

MACHIDA, H., and KIHO, Y. (1970). *In vivo* uncoating of tobacco mosaic virus. II. Complete uncoating to TMV-RNA. *Japan. J. Microbiol.* **14,** 441-449.

PELCHER, L. E., and HALASA, M. C. (1979). Factors influencing the production of intermediate particles during alkaline degradation of tobacco mosaic virus: Time, pH, salt concentration, and temperature. *J. Virol.* **29,** 431-437.

PELHAM, H. R. B. (1978). Leaky UAG termination codon in tobacco mosaic virus RNA. *Nature (London)* **272,** 469-471.

PELHAM, H. R. B., and JACKSON, R. J. (1976). An efficient mRNA dependent translation system from reticulocyte lysates. *Eur. J. Biochem.* **67,** 247-256.

PERHAM, R. N. (1969). Sucrose density-gradient analysis of the alkaline degradation of tobacco mosaic virus. *J. Mol. Biol.* **45,** 439-441.

PERHAM, R. N., and WILSON, T. M. A. (1976). The polarity of stripping of coat protein subunits from the RNA in tobacco mosaic virus under alkaline conditions. *FEBS Lett.* **62,** 11-15.

PERHAM, R. N., and WILSON, T. M. A. (1978). The characterization of intermediates formed during the disassembly of tobacco mosaic virus at alkaline pH. *Virology* **84,** 293-302.

RICHARDS, K., JONARD, G., GUILLEY, H., and KEITH, G. (1977). Leader sequence of 71 nucleotides devoid of G in tobacco mosaic virus RNA. *Nature (London)* **267,** 548-550.

SHAW, J. G. (1984). Early events in plant virus infections. *In* "Molecular Plant Virology" (J. W. Davies, ed.), Vol. II, in press. CRC Press, Boca Raton, Fla.

SPIRIN, A. S. (1969). The Second Sir Hans Krebs Lecture: Informosomes. *Eur. J. Biochem.* **10,** 20–35.

STUBBS, G., and STAUFFACHER, C. (1981). Structure of the RNA in tobacco mosaic virus. *J. Mol. Biol.* **152,** 387–396.

WILSON, T. M. A., and GLOVER, J. F. (1983). The origin of multiple polypeptides of molecular weight below 110,000 encoded by tobacco mosaic virus RNA in the messenger-dependent rabbit reticulocyte lysate. *Biochim. Biophys. Acta* **739,** 35–41.

WILSON, T. M. A., LOMONOSSOFF, G. P., and GLOVER, J. F. (1981). Dimethyl sulphoxide (DMSO) disassembles tobacco mosaic virus predominantly from the 5′-end of the viral RNA. *J. Gen. Virol.* **53,** 225–234.

WILSON, T. M. A., PERHAM, R. N., FINCH, J. T., and BUTLER, P. J. G. (1976). Polarity of the RNA in the tobacco mosaic virus particle and the direction of protein stripping in sodium dodecyl sulphate. *FEBS Lett.* **64,** 285–289.

ZIMMERN, D. (1975). The 5′-end group of tobacco mosaic virus RNA is $m^7G^{5'}$ $ppp^{5'}$ Gp. *Nucleic Acids Res.* **2,** 1189–1201.

Avoiding the Triple Flip and Two Twists in Cloning TMV

William O. Dawson
Department of Plant Pathology, Citrus Research and Education Center, 700 Experiment Station Road, Lake Alfred, FL 33850, USA

After becoming a member of the Cell Interaction Group at the University of California, Riverside, in 1972, Gary Jones, a friend and fellow member of the group, convinced me that definition of the functions of a virus would require a genetic approach. Thus began a saga that has moved rather unevenly since then. We began with the technology of the time, isolating temperature-sensitive mutants from a mutagenized population of virus. Although we obtained many, easily isolated, stable mutants of TMV and a few devilishly difficult mutants of cowpea chlorotic mottle virus (CCMV), the gains in understanding the biology of these viruses were substantially less than the original vision. The major lesson was that there was an inverse relationship between the simplicity of a viral genome and ability to use this genetic approach. These viruses had too few identifiable phenotypes from which we could select mutants. By the late seventies, it was clearly time for a change.

One option was to switch to a different virus, a larger virus with more phenotypes. After considering the available plant viruses at the time for characteristics needed for a genetic system, such as a usable local lesion assay host, ability to mechanically transmit the virus, and size of the genome, there was one leading candidate, tomato spotted wilt virus. This was a virus with a large genome and, in fact, had been the subject of prior genetic studies. This was one of the first viruses to show recombination (actually segment exchange).

Then, one of the seminal events of virology occurred. Taniguchi *et al.* (1978) did the first reverse-genetics of an RNA virus. They copied Qβ RNA into a cDNA and replicated it in a plasmid in *Escherichia coli*. This plasmid was able to infect bacterial cells with virus. This occurred at the dawn of recombinant DNA technology that would allow *in vitro* manipulation of the viral genome, followed by examination of any changes in phenotype. The power of this approach was immediately obvious.

This was where we needed to go. George Grantham and I were tired of working through about 40 *mCi* of ^{32}P every two weeks trying to characterize *ts* mutants. However, we were still committed to unfinished experiments and there was no one available to begin the new direction. I had hoped to get funds for a postdoctoral fellow to begin this new approach, but funds were short. In the fall of 1980, a recent Berkeley graduate, Mark Young, showed up looking for a job. After a long conversation, I decided that he could be the 'postdoc' on this project. He reminded me that he had not been to graduate school. I said "That's OK, I know nothing about recombinant DNA."

As I remember, Mark's first chore was to purify poly A polymerase from *E. coli*. This was before you could buy it, and commercial primers were yet to be invented. By spring, Andy Jackson allowed Mark and me to take my only sabbatical, which was two weeks in his laboratory to learn recombinant DNA technology. We learned about microfuge tubes and micropipettes, and probably saved a year compared to learning in our own laboratory. However, we still had a lot to learn before we were capable of developing a reverse-genetics system for one of our plant viruses. Mark left for the mountains of Tibet and later graduate school and was replaced by Kay Pauling and later Dave Beck, while George Grantham and I continued to characterize *ts* mutants.

As we progressed toward learning recombinant DNA techniques sufficiently to reach our objectives, one major problem had not been resolved. Once we had full-genomic cDNA clones of our virus, how were we going to infect plants? The cDNA of Qβ was infectious in the bacterial cells in which the plasmid grew, but the plasmid host was also the virus host. However, during this time, it was found that cloned DNA of cauliflower mosaic virus, a DNA pararetrovirus, could infect plants as could the cDNA of the potato spindle tuber viroid. Also, the cDNA of poliovirus infected animal cells. However, it was shown that the 5'-capped structure of TMV was required for infectivity. If the cDNAs were not directly infectious in plants, our plan was to move the cDNA clones into yeast to produce *in vivo* capped transcripts. Yet, because of this concern, we were attempting to clone three different viruses. David Beck was working with CCMV, David Knorr with southern bean mosaic virus, and George Grantham and I had begun working with TMV.

Another major concern was the reported error rate in the replication of RNA viruses and the possibility of populations with very little infectious RNA. Although, with animal and bacterial viruses, infectivity systems were available that would approach an infection per virion, with plant systems, the level of infection was many logs lower. Estimations of the number of virions required to induce a single local lesion were near a million. With work, some researchers had reduced that number one log. Did this mean that most of the virions of a very infectious virus like TMV are not infectious? Did this mean that most of our cDNA clones would be uninfectious? If we did everything correctly, would we have to test hundreds of cDNA clones to find one that was infectious? Thus, our strategy was to make multiple independent cDNA clones and to attempt to make each from a single RNA molecule.

Two major events from other laboratories greatly influenced our work during this time. The first was the Goelet *et al.* (1982) paper that reported the complete sequence of TMV. This was the second complete sequence of a plant virus and the first of an RNA plant virus. At the time, this was a considerable technical achievement, because the quality of the reverse transcriptase enzyme was substantially less than present preparations. Making cDNA of the entire genome of TMV was not trivial. The Cambridge strategy was to make and sequence a large series of small overlapping cDNAs. This was one of the first reports using synthetic primers. The first primer was a 7-mer made by the old phosphodiester chemistry that was a sufficiently major effort to convince them to rely on a random priming strategy for other cDNA clones. The complete sequence was derived from more than 400 cDNAs, in which each nucleotide was represented in an average of 13 different clones. Not only did this give the complete sequence, but it also was a measure of sequence variation in the population. They found two sequence variants at the 5' end, but few other variants of sequence throughout the rest of the genome. With this achievement, Philip Goelet left TMVology for greater challenges in neurobiology.

Before the Goelet *et al.* sequence, we aligned clones by restriction mapping with no idea which end was which. With the sequence available, we hoped that we would at least be able to align our clones. However, with the expected error rates of RNA viruses, we wondered how many restriction endonuclease sites would be common. One of the most amazing events of my career was learning that our sequence was essentially identical to the Goelet *et al.* (1982) sequence. To this date, we occasionally find a nucleotide or restriction site difference, but the limited variation among clones is about the same as the difference from the published sequence. The Goelet *et al.* (1982) sequence reported that the 5' end of the genome was a variable region with two different sequences. Professor Okada's laboratory pointed out that the "long variant" exactly corresponded to the tomato mosaic virus sequence (Meshi *et al.*, 1983), and all of our cDNA clones almost exactly matched the "short variant" reported by Goelet *et al.* (1982).

The second major event that greatly affected our work was the infectious cDNA clone of brome mosaic virus (BMV) from Paul Ahlquist's laboratory (Ahlquist *et al.*, 1984). This paper told us how we were going to infect plants with cDNA clones. They engineered a lambda phage promoter to exactly initiate transcription *in vitro* at the beginning of an inserted sequence (Ahlquist and Janda, 1984). Since they knew that *E. coli* RNA polymerase with this promoter and a dinucleotide cap analogue would produce a 5'-capped transcript (Contreras *et al.*, 1982), this allowed *in vitro* duplication of the original virion RNA. The good thing was that we now had a workable strategy for infecting plants once we obtained a full-genomic cDNA clone. The bad thing was that since Paul was so successful, we dropped all of our procedures and tried to duplicate his. We spent several months after the BMV paper came out watching TMV cDNA clones change in recombination-positive bacteria before we went back to our original methods that worked for TMV.

After our long lag phase, we then decided to put all of our efforts onto one virus. George Grantham and I were having pretty good success cloning TMV, so the two Daves joined forces with us. One of the first things that they did was to convince me that we essentially had all of TMV cloned if we were willing to put pieces together. Previously I thought we needed to obtain a full-genomic cDNA clone from a single RNA, but they convinced me to allow them to start piecing. A few days later I came to work and asked them how it was going and they said "pretty good, we have the whole genome assembled except the 5' terminus." For us that was a monumental decision, because to this day, it still is a triple flip with two twists to clone all of TMV into a plasmid in one piece. And best of all, it worked. Instead of only one infectious sequence in a million, most of our TMV cDNA clones have been infectious. Within a few months, we had an infectious cDNA clone of TMV, and as they say, the rest is history.

References

Ahlquist, P., French, R., Janda, M., and S. Loesch-Fries. (1984). Multicomponent RNA plant virus infection derived from cloned viral cDNA. *Proc. Natl. Acad. Sci. USA* **81**:7066-7070.

Ahlquist, P. and M. Janda. (1984). cDNA cloning and *in vitro* transcription of the complete brome mosaic virus genome. *Mol. Cell. Biol.* **4**:2876-2882.

Contreras, R., Cheroutre, H., Degrave, W., and W. Fiers. (1982). Simple, efficient *in vitro* synthesis of capped RNA useful for direct expression of cloned eukaryotic genes. *Nucleic Acids Res.* **10**:6353-6362.

Goelet, P., Lomonossoff, G. P., Butler, P. J. G., Akan, M. E., Gait, M. J., and J. Karn. (1982). Nucleotide sequence of tobacco mosaic virus RNA. *Proc. Natl. Acad. Sci. USA* **79**:5818-5822.

Meshi, T., Ishikawa, M., Takamatsu, N., Ohno, T., and Y. Okada. (1983). The 5'-terminal sequence of TMV RNA. Question on the polymorphism found in vulgare strain. *FEBS Lett.* **162**:282-285.

Taniguchi, T., Palmieri, M., and C. Weissmann. (1978). Qβ DNA-containing hybrid plasmids giving rise to Qβ phage formation in the bacterial host. *Nature* **274**:223-228.

Biographical Sketch

Bill Dawson received his Ph.D. at the University of Georgia in plant pathology with Cedric Kuhn, with a research emphasis on factors affecting the biosynthesis of cowpea chlorotic mottle virus. His postdoctoral training was at the University of California-Berkeley in Plant Pathology with David Schlegel, where they set up synchronized viral replication in leaves of intact plants. He moved through the professorial ranks from 1972-1992 at UC Riverside. In 1992 he was appointed the Graves Biotechnology Chair and Eminent Scholar at the University of Florida.

Proc. Natl. Acad. Sci. USA
Vol. 79, pp. 5818–5822, October 1982
Biochemistry

Nucleotide sequence of tobacco mosaic virus RNA

(oligonucleotide primers/cDNA library/RNA polymorphism/eukaryotic ribosome binding/leaky amber termination)

P. GOELET, G. P. LOMONOSSOFF*, P. J. G. BUTLER, M. E. AKAM†, M. J. GAIT, AND J. KARN

Medical Research Council Laboratory of Molecular Biology, Hills Road, Cambridge, CB2 2QH, England

Communicated by Sydney Brenner, June 18, 1982

ABSTRACT Oligonucleotide primers have been used to generate a cDNA library covering the entire tobacco mosaic virus (TMV) RNA sequence. Analysis of these clones has enabled us to complete the viral RNA sequence and to study its variability within a viral population. The positive strand coding sequence starts 69 nucleotides from the 5′ end with a reading frame for a protein of M_r 125,941 and terminates with UAG. Readthrough of this terminator would give rise to a protein of M_r 183,253. Overlapping the terminal five codons of this readthrough reading frame is a second reading frame coding for a protein of M_r 29,987. This gene terminates two nucleotides before the initiator codon of the coat protein gene. Potential signal sequences responsible for the capping and synthesis of the coat protein and M_r 29,987 protein mRNAs have been identified. Similar sequences within these reading frames may be used in the expression of sets of proteins that share COOH-terminal sequences.

The packaged single-stranded RNA of tobacco mosaic virus (TMV) *Vulgare* is an active mRNA for a protein of $M_r \approx 130{,}000$ and its readthrough product of $M_r \approx 165{,}000$ (1, 2) but fails to direct the synthesis of several other viral products, including the viral coat protein. These are expressed during infection from a set of subgenomic mRNAs. The mRNA for the coat protein (1) is known to contain the terminal 692 residues of the genomic RNA (3), and the mRNA for a M_r 30,000 protein (4) appears to contain the terminal 1,500 residues of TMV RNA. A number of other 3′-coterminal and colinear subgenomic mRNAs that are packaged have been detected by hybridization probes (5). In order to determine the full coding capacity of TMV and to study its mode of gene expression and the possible role of the RNA in virus assembly, we have completed the sequence of TMV RNA (3, 6).

cDNA cloning and sequence analysis

Strategy. Short overlapping cDNA fragments representing the entire TMV RNA sequence were cloned in bacteriophage M13 (7, 8), their sequences were determined by the dideoxy chain-termination method (7), and overlaps were determined by computer methods (9). TMV proved to be a poor template for the synthesis of cDNA longer than 2,000 nucleotides, and we found that the efficiency of the second-strand reaction was highly sequence dependent. These problems were circumvented by using a series of synthetic oligodeoxynucleotides to prime synthesis at either random or specific sites along the molecule. Efficient cloning of molecules that were rendered double-stranded was achieved by cleaving the synthetic DNA with restriction endonucleases. The use of several restriction enzymes ensured that overlapping sequences were cloned.

Priming with Synthetic Oligonucleotides. Mixtures of four- to seven-residue oligonucleotides, synthesized by phosphodiester chemistry (10), were used as nonspecific primers on TMV RNA or on cDNA. Double-stranded cDNA to most of TMV genome could be synthesized by using these primer "cocktails." To direct the synthesis of double-stranded cDNA to the termini and other poorly sampled regions of TMV RNA, oligonucleotide primers of 13 to 17 residues were synthesized by the solid-phase phosphotriester method (11). These included an oligonucleotide [d(TGGGCCCCATCCG)] complementary to the 3′-terminal 13 nucleotides (3) and a 15-residue oligonucleotide [d(CTCGCTTATTACTTT)] complementary to a sequence 222–236 nucleotides from the 5′ terminus (6). Double-stranded cDNA to the 5′ end was prepared by back priming with an oligonucleotide [d(GTATTTTTACAACAATT)] corresponding to the 5′-terminal 17 nucleotides of TMV RNA (6). This was inserted into the *Bam*HI site of M13 MP7 (8) after blunt-end ligation to *Bam*HI linkers (7). The oligonucleotides d(GTCAACTTCCAAAGATT) and d(CTGAATACCCTCT) (complementary to nucleotides 1,813–1,829 and 3,159–3,172) were synthesized to obtain clones to the 5′ sides of regions of TMV RNA well represented by cDNA clones from previous "shotgun" experiments. cDNA priming was with 10- to 100-fold molar excess of oligonucleotide over TMV RNA template (usually 5 μg of TMV RNA in a 25-μl reaction mixture) and standard incubation conditions for reverse transcription were used (12): 42°C and 60 min, with a 30°C and 15-min preincubation for the short oligonucleotide cocktails.

Second-Strand Synthesis, Cloning, and Assembly of the Sequence. cDNA freed from RNA by alkaline hydrolysis (100 mM NaOH, 1 mM EDTA for 15 min at 70°C) was used as a template for second-strand synthesis primed by "flip back" or added oligonucleotide primers. The standard reaction used Klenow DNA polymerase and incubation was in 10 mM Tris·HCl, pH 7.4/10 mM $MgCl_2$/10 mM dithiothreitol/100 mM NaCl/50 μM each deoxynucleoside triphosphate at 37°C for 30 min. In some experiments no attempt was made to purify the cDNA. Oligonucleotides generated in the first-strand reaction were used to prime second-strand synthesis on TMV cDNA by using the conditions of Wickens *et al.* (12) or the conditions described above after melting and annealing desalted products of the first-strand reaction (alkali treatment of the first-strand reaction product was shown by these approaches not to cause deamination of cytidine residues). These double-stranded cDNAs were digested by restriction endonucleases (*Sau*3a, *Taq* I, *Msp* I, *Hae* III, *Alu* I, *Rsa* I, *Tha* I, *Hin*fI, *Eco*RI, *Bgl* II, and *Hin*dIII) and inserted into appropriately linearized M13 vector replicative form DNAs (8) by direct ligation or with *Bam*HI linkers. Recombinant molecules were isolated by transfection of *Escherichia coli* JM101 and detected by *lac* complementation assay

The publication costs of this article were defrayed in part by page charge payment. This article must therefore be hereby marked "*advertisement*" in accordance with 18 U. S. C. §1734 solely to indicate this fact.

Abbreviation: TMV, tobacco mosaic virus.

* Present address: John Innes Institute, Colney Lane, Norwich, NR4 7UH, England.

† Present address: Dept. of Genetics, Univ. of Cambridge, Downing St., Cambridge, CB2 3EH, England.

(8) and plaque hybridization to ^{32}P-labeled TMV cDNA and kinase-labeled TMV RNA oligonucleotides (5). The recombinants were this plaque purified and grown up in liquid culture for storage in phage stocks. Template preparation and DNA sequence analysis were as described by Sanger et al. (7). Inserts containing an internal restriction site were treated as multiple ligations and were entered into the data base (9) as separate events. As the sequence approached completion, overlapping recombinants were selected by directed cDNA priming (see above), specific restriction enzyme digestions, and plaque hybridization to M13 clones representing the termini of large blocks of sequence.

Organization of the TMV genome

TMV Strains Are Polymorphic. The sequence presented in Fig. 1 is based on the sequences of more than 400 independently derived cDNA clones. Each nucleotide was represented on average in 13 independently isolated clones, including ones oriented in both the positive and negative senses. Only the 5'-terminal capped G and the last eight nucleotides were not determined by this method. At certain positions in the sequence, notably at the 5' end, polymorphisms in the sampled viral population were detected (Fig. 1). Most of the changes involve the third position of codons and so do not alter the amino acid sequence. The 5' end of TMV RNA was represented by two substantially different variants, which differ in length by three nucleotides. The amino acid sequences coded by these two variants differ at only four residues, and these changes are conservative—Glu to Asp at 108 and 177 and Thr to Ser at 93 and 129. These sequence polymorphisms cannot be due to errors in the *in vitro* reverse transcription of TMV RNA, because the changes observed consistently maintain the phase and nature of the coded amino acid sequence and clones with the altered sequences were reproducibly obtained at a level of 40–60%, suggesting a stability of the population with no selection for either variant.

Comparisons between the complete sequence presented above and partial sequences of TMV RNA obtained in other laboratories by direct RNA sequencing methods reveal similar patterns of sequence variability. The 3'-terminal 1,000 nucleotides of TMV RNA obtained by Guilley et al. (3) differ from our sequence at 14 residues, without altering the coded amino acid sequence. The published 5'-end sequence of TMV RNA (6) is most closely related to the shorter variant we present in Fig. 1, but 13 positions within the coding sequence are altered, leading to a few conservative amino acid substitutions. Within related stocks of TMV over a period of 14 years, a number of variations have become established around the nucleation region. A significant alteration is the substitution of a C (residue 5,481) for a U (3, 13) within the small loop of the origin of assembly hairpin sequence. These two kinds of variability reflect the low fidelity of RNA-dependent RNA replication (14).

TMV Has Only Three Open Reading Frames. Coding sequences on the infectious positive strand consist of only three major reading frames. The first methionine begins an open reading frame for a protein of M_r 125,941, which terminates at an amber codon at residue 3,417. An in-phase coding sequence extends beyond this terminator for another 1,497 nucleotides, allowing readthrough synthesis of a protein of M_r 183,253. These sequence data are consistent with the observations that cell-free translation of TMV leads to the synthesis of a protein of $M_r \approx 130{,}000$ and its readthrough product of $M_r \approx 165{,}000$, whose translation may be enhanced by addition of an amber suppressor tRNA (2). The end of the reading frame for the M_r 183,253 protein overlaps by five codons with a reading frame, in a second phase, that codes for a protein of M_r 29,987 [the M_r 30,000 protein described by Beachy and Zaitlin (4)], which terminates with a UAA codon two nucleotides before the initiator codon for the coat protein (3). The complementary sequence (negative strand) shows only two regions that could potentially code for proteins of $M_r \approx 15{,}000$, but these appear unlikely to correspond to expressed genes because they lack appropriate signal sequences for mRNA synthesis (see below) and because proteins of these sizes have not yet been detected in tissues infected by TMV.

Ribosome Binding and the 5' Ends of Subgenomic mRNAs. Protein synthesis in eukaryotes is initiated almost invariably at the AUG closest to the 5' end of a mRNA and does not reinitiate internally. The 5' noncoding regions of eukaryotic mRNAs tend to be low in G, and purines are preferred at positions −3 and +4 surrounding the initiator codon, the canonical sequence being $^{A}_{G}$---AUGG (15). In TMV and other polycistronic plant viral RNAs (16, 17) only the first AUG codon is active. The proposed initiation codon for the M_r 125,941 protein and its readthrough product, at residue 69, follows a sequence lacking G residues and shows purines in the preferred positions.

Synthesis of a separate coat protein mRNA and other subgenomic mRNAs during TMV infection exposes initiation codons and "ribosome binding sites" at the 5' end of these mRNAs. The coat protein mRNA begins with a capped G (3) corresponding to residue 5,703 of the genomic sequence. Between this G and the coat protein initiator codon at residue 5,712, which shows appropriate context nucleotides, is a short U+A-rich sequence. A similar U+A-rich sequence is seen near the presumed initiator codon for the M_r 29,987 protein (residue 4,903, Fig. 1). By analogy to the coat protein mRNA the G at residue 4,895 should be capped in the M_r 29,987 protein mRNA. These U+A-rich sequences, which we call Butler boxes, are not common in eukaryotic messengers in general but appear to be a regular feature of subgenomic mRNA encoded by plant viruses, and are similar to the U+A-rich sequences typically found at the 5' ends of their genomic RNAs [(16, 17); see legend to Fig. 1].

COOH-Terminally Overlapped Proteins. Hybridization and translation experiments have demonstrated that in addition to the coat protein and M_r 29,987 protein mRNAs of at least five other 3'-coterminal mRNAs coding for proteins of M_r between ≈20,000 and ≈90,000 are synthesized during infection. These mRNAs are packaged, because their sequences extend beyond the origin of assembly (13). T. Hunter, R. J. Jackson, and D. Zimmern (personal communication) have analyzed a set of proteins of M_r 30,000, 29,000, and 23,000 that are synthesized *in vitro* in response to packaged subgenomic mRNAs. Comparison of their peptide maps with our sequence data shows that these proteins have overlapping COOH-terminal sequences and initiate at methionine codons within the M_r 29,987 gene at the sites indicated in Fig. 1. At both sites the initiator methionine codon is preceded by the sequence GACAAAA. This sequence is likely to correspond to the 5' ends of the mRNAs specifying these proteins, but shows little homology to the Butler box.

We have examined the entire TMV sequence for other possible sites of internal initiation by looking for potential ribosome-binding sites closely preceded by either a Butler box or a sequence related to the flanking sequences for the M_r 27,875 and 19,478 proteins. The most promising of these are indicated in Fig. 1. Further sequence evidence at the protein level and on the subgenomic mRNAs is clearly required before the use of these putative sites in protein synthesis is established, but two independent observations suggest that at least some of these sites are active. Stimulation of TMV readthrough translation by coinjection of TMV RNA and suppressor tRNAs into *Xenopus* oocytes (ref. 18 and E. Kubli and M. Bienz, personal com-

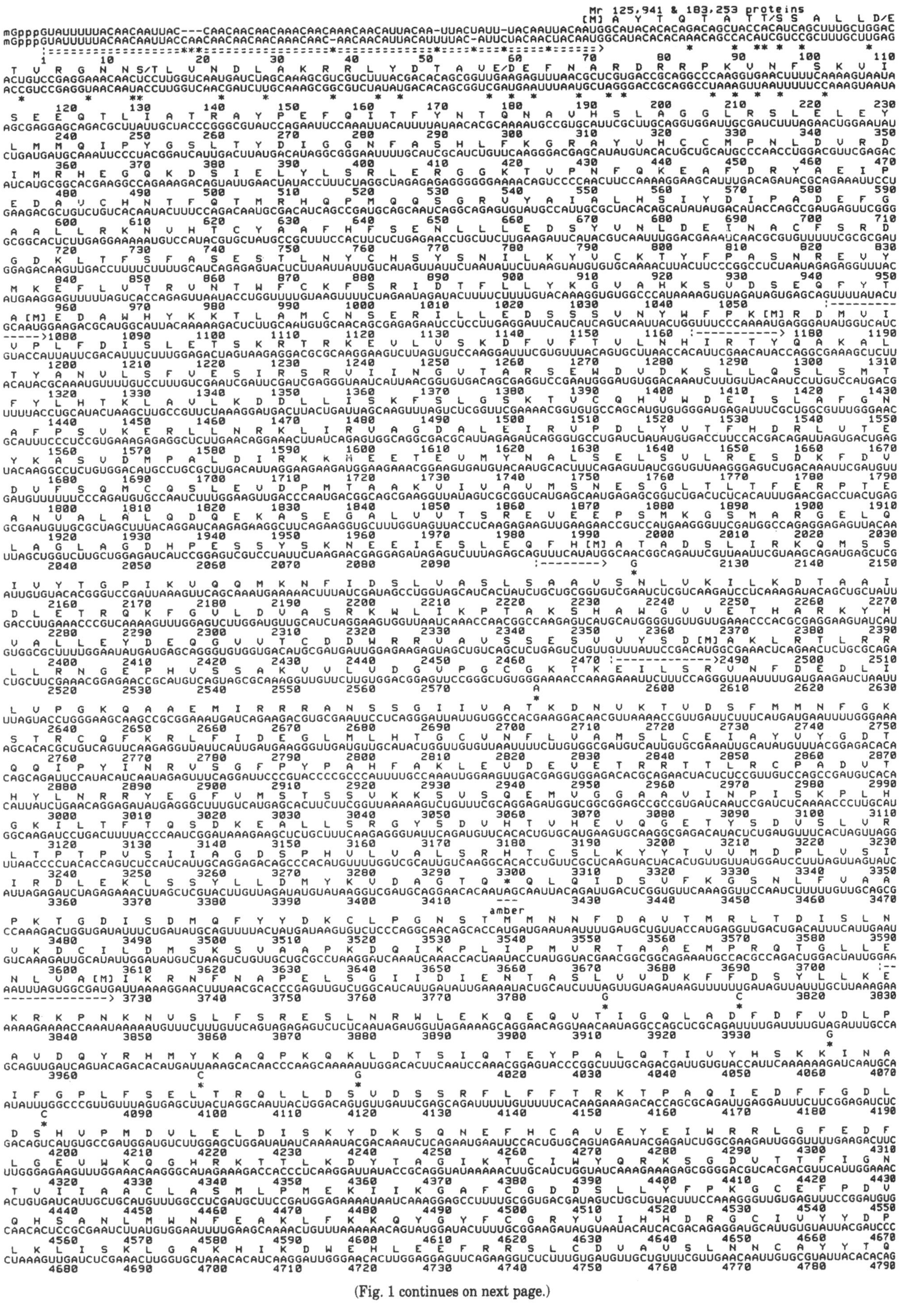

(Fig. 1 continues on next page.)

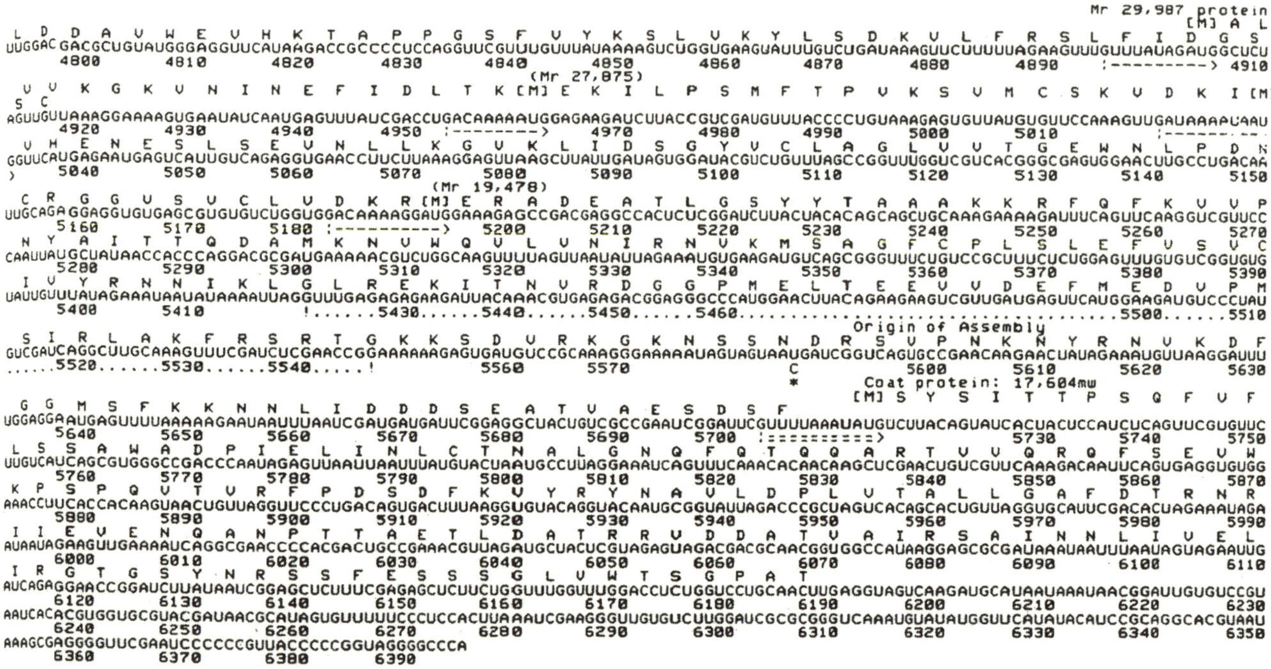

FIG. 1. Nucleotide and coding sequences for TMV. The standard one-letter amino acid code is used. Variants in the nucleotide sequence are shown on separate lines and are highlighted by an underlying *, and variations in the coding sequence are indicated in the overlying amino acid sequence—e.g., T/S at residue 93. At the 5′ end two variant sequences, whose lengths differ by three nucleotides in the noncoding sequence, are aligned with padding characters (−) in order to maximize their homologies. Residue numbers have been set to the uppermost variant because it is most similar to the sequence obtained for the 5′ end by Jonard *et al.* (6). The origin of assembly [defined by molar protection (13)] is marked by ·····; its hairpin loop center (13) is at residue 5,478. The leaky amber terminator (marked by an *) is underlined at residue 3,417. Known starts to mRNAs are underlined by :====>; potential starts by :--->. Initiator methionine residues are placed in brackets, [M]. Molecular weights of known proteins are presented above the initiator methionine; molecular weights of identified overlapped proteins are enclosed in brackets. The molecular weights of proteins corresponding to the potential starts are: residue 1,061, 87,438; residue 1,163, 83,467; residue 2,102, 48,417; residue 2,472, 34,554; residue 3,708, 46,218; and residue 5,020, 25,284. Note the similarities between the indicated U+A-rich sequences at the 5′ ends of the subgenomic mRNAs ("Butler boxes"), the 5′ end of TMV RNA, and the following plant viral RNA sequences.

Brome mosaic virus RNA 3

mGpppGUAAAAUACCAACUAAUUCUCGUUCGAUUCCGGCGAACAUUCUAUUUUACCAACAUCGGUUUUUUCAGUAGUGAUACUGUU UUUGUUCCCGAUGUCU.

Brome mosaic virus RNA 4 (mRNA for coat protein gene, which is silent in RNA 3 of this virus)

mGpppGUAUUAAUAAUGUCG.

Alfalfa mosaic virus RNA 3

mGpppGUUUUCAUCUUACACACGCUUGUGCAAGAUAGUUAAUCAUUCCAAUUCAACUCAAUUAACGUUUUUUACAGUGUAAUUCG UACUUUUCGUAAGUAAGUUUCUGUAAAAGCGUUUCUGUUUUAAUUUGGCCUAACACGUAAUUCGUACUCUUCGUGAGUAAGUUG UGUUAGCCAUACCUAUCCUUUAAAUUCCUGUCAAUUUAAAAAGAAAAUCAUUCCCAUUCGCGUAAUUCGUACUCUUCGUGAGUA AGUUGCAAAUGGAG.

Alfalfa mosaic virus RNA 4 (mRNA for coat protein gene, which is silent in RNA 3 of this virus)

mGpppGUUUUUAUUUUUAAUUUUCUUUCAAAUACUUCCAUCAUGAGU.

munication) induces a minor polypeptide of M_r 140,000. This would be consistent with initiation of protein synthesis at nucleotide 1,074 or 1,173 within the M_r 125,941 reading frame, giving rise to a protein of M_r either 87,438 or 83,467 and its readthrough product of M_r either 144,750 or 140,779. Partial proteolytic digestion of [^{35}S]fMet-tRNA-labeled translation products of packaged TMV RNA has shown minor NH_2-terminal peptides corresponding to the M_r 29,987 protein family and an NH_2-terminal peptide corresponding to a previously unidentified protein of M_r 45,000 (ref. 19, and H. R. B. Pelham personal communcation).

The Leaky Amber Terminator. Both amber suppressor tRNAs and wild-type tyrosine tRNAs from a number of species do promote readthrough translation of M_r 183,253 protein (18). It seems likely that $tRNA^{Tyr}$ is responsible for the 2–5% leaky termination of the TMV M_r 125,941 protein seen in infected tobacco and in cell-free systems. We have found no evidence for unusual secondary structure in the vicinity of the amber terminator, and this codon is not surrounded by nucleotides that could form additional base pairs with nucleotides surrounding the $tRNA^{Tyr}$ anticodon (M. Bienz, E. Kubli, and M. Altwegg, personal communication). Eukaryotic terminators show context nucleotide preferences (20, 21) that are met by the sequences surrounding the strong terminators of TMV but not by the leaky UAG terminator. It is followed by C, the most rarely used adjoining nucleotide (21). The leaky UAG terminator of the murine leukemia virus *gag* gene (22), does, however, obey these context rules. The use of amber codons to permit readthrough in RNA viruses contrasts with their relatively infrequent use in cellular messages (21) and may indicate that UAG is poorly recognized by eukaryotic release factor, sometimes allowing suppression by wild-type tRNAs, as found for UAG in *E. coli* (23).

RNA Packaging. The bidirectional assembly of TMV RNA with a coat protein disk preparation occurs in a "quantized" fashion on the 5′ side of the origin of assembly (24) but continuously on the 3′ side (25). One-dimensional Fourier analysis was used to detect any repeats around 50 or 100 bases apart that could be responsible for this pattern of assembly. There are no significant repeats (i.e., at the 0.5% level) at either 50- or 100-nucleotide intervals of single bases, purine- or pyrimidine-rich regions, or $^{G}_{A}$ di- or trinucleotides. In contrast, because of the amino acid compositions of the coded proteins, there is a strong tendency for G to be repeated with a period of 3 as the first codon base within each reading frame. Moreover, hairpin loops do not occur at intervals of 50 or 100 bases. The banding pattern seen during elongation in the 5′ direction with a disk preparation is therefore due not to any feature of the sequence but probably to the package size of protein added.

Comments

The TMV genome is composed of three closely packed open reading frames but the total coding capacity of the virus is substantially higher. Besides a sharing of NH_2-terminal sequences (2), there are also COOH-terminally overlapped proteins (T. Hunter, R. J. Jackson, and D. Zimmern, personal communication) involving sets of subgenomic mRNAs. The genomic and subgenomic RNAs of TMV are 3′-coterminal and colinear, and during infection complements to these RNAs are synthesized (5). We have suggested that incomplete transcription of the viral RNA into negative strands produces templates for the synthesis of the positive strand mRNAs. Butler box sequences found at the 5′ end of TMV RNA and the coat protein and M_r 29,987 protein genes, and at several other internal sites, may be signal sequences in this process. Many plant viruses are known to have similar readthrough products of their early genes (26) and it remains to be seen whether they also share the mechanism now described for generating COOH-terminally overlapped proteins.

The sequencing techniques employed in this work have sampled the viral population and revealed polymorphisms in the TMV sequence. The 5′ noncoding region shows the greatest variation, but no variants have been detected in the 3′ noncoding region. They must therefore be under different selective pressures, and this may be relevant to models of TMV replication.

One striking feature is the apparent lack of features associated with the virus assembly, with the exception of the origin of assembly. Thus there is no sign of sequence repeats that match with structural features in the virus, but rather the packaging appears to be determined by the protein alone, independent of the RNA sequence, allowing selection for other functions to dominate the genome.

Our cloning has generated a fully characterized phage collection that can now be used as a source of single-stranded DNAs for hybridization experiments and directed mutagenesis.

We thank Sydney Brenner, Don Northcote, and Leslie Barnett for their support and encouragement; Hans Matthes, Mohinder Singh, and Rodger Staden for assistance with oligonucleotide synthesis and computer methods; David Zimmern, Tony Hunter, Mariann Bienz, Erick Kubli, Tim Hunt, and Hugh Pelham for helpful discussions; and Panos Antoniou and Prospero Bennedetto for food for thought.

1. Hunter, T. R., Hunt, T., Knowland, J. & Zimmern, D. (1976) *Nature (London)* **260**, 759–764.
2. Pelham, H. R. B. (1978) *Nature (London)* **272**, 469–471.
3. Guilley, H., Jonard, G., Kukla, B. & Richards, K. E. (1979) *Nucleic Acids Res.* **6**, 1287–1307.
4. Beachy, R. N. & Zaitlin, M. (1977) *Virology* **81**, 160–169.
5. Goelet, P. & Karn, J. (1982) *J. Mol. Biol.* **154**, 541–550.
6. Jonard, G., Richards, K. E., Mohier, E. & Gerlinger, P. (1978) *Eur. J. Biochem.* **84**, 521–531.
7. Sanger, F., Coulson, A. R., Barrell, B. G., Smith, A. J. H. & Roe, B. A. (1980) *J. Mol. Biol.* **143**, 161–178.
8. Messing, J., Crea, R. & Seeberg, P. H. (1980) *Nucleic Acids Res.* **9**, 309–321.
9. Staden, R. (1980) *Nucleic Acids Res.* **8**, 3673–3694.
10. Gait, M. J. & Sheppard, R. C. (1977) *Nucleic Acids Res.* **4**, 1135–1158.
11. Duckworth, M. L., Gait, M. J., Goelet, P., Hong, G. F., Singh, M. & Titmas, R. C. (1981) *Nucleic Acids Res.* **7**, 1691–1706.
12. Wickens, M. P., Buell, G. N. & Schimke, R. T. (1978) *J. Biol. Chem.* **253**, 2483–2495.
13. Zimmern, D. (1977) *Cell* **11**, 463–482.
14. Domingo, E., Sabo, M., Taniguchi, T. & Weissmann, C. (1978) *Cell* **13**, 735–744.
15. Kozak, M. (1981) *Nucleic Acids Res.* **9**, 5233–5252.
16. Ahlquist, P., Luckow, V. & Kaesberg, P. (1981) *J. Mol. Biol.* **153**, 23–38.
17. Koper-Zwarthoff, E. C., Brederode, F. T., Veeneman, G., van Boom, J. H. & Bol, J. F. (1980) *Nucleic Acids Res.* **8**, 5635–5647.
18. Bienz, M. & Kubli, E. (1981) *Nature (London)* **294**, 188–190.
19. Pelham, H. R. B. (1979) *FEBS Lett.* **100**, 195–199.
20. Beaudet, A. L. & Caskey, C. T. (1971) *Proc. Natl. Acad. Sci. USA* **68**, 619–624.
21. Kohli, J. & Grosjean, H. (1981) *Mol. Gen. Genet.* **182**, 430–439.
22. Schinnick, T. M., Lerner, R. A. & Sutcliffe, J. G. (1981) *Nature (London)* **293**, 543–548.
23. Engelberg-Kulka, H. (1981) *Nucleic Acids Res.* **9**, 983–991.
24. Butler, P. J. G. & Lomonossoff, G. P. (1978) *J. Mol. Biol.* **126**, 877–882.
25. Lomonossoff, G. P. & Butler, P. J. G. (1980) *FEBS Lett.* **113**, 271–274.
26. Van Tol, R. G. L., Van Gemeren, R. & Van Vloten-Doting, L. (1980) *FEBS Lett.* **118**, 67–71.

Proc. Natl. Acad. Sci. USA
Vol. 83, pp. 1832–1836, March 1986
Genetics

cDNA cloning of the complete genome of tobacco mosaic virus and production of infectious transcripts

(*in vitro* transcription/sequence variation/population homology)

WILLIAM O. DAWSON, DAVID L. BECK, DAVID A. KNORR, AND GEORGE L. GRANTHAM

Department of Plant Pathology, University of California, Riverside, CA 92521

Communicated by Myron K. Brakke, November 18, 1985

ABSTRACT The entire genome of tobacco mosaic virus (TMV) was copied into a series of subgenomic cDNA clones. cDNA sequences of the 5′ and 3′ ends of TMV were cloned separately. A synthetic oligonucleotide primer was used to generate a *Pst* I site at the 5′ terminus, whereas a different primer was used to generate an *Nde* I site at the 3′ terminus. This strategy permitted removal of non-TMV sequences from cloned cDNA inserts by treatment with exonuclease VII following restriction endonuclease cleavage. *Pst* I linkers were added to TMV 3′ terminal cDNAs. Subgenomic cDNA fragments were ligated together into several independent full-genomic constructions from which TMV cDNA sequences could be cleanly excised as a single fragment by *Pst* I digestion. Full-genomic TMV cDNA was ligated immediately downstream from the λ phage promoter from pPM1 and transcribed *in vitro* with *Escherichia coli* RNA polymerase. RNA transcripts from three of four full-genomic cDNA constructions were infectious, even though they contained 6 non-TMV nucleotides at the 3′ end. Transcripts from a construction with 6 extra nucleotides at the 5′ end also were infectious. Progeny virus from plants infected with cDNA transcripts appeared identical to the parental virus. Restriction maps of independent cDNA clones of the same regions of the genome were identical to each other and as predicted from the reported nucleotide sequence of TMV. Also, sequences of the 200 nucleotides proximal to the 5′ termini of four independent cDNA clones were identical to each other and to published sequences, suggesting that independent isolates of TMV may have remarkably similar sequences.

Tobacco mosaic virus (TMV) is a virus with worldwide distribution that infects ≈200 plant species (1). Numerous strains of TMV cause serious losses in tobacco, tomato, and other crop plants. The virion is a rod-shaped particle 18 × 300 nm. The genome of TMV resides in a single strand of messenger-sense RNA encoding at least four proteins, two of which are translated from the same reading frame and two from processed mRNA fragments.

Historically, TMV has contributed to the understanding of the genetics of RNA viruses, including the first studies on biological variability, mutability and protein sequence variation of viral genomes (2–4). TMV continues to offer numerous advantages as a genetic system. Variants of this virus express a large number of easily scored phenotypic characters useful in studying specific viral functions. Isolates of TMV that vary in symptomology, host range, elicitation of host defense mechanisms, cross-protection, encapsidation, cell-to-cell and long-distance movement within the host, and several functions of RNA replication have been described. TMV strains infecting tobacco and tomato have been most intensely investigated. Of these, two viral strains and a variant have been fully sequenced (5–7). However, in spite of our understanding of the general organization and replication of this virus, only a few viral functions have been assigned to specific sequences. The inability to make site-specific mutations and to map random mutations in the RNA genome makes traditional genetic studies of limited use. Also, studies on the variation of any phenotypic character are hampered by the inability to obtain a uniform population of virus.

To overcome these limitations inherent in the genetic systems of RNA viruses, we have artificially extended the life cycle of TMV to include a double-stranded DNA phase. This approach has been utilized for several RNA pathogens, including bacteriophage Qβ (8), poliovirus (9, 10), potato spindle tuber viroid (11), and brome mosaic virus [BMV (12)]. In the DNA form, viral sequences can be precisely altered by using recombinant DNA technology. The entire genome, even that of variants carrying lethal mutations, can be maintained in bacterial plasmids and later transcribed into full-genomic length RNAs, thus providing a uniformity of inoculum not previously available. In addition, viral functions may be examined at cellular, cultured tissue, and whole plant levels in several hosts. These hosts, especially tobacco, are also readily transformed by means of Ti plasmid derivatives, and whole plants with inserted viral sequences can be easily regenerated. Therefore, functions of isolated genes or gene products may be examined in either cells or whole plants.

We report here on the construction of a library of subgenomic cDNA clones of TMV and from these the construction of functional full-genomic clones that can be easily manipulated. Transcripts of the full-genomic cDNAs ligated behind a promoter were infectious when inoculated onto test plants. Thus, an appropriate system for assigning functions to specific TMV sequences has been established.

MATERIALS AND METHODS

cDNA Synthesis. TMV RNA (strain U1) was extracted from purified virions with phenol/sodium lauryl sulfate. The RNA then was treated with methylmercuric hydroxide (13) and extracted with butanol/cetyltrimethylammonium bromide (14). The 3′ end of TMV RNA was polyadenylylated to an average length of 30 adenosines per 3′ end by the procedure of Sippel (15). First-strand cDNA synthesis was essentially by the procedure of Murray *et al.* (16) using 4 mM pyrophosphate, oligo$(dT)_{12-18}$ or 3′-specific synthetic primer, 5′-d(CATATGGGCCCC-OH) (Beckman Research Institute, Duarte, CA), no RNase inhibitor, and avian myoblastosis virus reverse transcriptase (Molecular Genetic Resources, Tampa, FL). Second-strand synthesis utilized DNA polymerase I (0.1 unit/μl) and RNase A (0.2 pg/μl) in 50 mM Tris·HCl (pH 7.2), 10 mM $MgSO_4$, 2 mM dithiothreitol, 80 μM dNTPs, and bovine serum albumin at 50 ng/μl in a total

The publication costs of this article were defrayed in part by page charge payment. This article must therefore be hereby marked "*advertisement*" in accordance with 18 U.S.C. §1734 solely to indicate this fact.

Abbreviations: TMV, tobacco mosaic virus; BMV, brome mosaic virus; ExoVII, exonuclease VII.

volume of 100 μl with 1–5 μg of RNA·DNA hybrid from the first-strand reaction. The reaction mixture was incubated 16 hr at 12°C. To select for cDNA to the 5′ end of the TMV genome, RNA was alkaline hydrolyzed after first-strand cDNA synthesis, and second-strand synthesis with Klenow fragment of DNA polymerase I (17) was primed with a 5′-specific synthetic primer, 5′-d(CTGCAGTATTTTTA-OH) (Beckman Research Institute). Double-stranded cDNA was blunt-end ligated to 5′-phosphorylated *Pst* I or *Bam*HI linkers (Pharmacia) followed by ligation into the *Pst* I or *Bam*HI site of pBR322. The recombinant plasmid was transformed into competent cells of *Escherichia coli* strain DHI prepared by the procedure of Hanahan (18). Nucleases were purchased from Bethesda Research Laboratories, Pharmacia, and International Biotechnologies.

***In Vitro* Transcription.** The transcription reaction mixture contained 4 μg of *Pst* I-digested plasmid DNA, 40 mM Tris·HCl (pH 7.9), 10 mM $MgCl_2$, 150 mM KCl, 0.1 mM EDTA, 2 mM dithiothreitol, 0.05% bovine serum albumin, 150 μM (each) ATP, CTP, and UTP, 150 μM m^7GpppG (Pharmacia), 100 units of RNasin (Promega Biotec, Madison, WI), and 1 unit of *E. coli* RNA polymerase (Promega Biotec) in a total volume of 92.5 μl. After 3 min at 37°C, GTP was added to 30 μM and after 15 min GTP was increased to 150 μM, in a final reaction volume of 100 μl. After 45 min an equal volume of cold RNA assay buffer (1% sodium pyrophosphate, pH 9/1% bentonite/1% Celite) was added, and pinto bean (*Phaseolus vulgaris* L.) and tobacco (*Nicotiana tabacum* L. var. Xanthi) leaves were immediately inoculated with the mixture.

Sequencing. DNA was sequenced by the method of Maxam and Gilbert (19). The 3′ terminal nucleotide sequence of RNAs was determined by direct RNA sequencing (20, 21). The sequence of the region adjacent to the 5′ end of TMV genomic RNA was determined by the dideoxy chain-terminating method (22) using reverse transcriptase and a synthetic primer that initiated cDNA synthesis at nucleotide 44 of TMV.

RESULTS

Construction of a Complete cDNA Copy of the TMV Genome. A series of subgenomic cDNA clones that collectively represent the entire genome of TMV was produced. An objective was to clone numerous independent cDNAs of each region of the viral genome so that independent full-genomic constructions could be made without being dependent on any one subgenomic cDNA sequence. For convenience in ligating the subgenomic clones into a complete copy of the TMV genome, only inserts ≈2 kilobases or larger were utilized. From cDNA produced by oligo(dT) priming, multiple independent clones were obtained of all regions of the virus genome except 200 nucleotides at the 5′ end. Fig. 1 shows

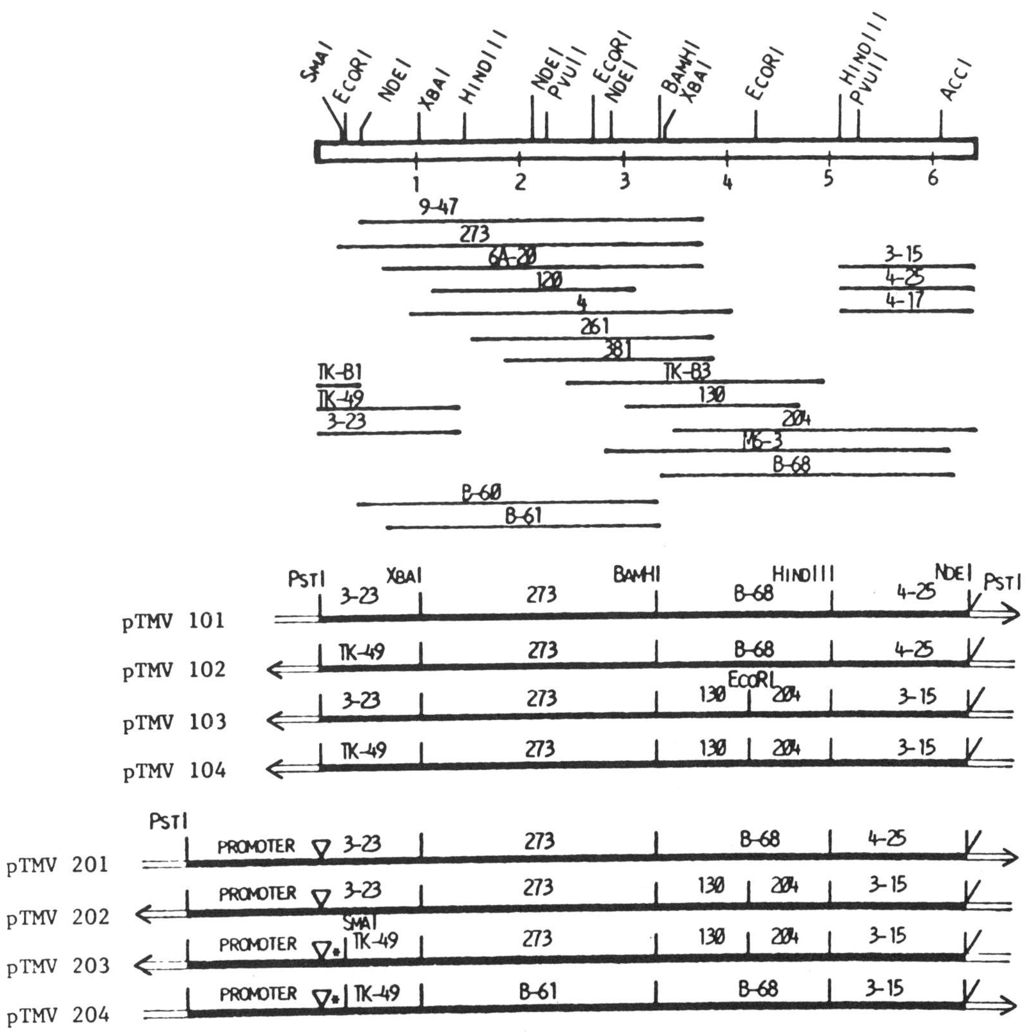

FIG. 1. Maps of TMV cDNA clones. Upper bar shows a partial restriction map of the complete cDNA of the TMV genome, with numbers under the bar indicating each kilobase. Subgenomic cDNA sequences are shown as fragments relative to the map above. Full-genomic cDNA clones are shown in pBR322 (unfilled lines) with arrows at the *Pst* I site proximal to the *Eco*RI site to show orientation. ▽, Junction of promoter to 5′ TMV sequence shown in Fig. 2*D*. *, Clone 3-23.

several of the characterized cDNA inserts that subsequently were used to construct full-genomic cDNA inserts.

Another objective was to produce a full-genomic construction that could be excised totally from plasmid vectors and contained no additional sequences. Thus, cDNA at both ends had to be engineered such that non-TMV sequences could be removed. 5′-end cDNA clones were designed with an adjacent *Pst* I site (which does not occur in the TMV sequence) so that after digestion with *Pst* I and exonuclease VII (ExoVII) (or other single-strand-specific exonuclease) only TMV sequences remained (Fig. 2*A*). To produce these clones, second-strand cDNA synthesis was primed with the synthetic 5′-specific oligomer, and the *Pst* I (1) to *Hin*dIII [nucleotide 1446 of Goelet *et al.* (5)] cDNA fragments were ligated into *Pst* I–*Hin*dIII-digested pUC19, resulting in clones TK-49 and 3-23 (Fig. 1).

cDNA clones containing the 3′ terminal sequence of TMV RNA were made by priming first-strand synthesis with the 3′-specific synthetic oligomer, resulting in an *Nde* I site at the 3′ end (Fig. 2*B*). However, this site is not unique because there are three other *Nde* I sites in the 5′ half of the cDNA

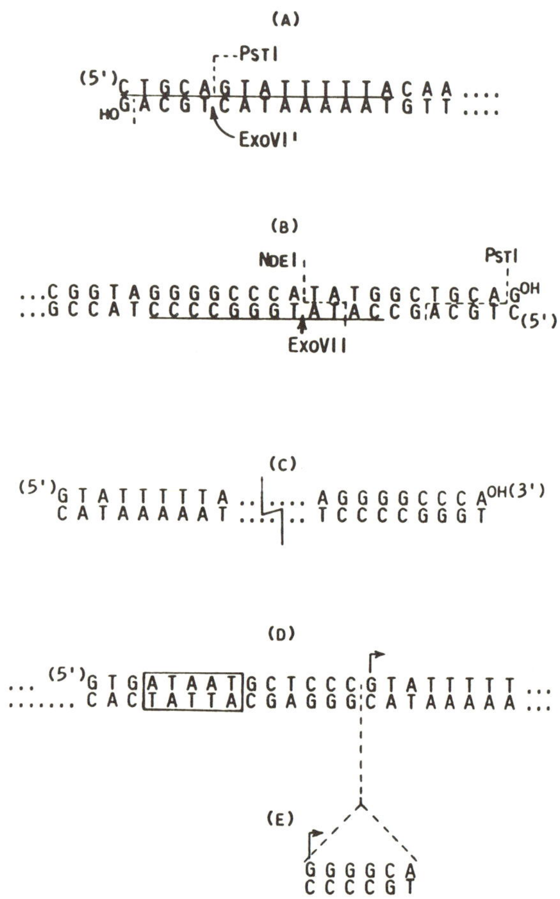

FIG. 2. Sequences of TMV cDNA constructs. In each construct the upper sequence represents the genomic-sense strand. In *A* and *B*, restriction endonuclease cleavage sites are indicated by dashed lines, and arrows indicate the point to which a protruding strand is removed by ExoVII. The bent arrows in *D* and *E* indicate the first nucleotide synthesized during transcription. (*A* and *B*) Sequence of the 5′ end (*A*) and 3′ end (*B*) with the synthetic primers underlined. (*C*) Nucleotide sequences produced at the termini when the 5′ *Pst* I–ExoVII-, *Bam*HI-digested fragment of pTMV103 is ligated *in vitro* at the *Bam*HI site with the 3′ *Bam*HI-, *Nde* I–ExoVII-digested 3′ fragment. (*D*) Sequence of pTMV201, pTMV202, pTMV203, and pTMV204 at the promoter–TMV junction. Box indicates *E. coli* promoter consensus sequence (*E*). Additional nucleotides of the insertion mutant pTMV205 are as indicated by dashed lines.

genome of TMV. Therefore, a *Pst* I linker was added after second-strand synthesis. *Hin*dIII (5080)–*Pst* I (6400 + 6) 3′ fragments of cDNA were ligated into *Pst* I–*Hin*dIII-digested pUC18, resulting in clones 3-15, 4-25, and 4-17 (Fig. 1). A full-length genomic construct with the engineered 5′ and 3′ ends could then be excised as a single fragment by *Pst* I.

Several full-length constructions of TMV cDNA were made by ligating together cloned cDNA restriction fragments representing all regions of the genome. Examples are plasmids pTMV101, pTMV102, pTMV103, and pTMV104 in Fig. 1. Constructions were made progressively by first inserting into pUC18 a cDNA containing the 5′ end to the *Bam*HI (3332) site and inserting into pUC19 a cDNA containing the *Bam*HI (3332) site to 3′ end. These two inserts, excised by digesting the plasmid DNA with *Pst* I and *Bam*HI, were then ligated together into the *Pst* I site of pBR322 (23).

Constructs of cDNA also were produced that contain only the sequence of TMV (Fig. 2*C*). 5′ cDNA fragments [*Pst* I–ExoVII-treated 5′ end (1) to *Bam*HI (3332)] were ligated *in vitro* to 3′ cDNA fragments [*Bam*HI (3332) to *Nde* I–ExoVII-treated 3′ end (6400)].

Restriction Map and Sequence. cDNA inserts depicted in Fig. 1 have been mapped by using 21 different restriction endonucleases. Restriction maps of each of these cDNA clones and others that we have characterized conform exactly to the restriction map predicted from the nucleotide sequence of TMV published by Goelet *et al.* (5). Additionally, we have determined the nucleotide sequence of four independent cDNA clones through the 5′ "variable region" (5), which occurs within the first 221 nucleotides adjacent to the 5′ terminus. The nucleotide sequences (data not shown) were identical to each other through this area and exactly identical to the "short" variant reported by Goelet *et al.* (5).

Infectivity. The major objective for producing full-genomic cDNA clones of TMV was to artificially extend the life cycle of TMV through a DNA phase. This requires that plants be infected with TMV in some manner utilizing the cDNA. Since the cDNAs of poliovirus (9, 10) and potato spindle tuber viroid (11) were infectious, we tested the infectivity of TMV cDNA. Plants were inoculated with the excised cDNA of the TMV genome and with uncut plasmids containing the complete TMV cDNA, both in double-stranded and single-stranded forms and both with and without the λ promoter described below. None became infected, including plants inoculated with cDNA whose transcripts are shown to be infectious in the next section (Table 1). Also, total RNA from *E. coli* cells with plasmids containing cDNA of the complete

Table 1. Assays of infectivity of TMV cDNA

Sample	Lesions per eight pinto bean leaves	Infected Xanthi/total
pTMV204		
Pst I-cut	0	0/5
Uncut	0	0/5
pTMV101		
Pst I-cut	0	0/5
Uncut	0	0/5
pTMV103		
Pst I-cut	0	0/5
Uncut	0	0/5
In vitro-ligated complete TMV cDNA	0	0/5
Total RNA of *E. coli* with pTMV204	0	0/5
TMV RNA, 1 ng/μl	728	5/5

Five micrograms of DNA in 100 μl was assayed by mechanical inoculation of pinto bean and Xanthi tobacco.

TMV genome with or without the λ promoter was not infectious.

Ahlquist *et al.* (12) infected plants with BMV RNA produced *in vitro* using the phage λ P_R promoter to direct transcription of BMV cDNA by *E. coli* RNA polymerase. The promoter was engineered into plasmid pPM1 so that transcription of any DNA ligated into the *Sma* I site initiates with the first purine of the inserted sequence (24). We utilized the promoter construction in pPM1 (a gift from Paul Ahlquist) to direct synthesis of an RNA transcript of TMV cDNA, beginning with the first base of the TMV sequence. To add the pPM1 promoter sequence to TMV cDNA, a 270-base-pair fragment, prepared from clone 3-23 by blunt-ending the 5′ *Pst* I site with ExoVII followed by *Eco*RI cleavage, was ligated into *Sma* I-*Eco*RI-digested pPM1 (Fig. 2*D*). From this construct, a 1.2-kilobase fragment containing the entire promoter region (1010 nucleotides) and the 5′ 256 nucleotides of TMV cDNA was removed by cleavage with *Pst* I and *Xcy* I. This fragment was then ligated to *Xcy* I (256)–*Pst* I (6400 + 6) TMV cDNA constructs and inserted into the *Pst* I site of pBR322, giving the recombinant plasmids pTMV201, pTMV202, pTMV203, and pTMV204 (Fig. 1).

Plasmids pTMV201–pTMV204 were cut with *Pst* I to yield two fragments (full-genomic cDNA of TMV ligated to the λ promoter and linear pBR322), which then were transcribed with *E. coli* RNA polymerase in the presence of m^7GpppG. Contreras *et al.* (25) have shown that synthetic RNA cap structures can be incorporated efficiently during transcription by *E. coli* RNA polymerase. Full-length transcripts should contain the exact sequence of TMV plus 6 additional nucleotides added to the 3′ end of the RNA. Transcripts of three of the four different constructions tested were infectious (Table 2).

TMV cDNA with no additional 3′ nucleotides was prepared by *in vitro* ligation of the *Pst* I (−1010) to *Bam*HI (3332) λ promoter-containing fragment from pTMV204 with the 3′ terminal *Bam*HI (3332) to *Nde* I-ExoVII (6400)-treated fragment from pTMV204. *In vitro* transcription of this cDNA also yielded infectious TMV RNA.

Several controls demonstrated that infectious RNA resulted from the transcription reaction and not from contamination. Infectivity of TMV cDNA transcripts resulted only when a synthetic cap structure was included in the transcription reaction. Use of the unmethylated cap analogue GpppG resulted in almost as many lesions as use of m^7GpppG (Table 2). RNase treatment destroyed infectivity after, but not before, the transcription reaction. DNase treatment before, but not after, transcription destroyed infectivity. As shown above, transcripts terminating at the *Pst* I site, which extended the 3′ terminus by 6 nucleotides, were infectious. The cDNA clones linearized with either *Acc* I, to give a transcript shortened by 346 nucleotides, or *Eco*RV, to give a transcript 962 nucleotides longer than TMV, were not infectious (Table 2).

We could not compare the specific infectivity of the *in vitro* transcripts to that of parental RNA because the size of *in vitro* transcripts was not homogeneous (data not presented), making quantification of yield of full-length TMV transcripts difficult. However, transcripts of pTMV201, pTMV203, and pTMV204 reproducibly infected plants (Table 2), with transcripts of pTMV204 exhibiting the highest infectivity. In 12 different experiments, pTMV204 transcripts infected every plant inoculated and gave >200 lesions per pinto bean leaf in several experiments.

Transcripts of pTMV202 were not infectious in six different experiments in which transcripts from other constructions were infectious. pTMV202 differs from pTMV203 only in nucleotides 256–1446 but shares nucleotides 256–3332 with pTMV201 (Fig. 1). Therefore, the alteration that reduces or destroys the infectivity of transcripts of pTMV202 is not known.

Table 2. Infectivity of transcripts from TMV cDNA produced under different conditions

Sample	Lesions per eight pinto bean leaves	Infected Xanthi/total
Experiment 1		
pTMV201	4	5/6
pTMV202	0	0/6
pTMV203	7	1/6
pTMV204	320	6/6
+ GpppG	227	6/6
Acc I-cut, not *Pst* I-cut	0	0/6
*Eco*RV-cut, not *Pst* I-cut	0	0/6
pTMV101	0	0/6
pTMV104	0	0/6
TMV RNA, 100 pg/μl	73	6/6
Experiment 2		
pTMV204		
Complete	88	
Without cap	0	
0 time reaction	0	
RNase A, 0.12 μg/ml		
Before transcription	23	
After transcription	0	
DNase, 0.6 μg/ml		
Before transcription	0	
After transcription	31	
Experiment 3		
In vitro-ligated complete sequence (≈500 ng)	4	
pTMV204 (500 ng)	6	
Experiment 4		
pTMV204	612	
pTMV205	2	

Unless stated otherwise, 4 μg of *Pst* I-cut plasmid was incubated with the complete transcription reaction mixture including m^7GpppG.

Characteristics of the Progeny Virus. TMV RNA produced *in vitro* caused systemic mosaic symptoms on Xanthi tobacco leaves and local lesions on pinto bean, Xanthi-nc tobacco, and *Nicotiana glutinosa* leaves. All of these symptoms were identical to those caused by the parental virus. Virus transferred from primary local lesions induced by the transcripts produced symptoms identical to those caused by parental virus in each of these hosts. No differences were observed when progeny virus isolated from plants infected with TMV cDNA transcripts was compared to the parental virus in relation to host range and symptoms on 12 different plant species (data not presented).

Progeny virus isolated from plants infected by cDNA transcripts appeared identical to the original parental virus by electron microscopy and serology in double-diffusion tests. RNA isolated from the progeny virus was identical to that of the parental by RNA transfer hybridization analysis. The 3′ terminal nucleotide sequence of RNA isolated from progeny virus of pTMV203 and pTMV204 was identical to that of the parental virus RNA (data not shown). No differences were observed in specific infectivities of virions or virion RNA isolated from plants infected either with *in vitro* transcripts or with parental virus (data not presented).

Addition of Nucleotides to the 5′ End. Constructions were made that caused 6 extra nucleotides to be inserted at the 5′ end of TMV transcripts. pTMV205 was constructed with the same sequence as pTMV204 with the insertion of 6 extra nucleotides between the λ promoter from pPM1 and the 5′

terminus of TMV cDNA (Fig. 2*E*). Transcripts of pTMV205 were infectious at a low level (Table 2). Biological characteristics of the progeny virus were identical to those of the parental virus. The nucleotide sequence at the 5′ end of viral RNA isolated from plants infected with pTMV205 transcripts did not contain the additional nucleotides but was identical to that of the parental virus RNA (data not presented).

DISCUSSION

We have constructed several independent cDNA clones of the complete genome of TMV from a library of subgenomic cDNA clones. Our strategy was designed to produce full-genomic constructions that could be excised cleanly as single fragments, allowing movement to other vectors or the addition of promoters to allow transcription of the entire genome. Infectious transcripts were produced from three of four full-genomic TMV cDNA constructions.

The published sequence of TMV (5) greatly facilitated the characterization of clones and the design of primers. However, because of the suggested sequence variation in the virus population (5) and because the relationship between the isolate sequenced and ours was unknown, we did not know whether our isolate would conform to the published sequence. Yet, each characterized cDNA clone exactly matched the predicted restriction map. Also, the nucleotide sequence of the 5′ variable region of four of the cDNA clones corresponded exactly to the "short variant" of Goelet *et al.* (5). Since the nucleotide sequence of the "long variant" of Goelet *et al.* (5) corresponds to the sequence for the tomato strain of TMV (6) and our independent isolate of TMV is so similar to the published sequences, the 5′ region of TMV appears not to be polymorphic.

A concern as we undertook this project was heterogeneity of the original virus population. Since estimates of the number of virions required to initiate an infection have been reported to be far in excess of one (26), it was not possible to estimate the proportion of infectious RNA molecules in the initial population. Thus, there was the possibility of cloning a noninfectious RNA. In addition, the frequency of variants in a TMV population is extremely high (2, 27), further increasing the possibility of generating cDNAs from other than wild-type starting material. For these reasons we produced multiple, independent cDNA clones of each region of the viral genome to increase the probability of generating a composite full-length construct corresponding to an infectious parental genotype. This approach also permits a survey of the original viral RNA population by comparison of different cDNA clones of the same region. Three of four full-genomic constructions allowed production of infectious transcripts, and all of the cDNA clones that were sequenced and/or extensively analyzed with restriction endonucleases fit a single map, implying that most of the RNAs from the original virus population were similar and infectious.

Although it was possible to produce an infectious RNA transcript consisting of the exact TMV sequence without additional nucleotides, this proved to be unnecessary and inconvenient. Transcripts with additional bases at the 3′ end also were infectious, and virus isolated from infected plants appeared to be identical to the parental U1 TMV, both physically and biologically. However, addition of extra nucleotides to the 5′ terminus of the transcripts reduced infectivity. Yet, after replication, the progeny virus regained levels of infectivity equal to that of the parental virus, and the extra nucleotides were not retained. This extends earlier observations with poliovirus (9) and BMV (12) demonstrating the ability of enzymes involved in replication of RNA viral genomes to recognize and/or replicate only viral sequences.

The principal objective of this work was to extend the life cycle of TMV through a DNA phase and thereby generate an experimental system for extending studies of the molecular genetics of TMV. Accomplishment of this objective now allows accurate mapping of viral functions, production of homogeneous virus populations, and confirmation of the biological activity of viral cDNAs prior to transformation into plant hosts. Viral sequences from diverse sources may be recombined to create unique genomes for studying regulation and biological functions of viral genes. Such studies may provide new strategies for controlling plant viruses.

We thank A. O. Jackson for generously inviting us into his laboratory and introducing us to recombinant DNA techniques, Kay Pauling and Mark Young for efforts on the initial stages of this project, Carol Boyd for technical assistance, BettyAnn Merrill for typing, and Gary Jones, Noel Keen, Andrew Jackson, and Paul Ahlquist for useful comments on the manuscript. This research was funded in part by Grants PCM 7921664 from the National Science Foundation and 85-CRCR-1-1795 from the U.S. Department of Agriculture.

1. Zaitlin, M. & Israel, H. W. (1975) *Descriptions of Plant Viruses* (Commonwealth Mycological Institute/Association of Applied Biologists, Kew, U.K.), No. 151.
2. Kunkel, L. O. (1947) *Annu. Rev. Microbiol.* **1,** 85–100.
3. Mundry, K. W. & Gierer, A. (1958) *Z. Vererbungsl.* **89,** 614–630.
4. Wittman, H. G. (1960) *Virology* **12,** 609–612.
5. Goelet, P., Lomonossoff, P., Butler, P. J. G., Akam, M. E., Gait, M. J. & Karn, J. (1982) *Proc. Natl. Acad. Sci. USA* **79,** 5818–5822.
6. Meshi, T., Ishikawa, M., Takamatsu, N., Ohno, T. & Okada, Y. (1983) *FEBS Lett.* **162,** 282–285.
7. Nishiguchi, M., Kituchi, S., Kiho, Y., Ohno, T., Meshi, T. & Okada, Y. (1985) *Nucleic Acids Res.* **13,** 5585–5590.
8. Taniguchi, T., Palmieri, M. & Weissmann, C. (1978) *Nature (London)* **274,** 223–228.
9. Racaniello, V. R. & Baltimore, D. (1981) *Science* **214,** 916–919.
10. Semler, B. L., Dorner, A. J. & Wimmer, E. (1984) *Nucleic Acids Res.* **12,** 5123–5144.
11. Cress, D. E., Kiefer, M. C. & Owens, R. A. (1983) *Nucleic Acids Res.* **11,** 6821–6835.
12. Ahlquist, P., French, R., Janda, M. & Loesch-Fries, S. (1984) *Proc. Natl. Acad. Sci. USA* **81,** 7066–7070.
13. Maniatis, T., Fritsch, E. F. & Sambrook, J. (1982) *Molecular Cloning: A Laboratory Manual* (Cold Spring Harbor Laboratory, Cold Spring Harbor, NY).
14. Langridge, J., Langridge, P. & Bergquist, P. L. (1980) *Anal. Biochem.* **103,** 264–271.
15. Sippel, A. E. (1973) *Eur. J. Biochem.* **37,** 31–40.
16. Murray, M. G., Hoffman, L. M. & Jarvis, N. P. (1983) *Plant Mol. Biol.* **2,** 75–84.
17. Bothwell, A. L., Paskind, M., Imonishi-Kari, T., Rajewsky, K. & Baltimore, D. (1981) *Cell* **24,** 625–637.
18. Hanahan, D. (1983) *J. Mol. Biol.* **66,** 557–580.
19. Maxam, A. & Gilbert, W. (1980) *Methods Enzymol.* **65,** 497–559.
20. England, T. & Uhlenbeck, O. (1978) *Nature (London)* **275,** 560–561.
21. Donnis-Keller, H. (1980) *Nucleic Acids Res.* **8,** 3133–3142.
22. Zimmern, D. & Kaesberg, P. (1978) *Proc. Natl. Acad. Sci. USA* **75,** 4257–4261.
23. Crouse, G. F., Frischauf, A. & Lebrach, H. (1983) *Methods Enzymol.* **101,** 78–89.
24. Ahlquist, P. & Janda, M. (1984) *Mol. Cell. Biol.* **4,** 2876–2882.
25. Contreras, R., Cheroutre, H., Degrave, W. & Fiers, W. (1982) *Nucleic Acids Res.* **10,** 6353–6362.
26. Steere, R. (1955) *Phytopathology* **45,** 196–208.
27. Dawson, W. O. & Jones, G. E. (1976) *Mol. Gen. Genet.* **145,** 307–309.

Coat Protein Mediated Resistance

Roger N. Beachy
Donald Danforth Plant Science Center and Department of Biology, Washington University, St. Louis, Missouri

Control of virus diseases has historically been accomplished in a limited number of crops by deploying genes for resistance (e.g., the N locus of tobacco; the Tm1, Tm2 resistance loci in tomato). In other cases disease control requires chemical control of insect vectors, and physical removal of infected plants, including crop and weed hosts. A limited number of crops are protected from severe virus infections by pre-infecting the plants with mild strains of the problem virus, a practice known as cross protection. (A commentary on cross protection by Robert Fulton is provided in this volume.) Practical applications of cross protection are limited because it is difficult to identify, or generate, mild strains that provide a sufficient level of protection while having acceptable affects on yield. Similarly, the few examples of characterized single genes for virus resistance have limited the success of plant breeding for virus resistance.

As a graduate student in the early 1970s in the laboratory of the late Harry H. Murakishi, Department of Botany and Plant Pathology, Michigan State University, I became interested in the nature of disease resistance and cross protection for control of virus diseases. My Ph.D. thesis project involved developing cell cultures of tobacco plants that contain the N-gene, demonstrating that such cultures produced necrotic lesions upon infection by TMV and were useful for studies of virus replication (Beachy and Murakishi, 1971a). These studies expanded my interests in virology, and led me to apply for a postdoctoral position with Milton Zaitlin to study the genome of TMV and how its genes were expressed.

During this period, studies of the mechanisms of cross protection to control TMV were being conducted by Robert Fulton and colleagues. These workers suggested that cross protection was caused when coat protein (CP) from the protecting strain interfered with infection by a second, related strain, perhaps by blocking its disassembly upon infection. Zaitlin (1976) used a TMV mutant that did not produce soluble coat protein and demonstrated cross protection; it was concluded that re-assembly of the challenge virus by the protecting strain was not required. These studies raised the possibility that cross protection was caused by an effect of viral RNA [either (+)-sense or (-)-sense RNA sequences] rather than CP. These and other studies of cross protection would be instrumental in our subsequent studies of transgenic plants.

During the late '70s and early '80s, researchers at several locations were conducting experiments to determine how the soil microbe *Agrobacterium tumefaciens* transferred tumor-inducing DNA to its hosts. It was correctly anticipated that these studies would eventually enable scientists to develop plants that contain foreign gene sequences. As a member of the faculty in the Department of Biology at Washington University, St. Louis, I was fortunate to interact with Mary Dell Chilton, who was developing methods to transfer genes to plant cells. Josef (Jeff) Schell and Marc van Montague from the University of Ghent (Belgium) were also working to transform plant cells. At Monsanto Company, St. Louis, MO, a new group in plant biology under the direction of Ernest Jaworski was established, with a goal to perfect procedures to regenerate cultured cells to plants, and to develop 'genetically engineered plants'.

In 1981 I proposed a research project to determine whether cross protection against TMV could be mimicked in transgenic plants that carried genes encoding virus coat protein (CP) or viral RNA. With financial support from Monsanto Company and in collaboration with Monsanto scientists Robert T. Fraley, Steve G. Rogers, and Robert Horsch, in 1982 we established a small team to prepare cloned cDNAs that contained segments of the TMV genome. Having received a preprint of the Goelet *et al.* (1982) paper (described in this volume) that provided the sequence of TMV RNA, we began to develop cloned cDNAs for the CP, the 30 kDa movement protein, and fragments of the replicase. The team at Monsanto developed gene expression cassettes into which cDNAs were ligated, and applied *Agrobacterium*-mediated transformation to produce transgenic tobacco plants that contained the new gene constructs. Plants were returned to Washington University for molecular characterization and challenge inoculation with TMV and other tobamoviruses.

The first sets of transgenic plants (produced in 1983-84) contained genes that encoded (+) sense RNA and would produce CP, and others that produced the analogous (-) sense sequences. The genes were under control of the 19S promoter from cauliflower mosaic virus (CaMV), considered at that time to give higher levels of gene expression than other promoters. However, these plants contained disappointingly low levels of gene transcript and no CP was detected by western blot assays. By early 1985 a set of transgenic plants carrying the CP gene that contained the more active CaMV 35S promoter were tested, and by mid-year we had gathered convincing evidence of gene expression and accumulation of CP. By fall, Patricia Powell Abel, a graduate student, and Rick Nelson, a postdoc, completed convincing studies which showed that R0 parental plants, as well as R1 progeny plants, that contained CP were more resistant to systemic infection by TMV than progeny that did not produce CP. Furthermore, plant lines that contained higher levels of CP were more resistant than those with lower levels of CP. These observations were published in 1986 in the *Science* article reproduced here. The phenomenon was initially referred to as 'genetically engineered cross protection' but was later renamed 'coat protein mediated resistance' (CP-MR) to more accurately describe the gene and mode of action.

Other research groups, some working in parallel with our group and others subsequent to us, tested the prediction that a similar approach would work to develop CP-MR in other plants and against other viruses. Studies showed that the TMV coat protein gave resistance to TMV in tomato (Nelson *et al*., 1988), and CP genes from alfalfa mosaic virus (Loesch-Fries *et al*., 1987; Tumer *et al*., 1987) and potato virus X (Hemenway *et al* 1988) provided resistance to the respective viruses in tobacco plants. Subsequently, other researchers attempted to develop resistance to viruses in other taxonomic groups with varying success and, in some cases, a notable lack of success. With time, it became clear that there are several strategies for developing resistant plants, some of which are mediated by protein and others mediated by RNA. A landmark paper by Bill Dougherty and colleagues (Lindbo and Dougherty, 1992) demonstrated RNA-mediated resistance to tobacco etch potyvirus (i. e., protein synthesis is not required for resistance), an observation that can arguably be described as having led to a greater understanding of RNA-mediated suppression of gene expression in plants (Baulcombe, 1996).

Since 1986 our group has largely retained its focus and interest in describing the mechanisms and implications of CP-MR in the model TMV-tobacco system, with the goal to increase the efficacy and breadth of resistance. We recently demonstrated that mutants of the CP that create stronger CP-CP interactions than does wild type CP confer greater levels of CP-MR than wild type CP. Mutants of CP that have low levels of CP-CP interaction confer little or no resistance (Bendahame *et al*., 1997). Our current model proposes that increasing H-bonds between adjacent CP molecules raises the level of CP-MR, regardless of whether or not the CP can package with viral RNA. The model also proposes that increasing other interactions, for example those formed by hydrophobic interactions, will provide broad protection against distantly related tobamoviruses. These studies also revealed that certain mutants of CP can reduce virus replication and local spread of infection, and suggest that CP can have a variety of regulatory and functional roles in replication.

It has been personally gratifying to see the generally positive response that the discovery of coat protein mediated resistance has had on plant pathology, virology and biotechnology. Examples of CP-MR, including the results of field studies, continue to be reported in the literature. It also has been 'interesting' to receive the criticisms of both scientists and non-scientists who either oppose the use of such technologies in agriculture, or are critical of how its discovery was commercialized. Nevertheless, CP-MR and other forms of pathogen derived resistance, have found their way into commercial agriculture: squash plants that carry multiple coat protein genes are resistant to as many as four different viruses; transgenic papaya plants that carry the CP gene for papaya ringspot virus have restored papaya production to certain areas of Hawaii; and potato plants with CP-MR to potato virus X, and other examples are expected to be commercialized in the next several years. All in all, CP-MR is a good example of how knowledge gathered through many years of biological observations, molecular biology, biotechnology, and plant breeding can be employed to benefit agriculture.

References

Baulcombe, D. C. (1996). Mechanisms of pathogen-derived resistance to viruses in transgenic plants. *Plant Cell* **8**:1833-1844.

Beachy, R. N. and H. H. Murakishi. (1971a). Local lesion formation in tobacco tissue culture. *Phytopathology* **61**:877-878.

Bendahmane, M., Fitchen, J. H., Zhang, G., and R. N. Beachy. (1997). Studies of coat protein mediated resistance to TMV: correlation between capacity of mutant coat protein assembly and CP-MR. *J. Virol.* **71**:7942-7950.

Goelet, P., Lomonossoff, G. P., Butler, P. J. G., Akam, M. E., Gait, M. J., and J. Karn. (1982). Nucleotide sequence of tobacco mosaic virus RNA. *Proc. Natl. Acad. Sci. USA* **79**:5818-5822.

Hemenway, C., Fang, R. F., Kaniewski, W. K., Chua, N. H., and N. E. Tumer. (1988). Analysis of the mechanism of protection in transgenic plants expressing the potato virus X coat protein or its antisense RNA. *EMBO J.* **7**:1273-1280.

Lindbo, J. A. and W. G. Dougherty. (1992). Pathogen-derived resistance to a potyvirus: Immune and resistant phenotypes in transgenic tobacco expressing altered forms of a potyvirus coat protein nucleotide sequence. *Mol. Plant-Microbe Interact.* **5**:144-153.

Loesch-Fries, L. S., Merlo, D., Zinnen, T., Burhop, L., Hill, K., Krahn, K., Jarvis, N., Nelson, S., and E. Halk. (1987). Expression of alfalfa mosaic virus RNA 4 in transgenic plants confers virus resistance. *EMBO J.* **6**:1845-1851.

Nelson, R. S., McCormick, S. M., Delannay, X., Dube, P., Layton, J., Anderson, E. J., Kaniewska, M., Proksch, R. K., Horsch, R. B., Rogers, S. G., Fraley, R. T., and R. N. Beachy. (1988). Virus tolerance, plant growth, and field performance of transgenic tomato plants expressing coat protein from tobacco mosaic virus. *Biotechnology* **6**:403-409.

Tumer, N. E., O'Connell, K. M., Nelson, R. S., Sanders, P. R., Beachy, R. N., Fraley, R. T., and D. M. Shah. (1987). Expression of alfalfa mosaic virus coat protein gene confers cross-protection in transgenic tobacco and tomato plants. *EMBO J.* **6**:1181-1188.

Zaitlin, M. (1976). Viral cross protection: more understanding is needed. *Phytopathology* **66**:382-383.

Biographical Sketch

Roger Beachy received a B. A. degree in education from Goshen College, Goshen, Indiana, in 1966. He began graduate school in the fall of 1968 in the Department of Botany and Plant Pathology, and conducted research for the Ph.D. in the laboratory of Prof. Harry H. Murakishi. From 1973 to 1976, he was a postdoc and NIH fellow at Cornell University with Prof. Milton Zaitlin, and from 1976 to 1978 he was with Dr. John Thompson at the USDA Laboratory of Nutrition at Cornell University. From 1978 to 1991, he held positions of Assistant, Associate, and Full Professor, Department of Biology at Washington University, St. Louis, MO. He also served as Director of the Center for Plant Science and Biotechnology from 1987 to 1991. From July, 1991 he was Professor and Head of the Division of Plant Biology at The Scripps Research Institute, La Jolla, CA. In January, 1999 he became President and Director of the Donald Danforth Plant Science Center, St. Louis, MO. Among his honors and awards are: NIH-NRSA, postdoctoral fellow, 1975; Fellow, AAAS, 1987; Ruth Allen Award from the American Phytopathology Society, 1990; Elected to the National Academy of Sciences, 1997; Headliner of the Year Award, San Diego Press Club.

Delay of Disease Development in Transgenic Plants That Express the Tobacco Mosaic Virus Coat Protein Gene

PATRICIA POWELL ABEL, RICHARD S. NELSON, BARUN DE, NANCY HOFFMANN, STEPHEN G. ROGERS, ROBERT T. FRALEY, ROGER N. BEACHY

A chimeric gene containing a cloned cDNA of the coat protein (CP) gene of tobacco mosaic virus (TMV) was introduced into tobacco cells on a Ti plasmid of *Agrobacterium tumefaciens* from which tumor inducing genes had been removed. Plants regenerated from transformed cells expressed TMV mRNA and CP as a nuclear trait. Seedlings from self-fertilized transgenic plants were inoculated with TMV and observed for development of disease symptoms. The seedlings that expressed the CP gene were delayed in symptom development and 10 to 60 percent of the transgenic plants failed to develop symptoms for the duration of the experiments. Increasing the concentration of TMV in the inoculum shortened the delay in appearance of symptoms. The results of these experiments indicate that plants can be genetically transformed for resistance to virus disease development.

FOR A NUMBER OF YEARS, AGRICULTURALISTS HAVE INOCULATED plants with mild strains of viruses or viroids to prevent more virulent strains from infecting the plant and causing severe disease symptoms. This practice, referred to as cross-protection, has been used to reduce yield losses in crops such as tomatoes, potatoes, and citrus, due to tomato mosaic virus (TMV), potato spindle tuber viroid, and citrus tristeza virus, respectively (*1*). In addition to cross-protection, there are several other types of plant resistance responses to pathogens. These responses are often nonspecific and have been categorized as induced resistance. Extensive discussions on the nature and utility of cross-protection and induced resistance have been presented (*2–4*). Successful cross-protection is most often judged by the ability of the first virus to suppress or delay disease symptoms caused by the superinfecting (challenge) virus. In some cases replication of the superinfecting virus is suppressed (*5*). In most cases symptoms are suppressed or delayed for a period of time, after which the severe strain overcomes the protection, and symptoms develop (*6*).

A number of models have been proposed to explain cross-protection. Gibbs (*7*) suggested that replication of the inducing virus depletes the host cell of a component needed for replication of the challenger. Another hypothesis proposes that capsid protein

P. P. Abel is a graduate student, R. S. Nelson is a postdoctoral associate, and R. N. Beachy is a professor in the Department of Biology, Washington University, St. Louis, MO 63130. N. Hoffmann is a research assistant and S. G. Rogers and R. T. Fraley are research scientists at Monsanto Company, 700 Chesterfield Village Parkway, St. Louis, MO 63107. Barun De is research scientist at the Center for Disease Control, CID-Influenza Branch, 1600 Clifton Road, Atlanta, GA 30333.

Fig. 1. Suppression of symptoms, due to viral infection, by genetically engineered plants. (A) Leaves from plants inoculated with TMV 10 days previously. Leaf at left is part of a transgenic plant containing and expressing the coat protein (CP) gene from TMV. Leaf at right is part of a transgenic plant not containing the CP gene. The leaf from the plant not containing the CP gene shows characteristic symptoms of a systemic infection by TMV (that is, light and dark green areas and leaf margin deformity). (B) Normal and transgenic plants 14 days after inoculation with TMV. Nontransformed *N. tabacum* cv. Xanthi plants (left) develop typical systemic symptoms of TMV infection during this period of time. The transgenic plant containing and expressing the TMV coat protein gene (right) has not developed symptoms.

produced by the first infection encapsidates the RNA of the challenging strain, thereby preventing its replication (*8*). Experiments with viroids (*9*) and with defective strains of tobacco mosaic virus (*10*), however, argue against the involvement of virus capsid protein in cross-protection. It has also been suggested that uncoating of the challenge virus is blocked by the inducing virus (*11*). Palukaitis and Zaitlin (*4*) proposed that sense or antisense RNA of the inducing virus anneals with RNA of the challenger virus thereby blocking replication of the challenge virus. From these examples it is evident that although cross-protection has been used to reduce economic losses caused by virus infection, the mechanism responsible for cross-protection has not been elucidated.

There are potential disadvantages to the widespread use of cross-protection in field agriculture. First, the mild strain (protecting strain) might undergo mutation to a highly virulent form which could be selected during replication, thereby leading to extensive crop losses rather than protection (*4*). Second, the protecting strain might act in synergism with a nonrelated virus resulting in a disease condition that is more severe than that caused by either virus alone. Such a condition occurs when TMV and cucumber mosaic virus concurrently infect tomato plants (*12*). Third, a protecting virus in one plant might be a severe pathogen in another, thereby making it unwise to spread the virus extensively in field situations. Fourth, the protecting strain may itself cause a small but significant loss in yield (*3*). Most or all of these objections could be overcome if cross-protection was engendered in plants as the result of expression of a single viral gene, rather than as a result of infection with an intact virus.

To further understand the mechanisms of cross-protection, and to produce plants that carry a heritable genetic trait for viral resistance, we began a series of experiments designed to express gene sequences of TMV as chimeric nuclear genes in transgenic tobacco plants. These plants would then be tested for the capacity to suppress symptom development upon infection by a severe strain of TMV. We now report the results of experiments showing that plants can be genetically engineered to suppress symptoms due to viral infection (Fig. 1).

Construction of the chimeric gene encoding the TMV–coat protein. The complete nucleotide sequence of the U_1 (common) strain of TMV has been determined by sequencing cloned complementary DNA's (cDNA's) derived from viral RNA (*13*). Nucleotides 70 to 4919 encode two polypeptides that are postulated to be subunits of the TMV replicase (*14*). The 130-kD polypeptide terminates at an amber codon which is infrequently suppressed to produce a 180-kD polypeptide which terminates at nucleotide 4919 (*15*). Both the 130-kD and 180-kD polypeptides have been identified in virus-infected cells (*16*). Nucleotides 4903 to 5709 encode a 30-kD polypeptide which has been implicated in facilitating the movement of virus from cell to cell (*17*). The coat or capsid protein (CP) is encoded within the nucleotides 5712 to 6190. The 3′ untranslated region of the RNA extends to nucleotide 6395. A double-stranded (ds) cDNA of the CP cistron of TMV-RNA was generated and cloned into the plasmid pUC9 (Fig. 2). The cDNA containing the viral coat protein (CP) coding sequence was excised from the pUC plasmid by restriction with Aha III [nucleotide 5707] (*13*) and with Bam HI. To confirm that the cloned cDNA contained an open reading frame encoding the TMV-CP, we cloned the ds-cDNA into a plasmid for in vitro transcription, and the transcript was translated in vitro in a wheat germ system. The results of these experiments (Fig. 3) demonstrate that the protein product of these reactions comigrates with TMV-CP during electrophoresis and is immunoreactive with antiserum to CP produced in rabbits.

The coat protein coding sequence was ultimately inserted into the expression cassette vector pMON316 (*18*) to create a chimeric gene containing the 35*S* promoter from cauliflower mosaic virus (CaMV) and a polyadenylation signal from the nopaline synthase gene (NOS) (Fig. 2). Leaf disks of *Nicotiana tabacum* cv. Xanthi were inoculated with *Agrobacterium tumefaciens* cells containing the pTM319::pTiB6S3-SE cointegrate plasmid, and transformed cells were selected for kanamycin resistance and regenerated into plants as described (*19*). A total of eight different transgenic plants were regenerated and analyzed for expression of the chimeric gene.

Quantitation of gene copy number and expression of the chimeric gene in transgenic plants. The number of chimeric TMV-CP genes present in transgenic plants was determined (Fig. 4A). The expected size of the Bam HI fragment containing the CP gene is approximately 2.3 kb. The copy number of the chimeric TMV-CP gene was one to five per diploid genome for six of the transgenic plants, and more than ten copies for two others (Fig. 4A). Plants having high copy numbers of the CP gene were shown to have multiple sites of insertion (*20*). Since there are greater numbers of copies of the TMV-CP gene in some plants than in others, but not equivalent differences in the amounts of RNA or protein (Fig. 4, B and C), it is probable that some copies of the gene are silent in plants with high numbers of genes.

RNA was isolated from transgenic plants and analyzed for TMV-related messenger RNA (mRNA) (Fig. 4B). As controls in these experiments, RNA isolated from nontransformed plants was included on each blot. The size of the chimeric mRNA is expected to equal the sum of the cDNA (697 nucleotides), the 5′ untranslated nucleotides (42 nucleotides), the approximately 160 nucleotides contributed by the NOS 3′ end, plus polyadenylate residues—altogether about 950 to 1000 nucleotides in length. This RNA would, therefore, be at least 200 nucleotides longer than the low molecular size component (LMC) RNA, the subgenomic RNA encoding the CP in TMV-infected cells (*21*). The RNA blot analyses confirmed that the transcript, found only in transgenic plants carrying pTM319, was larger than the LMC RNA (Fig. 4B). This transcript binds to oligo(dT) cellulose, confirming that it is polyadenylated (*20*). In addition to the CP mRNA of about 0.95 kb, a second RNA of about 2 kb containing sequences of TMV RNA and binding to oligo(dT) cellulose, was found. This RNA presumably results either from multiple sites of transcription initiation, or multiple sites of polyadenylation of the transcript. The reasons for the variability in the relative amounts of these two RNA's is unknown. There was no more than a two- to fivefold variability in mRNA levels between different transgenic plants.

The amount of TMV-CP that accumulated in transgenic plants

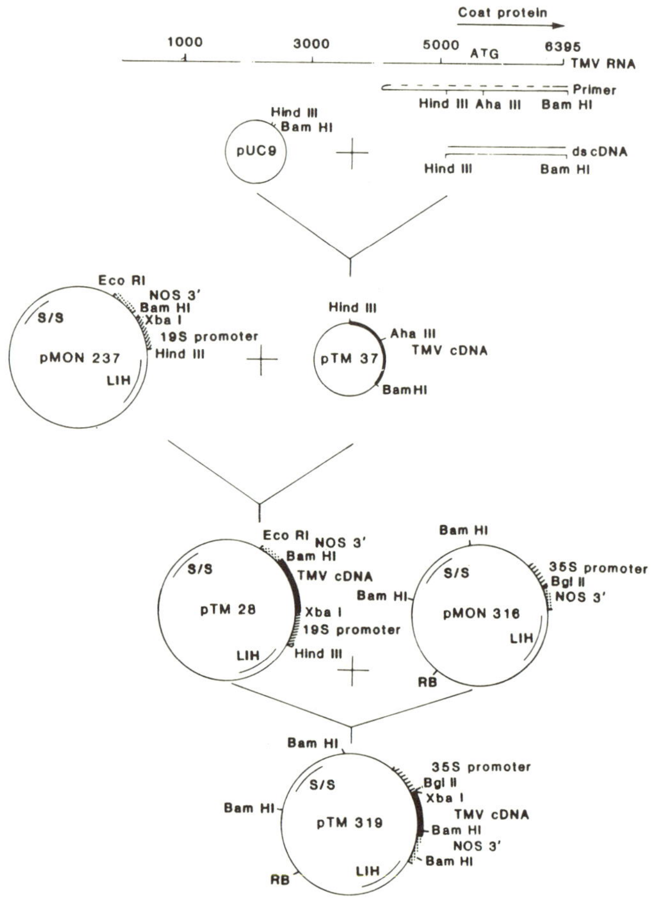

Fig. 2. Construction of the TMV-CP cDNA clone and the intermediate plasmids. A cDNA clone of the 3′ region of TMV RNA (containing nucleotides 5707 to 6395 and including the coat protein cistron) was generated (*29*) by reverse transcriptase with a 35-base oligonucleotide as primer. The primer was complementary to 26 nucleotides at the 3′ end of the viral RNA; the remaining nucleotides generated a Bam HI site. Second strand synthesis was done with the Klenow fragment of DNA polymerase, and the Hind III–Bam HI fragment of TMV cDNA was cloned into the plasmid pUC9 (*30*). The resultant plasmid, pTM37, was digested with Aha III (nucleotide 5707) and Bam HI to excise the CP sequences. This fragment was ligated into the vector pMON 237 (*31*) at the Xba I site which was made blunt, and the Bam HI site to produce pTM 28. The CP coding sequence was excised from pTM 28 by digestion with Xba I and Bam HI and transferred to another plasmid to acquire a Bgl II site at the 5′ end of the coding sequence and an Eco RI site at the 3′ end. The Bgl II–Eco RI fragment was inserted into the expression cassette vector pMON 316 between the CaMV 35*S* promoter (*18*) and the NOS 3′ untranslated region. The resultant plasmid pTM 319 was mated into *A. tumefaciens* strain GV 3111 carrying the disarmed pTi B6S3-SE plasmid (*32*). Recombination between the plasmid LIH region and the LIH region of the transferred DNA (T-DNA) was selected by screening for *A. tumefaciens* colonies resistant to spectinomycin and streptomycin (S/S on the plasmid). These colonies were subsequently used to transform *N. tabacum* cv. Xanthi leaf disks which were then regenerated into whole plants (*19*).

Table 1. Segregation of the TMV-CP gene in transgenic plants and chi-square (χ^2) analysis for a 3:1 expression ratio. Seeds were germinated in a greenhouse under natural light supplemented with artificial light (150 μE m^{-2} sec^{-1}) to produce 14-hour days and 10-hour nights. Seedlings were transplanted 20 days after planting. Leaf disks from the two youngest expanding leaves were sampled, frozen in liquid nitrogen, and ground in extraction buffer (0.035*M* potassium phosphate buffer, *p*H 7.5, 0.4*M* NaCl, 10 m*M* 2-mercaptoethanol). Tissue was ground in extraction buffer at a ratio of 1 ml of buffer to 1 gram of fresh leaf. After centrifugation of the crude extracts, supernatants were diluted 1:1 with Laemmli buffer mix (*37*) and boiled. Samples were analyzed for the presence of CP, after SDS-PAGE and by immunological reactions (*38*). *P*, probability of such a χ^2 value or larger for 3:1 ratio. CP, coat protein; T_2, second generation from transformation event.

Transformant	Number of T_2 seedlings		χ_1^2 (3:1)	*P*
	Expressing CP	Not expressing CP		
3646	42	17	0.36	0.5–0.7
3773	32	8	0.17	0.5–0.7
3404	41	17	0.58	0.3–0.5
3648	39	1	10.8	<0.005

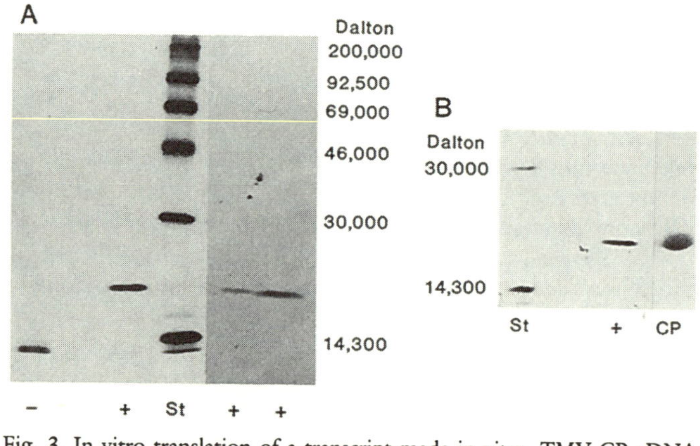

Fig. 3. In vitro translation of a transcript made in vitro. TMV-CP cDNA containing nucleotides 5707 to 6395 of TMV-RNA was ligated into the plasmids SP65 and SP64 (Promega Biotec), restricted with Bam HI or Eco RI, respectively, and used as the template for SP6 polymerase (*33*). ^3H-labeled uridine triphosphate (UTP) was used to label the RNA used as the template in a wheat germ extract in an in vitro translation reaction. Products of the translation reaction were labeled with [^3H]leucine. (A) Translation product when RNA in the positive sense (+) was used as template and when antisense (−) RNA was used. The center lane contains molecular size standards. The lanes at the right contain the translation products that were immunoprecipated with antiserum to TMV. (B) Electrophoretic mobility of the translation product (+) and authentic coat protein (CP). Authentic CP was stained with Coomassie brilliant blue.

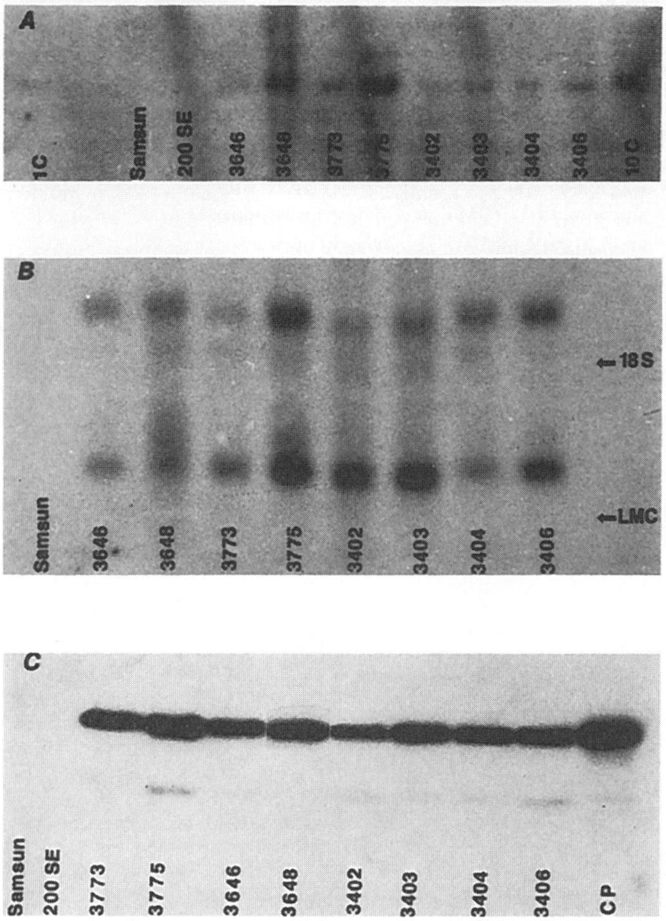

Fig. 4. Integration and expression of the chimeric CP gene in transgenic plants. (A) Quantitation of TMV-CP DNA sequences in transgenic plants. Genomic DNA was isolated from nontransformed tobacco plants and transgenic plants containing the vector alone, or the vector plus the CP gene. DNA (10 μg) was digested with Bam HI, separated on 1 percent agarose gels, transferred to nitrocellulose (*34*), and hybridized with ^{32}P-labeled transcripts of the cDNA (as in Fig. 4B). Reconstructions representing one copy (1C) and ten copies (10C) per genome equivalent of pTM319 are indicated. Lanes containing DNA from normal tobacco leaves (Samsun) are as indicated. Sample 200 SE shows DNA from transgenic plants containing only the vector. The remaining lanes show DNA from leaves of individual transgenic plants containing pTM 319 (Fig. 2). (B) Detection of the TMV coat protein related transcripts in tissue from transgenic plants. RNA was isolated from leaves by the method of Haffner *et al.* (*35*) with minor modifications. Total RNA was separated on a 1.0 percent agarose gel containing formaldehyde and transferred to nitrocellulose (*29*). The blot was hybridized with a ^{32}P-labeled in vitro transcription product complementary to CP mRNA. The transcript probe was prepared as described in Fig. 3 except that [^{32}P]GTP was used in place of [^{3}H]UTP. The first lane (Samsun) contains 6 μg of RNA from leaves of a normal tobacco plant (nontransformed). Successive lanes contain 6 μg of total RNA from leaves of eight individual transgenic plants. The positions of 18*S* ribosomal RNA (rRNA) and LMC, the CP mRNA present in TMV-infected plants (*21*), are indicated. The slight variability in the amount of RNA placed in different lanes is reflected in the amount of probe bound to the 18*S* rRNA. (C) Detection of coat protein in leaves of transgenic plants. Protein was isolated by powdering leaf samples in liquid nitrogen followed by grinding in 0.03*M* potassium phosphate buffer, *p*H 7.5, 0.4*M* NaCl, and 10 m*M* 2-mercaptoethanol. The protein concentration of clarified supernatants was determined by the method of Bradford (*36*). The supernatant was combined with Laemmli sample buffer (*37*) in a 1:1 ratio and boiled for 3 minutes. Total protein (50 μg) from eight individual transformants was separated on a 12.5 percent sodium dodecyl sulfate (SDS) polyacrylamide gel, transferred to nitrocellulose (*38*), and reacted first with rabbit antibody to TMV, then with ^{125}I-labeled donkey antiserum to rabbit serum (Amersham). Proteins extracted from normal Samsun tobacco leaves and transgenic plants containing the vector alone (200 SE) are indicated. The right lane contained 100 ng of purified TMV-CP. The identity of the protein migrating ahead of the CP is unknown.

was determined as described in Fig. 4C. The amount of antigen in each sample was determined by blotting 100 ng of purified TMV-CP on each filter. The amount of coat protein in transgenic plants was between 20 and 50 ng per 50 μg of leaf protein (Fig. 4C), or about 0.1 percent of the extracted leaf protein. Furthermore the amount of TMV-CP, as a percentage of cell protein, was approximately equivalent in leaves at the top, middle, and bottom of the transgenic plants (*22*).

Molecules of TMV-CP assemble to multimeric forms depending upon the concentration of the CP, *p*H, ionic strength, and temperature. Interaction of the monomeric and aggregated states of the CP with viral RNA in vitro has been extensively studied (*23*). To determine whether or not the CP produced in transgenic plants is assembled into multimeric forms, protein was extracted from leaves of these plants under conditions that favor maintenance of the 20*S* disk (*24*). Leaf protein was subjected to velocity sedimentation in sucrose gradients, with hemoglobin (4*S*) and catalase (11*S*) as marker proteins in accompanying parallel gradients. As shown in Fig. 5, TMV-CP was located only in fractions 1 to 5, indicating that the coat protein which accumulates in transgenic plants consists of monomers and multimers with sedimentation coefficients of about 4*S*. The lack of more complex multimers, such as the 20*S* disks, probably reflects the relatively low concentration (estimated to be no more than 0.1 percent of cell protein) of the monomers in leaf tissues.

Infection of transgenic plants with TMV. Experiments were initiated to determine whether transgenic plants could suppress symptoms caused by TMV infection, as in viral cross-protection. In our initial experiments rooted cuttings of transgenic plant 3646, and control plants, consisting of either plants that contained the vector without the CP gene, or nontransformed plants, were infected. Two or three leaves of the rooted cuttings were inoculated by rubbing them with a suspension containing 0.5 μg of TMV per milliliter of phosphate buffer and symptom development was monitored daily in the greenhouse. In experiments with 17 rooted cuttings, each of the 10 plants that did not contain the chimeric CP gene developed classical symptoms of TMV infection. Four of the plants that expressed the CP gene and developed symptoms did so from 2 to 14 days later than any of the controls. Three of the seven transgenic plants expressing the CP gene did not develop symptoms over a 20-day period of observation (*25*).

The limiting factor in experiments with rooted cuttings is in obtaining a sufficient number of uniform plants to carry out experiments that are statistically significant. Therefore all other experiments were done with seedling progeny of self-fertilized transgenic plants. If transgenic plants express the CP gene at a single locus, three of every four seedlings should contain TMV-CP. Seedlings were classified as "expressors" (E) or "nonexpressors" (NE) of TMV-CP on the basis of immunoblot analysis. Expressor and nonexpressor plants were indistinguishable from each other as judged by growth rates, physical appearance, and fertility. An example of the analysis of the seedling progeny of transgenic plant 3773 (Fig. 6) demonstrates that 32 of the 40 seedlings tested expressed the TMV-CP gene. Seedling progeny of transgenic plants 3646, 3773, and 3404 each segregate with a ratio consistent with expression of the TMV-CP gene at a single locus (Table 1). The segregation ratio of seedlings from plant 3648 did not fit the 3:1 expression model, suggesting that the CP gene was expressed from genes inserted at multiple loci. Plant 3648 contained multiple copies of the CP gene (Fig. 4A).

To determine whether transgenic seedlings could suppress symptoms caused by infection, plants were inoculated with TMV. Progeny that did not contain the CP gene and seedlings of nontransformed plants served as controls. In one experiment a total of 37 to 40 seedlings in the three- to four-leaf stage from each of four transgenic plant lines were inoculated with a suspension of TMV (0.25 μg/ml). These plants were held in a growth chamber under conditions that favored rapid appearance of symptoms. Under these conditions typical symptoms of TMV infection appeared in 3 to 4 days on all control seedlings. In contrast, seedlings of transgenic plants that expressed the TMV-CP were significantly delayed in symptom development compared with the controls (Fig. 7). The results of these experiments indicated that delay in symptom appearance is correlated with expression of the TMV-CP cDNA. In every experiment there were some inoculated transgenic seedlings that did not develop symptoms during the observation period. For example, in one experiment 47 percent of seedlings derived from transgenic plant 3404 that expressed TMV-CP did not exhibit symptoms 30 days after inoculation (*25*). In other examples, 4 to 28 percent of the CP-expressing seedlings did not develop symptoms within 12 days after inoculation (the duration of the experiment).

One of the features of cross-protection against TMV is that increasing the concentration of virus in the challenge infection reduces the length of time between the challenge infection and symptom development (*6*). Such an experiment was carried out with transgenic plants that expressed the TMV-CP gene. Seedlings from transgenic lines 3773 and 3404 were inoculated with increasing amounts of U_1-TMV and symptom appearance was monitored in a greenhouse (Fig. 8). The delay in appearance of symptoms in those plants expressing the CP gene was again apparent when compared with segregants that were not expressing CP. In addition, as the concentration of TMV in the inoculum was decreased from 2.0 μg/ml to 0.4 μg/ml, the delay in symptom appearance in the expressing plants increased from about 1 day to about 10 days as judged by comparison with nonexpressing controls. These results are similar to the cross-protection experiments reported for TMV in tomato (*6*). In our experiment, which was carried out in a greenhouse, a large percentage of the transgenic plants (up to 60 percent of plants

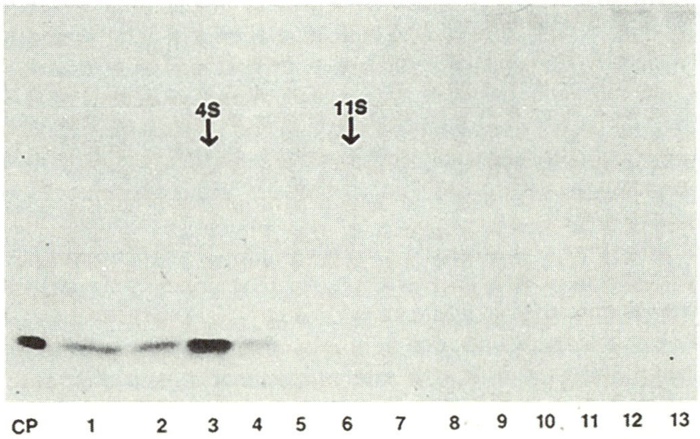

Fig. 5. Determination of the sedimentation coefficient of the coat protein in transformed plants. Protein was isolated from leaves of transgenic plants by grinding in 0.03*M* sodium phosphate, *p*H 7.0, at 20°C, conditions which favor maintenance of CP in higher states of aggregation (*24*). The protein was then placed on a linear sucrose gradient (5 to 30 percent) and subjected to centrifugation in a Beckman SW40 rotor at 35,000 rev/min for 16 hours. Fractions (0.8 ml) were collected, concentrated, boiled with Laemmli sample buffer, and subjected to immunoblot analysis as in Fig. 4. Catalase (11*S*) and hemoglobin (4*S*) were placed on a parallel gradient as standards to determine sedimentation coefficients.

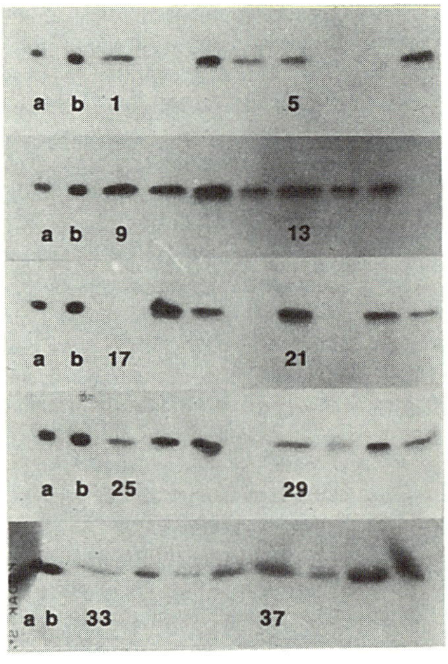

Fig. 6. Segregation of expression of the CP gene in seedlings. Forty micrograms of protein extracted from leaves of seedlings grown from self-fertilized transgenic plants (line 3773) were subjected to SDS–polyacrylamide gel electrophoresis (PAGE) and immunoblot analysis as in Fig. 4C. Of the 40 seedlings assayed, 8 did not express the coat protein gene, and 32 did. Lanes a and b contain 10 and 20 ng of TMV-CP, respectively.

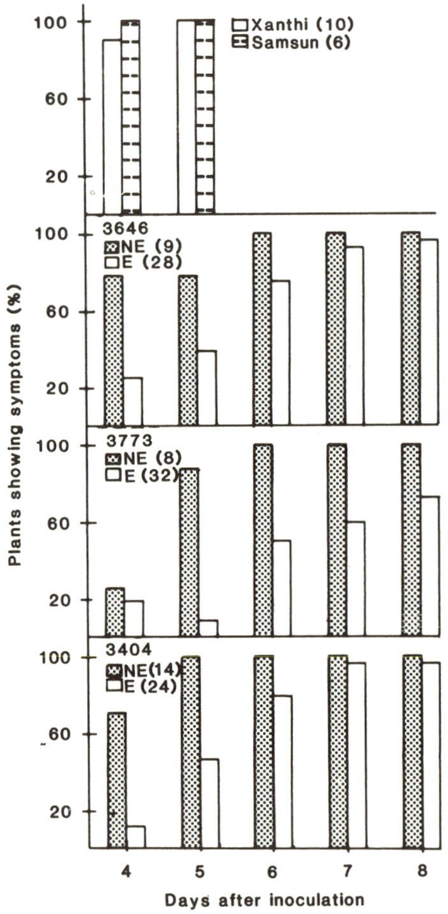

Fig. 7. The percentage of plants showing TMV infection on their leaves at successive days after inoculation with purified TMV. Seeds of each plant type, *N. tabacum* cv. Xanthi and cv. Samsun and transgenic lines 3646, 3773, 3404, and 3648, were germinated and assayed for CP (Table 1). Four days later the two youngest expanding leaves were inoculated with a suspension of purified TMV (0.25 μg/ml); carborundum (330 grit, Fisher Scientific) was used as an abrasive. The seedlings were then placed in growth chambers in a light-dark cycle of 14 hours of light and 10 hours of darkness; the temperature was 26°C for the light period and 19°C for the dark period. The relative humidity was approximately 50 percent and light flux density (PAR), supplied by fluorescent and incandescent lamps, was 340 μE m^{-2} sec^{-1}. Plants were scored as having positive symptoms when the first and second leaves above the inoculated leaves showed obvious vein clearing. CP analyses were not completed until all visual observations were recorded in order to avoid observation bias. NE, segregated seedlings from transgenic plants which were not expressing coat protein; E, segregated progeny from transgenic plants which were expressing coat protein. Numbers in parentheses indicate sample size for each plant type. Plants expressing CP were delayed in disease expression compared with respective nonexpressing plants for each plant line as determined by the Mann-Whitney *U* test ($\alpha = 0.005$). The test was conservative in that (i) no corrections were made for ties in scoring, and (ii) in the case of line 3773 (E progeny at 4 and 5 days after inoculation) once a plant was scored positive for symptom development it was never released from that category during calculations. A fourth transgenic plant line, 3648, had only one NE plant. The 39 E plants in line 3648 showed symptom development patterns similar to transgenic plant line 3773 (*28*).

inoculated with 0.4 μg of TMV per milliliter) did not develop symptoms during the 19-day period of the experiment (Fig. 8).

Toward an understanding for cross-protection. A double-stranded cDNA encoding the coat protein (CP) of TMV was ligated to the CaMV 35*S* promoter and NOS 3′ sequences, and introduced into tobacco cells on a Ti plasmid. Eight different transgenic plants regenerated from transformed cells produce CP to the level of about 0.1 percent of the extractable cellular protein. Bevan *et al.* (*26*) recently reported the results of similar experiments and provided data to indicate the TMV-CP accumulated to a level of about 0.002 percent of the cell protein (*27*). The reasons for the apparent differences in levels of expression of a similar gene in transgenic plants generated in different laboratories remains to be determined. It may reflect differences in the 35*S* promoter fragment, or the host plant genotype used in the different laboratories.

On inoculation with TMV, symptom development in transgenic plants expressing the CP gene was altered compared to control plants. First, not all transgenic plants developed disease symptoms. The results of virus replication experiments have indicated that plants that do not develop symptoms are not infected (*28*). As was expected, increasing the concentration of TMV in the inoculum increases the number of plants that become infected and develop symptoms (Fig. 8). Second, plants that developed symptoms did so 1 to 10 days later than the control plants, depending on the concentration of TMV in the inoculum (Figs. 7 and 8). Both of these observations are consistent with the hypothesis that the number of sites susceptible to infection are reduced in leaves of transgenic plants that express the CP gene compared with leaves of control plants.

Although Bevan *et al.* (*26*) reported the accumulation of TMV-CP in transgenic plants, they did not discuss virus infection of the plants. If the level of expression in these plants is indeed 50 to 100 times lower than in the plants described above, it may be possible to use the plants with lower CP levels to establish a threshold level of gene expression that is needed to give protection.

Delay in symptom development such as that demonstrated in Figs. 1, 7, and 8 is a measure of the degree of cross-protection in these plants (*3*). Whether or not the delay in symptom development in the transgenic plants is the result of a mechanism similar to that of classical cross-protection remains to be determined. However, identification of the mechanism (or mechanisms) of classical cross-protection will be technically difficult. Identifying the mechanism that gives rise to resistance in transgenic plants will be somewhat easier since it involves the expression of a single virus-related gene, rather than on replication of closely related viruses. Regardless of the precise mechanism involved in the virus disease resistance in these transgenic plants, the methods described here provide a way of producing virus-resistant plants that should complement those used in classical plant breeding.

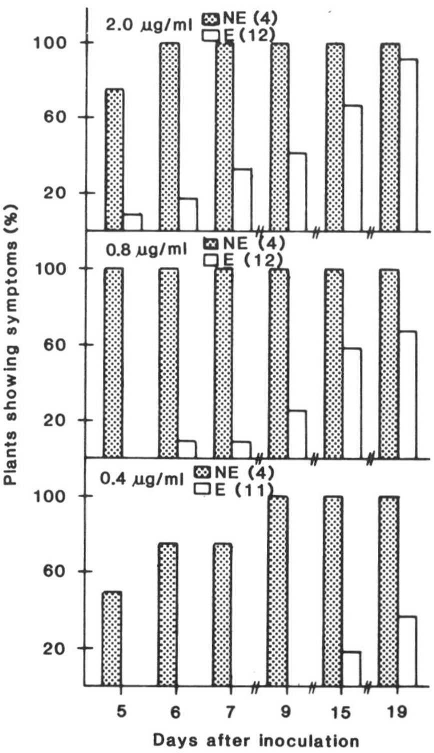

Fig. 8. The percentage of plants showing symptoms of TMV infection at succeeding days after inoculation with various concentrations of purified U_1-TMV. Seeds of transgenic lines 3773 and 3404 were germinated and assayed for CP (Table 1), or for the presence of nopaline (*39*). Approximately 4 days later, the two youngest expanding leaves of each plant were inoculated with suspensions containing purified TMV at 0.4, 0.8, or 2.0 μg/ml. The seedlings were kept in a greenhouse under conditions described in Table 1. Plants were scored as having positive symptoms when leaves above the inoculated leaves showed vein clearing that subsequently led to severe chlorosis. Equal numbers of plants from lines 3773 and 3404 were used except for the inoculum at 0.4 μg/ml, where one less 3773 plant was used. NE, segregated progeny from transgenic plants that did not express coat protein. E, segregated progeny from transgenic plants that expressed the coat protein gene. Numbers in parentheses indicate the sample size for each plant type.

REFERENCES AND NOTES

1. L. Broadbent, *Annu. Rev. Phytopathol.* **14**, 75 (1976); K. H. Fernow, *Phytopathology* **57**, 1347 (1967); A. S. Costa and G. W. Muller, *Plant Dis.* **64**, 538 (1980).
2. L. Sequeira, *Trends Biotechnol.* **2**, 25 (1984).
3. R. I. Hamilton, in *Plant Disease: An Advanced Treatise*, J. G. Horsfall and E. B. Cowling, Eds. (Academic Press, New York, 1980), vol. 5, p. 279.
4. P. Palukaitis and M. Zaitlin, in *Plant-Microbe Interactions: Molecular and Genetic Perspectives*, T. Kosuge and E. W. Nester, Eds. (Macmillan, New York, 1984), vol. 1, p. 420.
5. J. A. Dodds, S. Q. Lee, M. Tiffany, *Virology* **144**, 301 (1985).
6. A. C. Cassells and C. C. Herrick, *ibid.* **78**, 253 (1977).
7. A. J. Gibbs, *Adv. Virus Res.* **14**, 263 (1969).
8. G. A. de Zoeten and R. W. Fulton, *Phytopathology* **65**, 221 (1975).
9. C. L. Niblett, E. Dickson, K. H. Fernow, R. K. Horst, M. Zaitlin, *Virology* **91**, 198 (1978).
10. M. Zaitlin, *Phytopathology* **66**, 382 (1976).
11. J. L. Sherwood and R. W. Fulton, *Virology* **119**, 150 (1982).
12. C. Garces-Orejuela and G. S. Pound, *Phytopathology* **57**, 232 (1957).
13. P. Goelet *et al.*, *Proc. Natl. Acad. Sci. U.S.A.* **79**, 5818 (1982).
14. M. Zaitlin, C. T. Duda, M. A. Petti, *Virology* **53**, 300 (1973); T. R. Hunter, T. Hunt, J. Knowland, D. Zimmern, *Nature (London)* **260**, 759 (1976); R. Scalla, P. Romaine, A. Asselin, J. Rigaud, M. Zaitlin, *Virology* **91**, 182 (1978).
15. H. R. B. Pelham, *Nature (London)* **272**, 469 (1978).
16. F. Sakai and I. Takebe, *Virology* **62**, 426 (1974); G. Bruening, R. N. Beachy, R. Scalla, M. Zaitlin, *ibid.* **71**, 498 (1976).
17. R. N. Beachy and M. Zaitlin, *ibid.* **81**, 160 (1977); M. Nishiguchi, F. Motoyoshi, N. Oshima, *J. Gen. Virol.* **39**, 53 (1978); D. A. Leonard and M. Zaitlin, *Virology* **117**, 416 (1982).
18. S. G. Rogers *et al.*, in *Biotechnology in Plant Science: Relevance to Agriculture in the Nineteen Eighties*, M. Zaitlin, P. Day, A. Hollaender, Eds. (Academic Press, New York, 1986), p. 219.
19. R. B. Horsch *et al.*, *Science* **227**, 1229 (1985).
20. P. P. Abel and R. N. Beachy, unpublished results.
21. A. Siegel, V. Hari, I. Montgomery, K. Kolacz, *Virology* **73**, 363 (1976).
22. Plants used in these experiments were at the 12- to 15-leaf stage; P. P. Abel and R. N. Beachy, unpublished results.
23. L. Hirth and K. E. Richards, *Adv. Virus Res.* **26**, 145 (1981).
24. A. C. H. Dunham and A. Klug, *Nature (London) New Biol.* **229**, 42 (1971).
25. P. P. Abel, R. S. Nelson, R. N. Beachy, unpublished results.
26. M. W. Bevan, S. E. Mason, P. Goelet, *EMBO J.* **4**, 1921 (1985).
27. Based on the data of Bevan *et al.* (*26*) that the level of CP in their transgenic plant was 6000 times less than that of CP in infected cells, and the estimate that TMV accumulates to about 10 percent of cellular protein in chronic infections.
28. R. S. Nelson, P. P. Abel, R. N. Beachy, unpublished results.
29. T. Maniatis, E. F. Fritsch, J. Sambrook, *Molecular Cloning* (Cold Spring Harbor Laboratory, Cold Spring Harbor, NY, 1982).
30. J. Vieira and J. Messing, *Gene* **19**, 259 (1982).
31. pMON 237 is a derivative of pMON 200 (*19*) which contains the CaMV 19*S* promoter, a polylinker, and the NOS 3′ end; S. G. Rogers *et al.*, *Methods Enzymol.*, in press.
32. R. T. Fraley *et al.*, *Biotechnology* **3**, 629 (1985).
33. D. A. Melton *et al.*, *Nucleic Acids Res.* **12**, 7035 (1984).
34. E. M. Southern, *J. Mol. Biol.* **98**, 503 (1975).
35. M. H. Haffner, M. B. Chin, B. G. Lane, *Can. J. Biochem.* **56**, 729 (1978).
36. M. M. Bradford, *Anal. Biochem.* **72**, 248 (1976).
37. U. K. Laemmli, *Nature (London)* **227**, 680 (1970).
38. H. Towbin, T. Staelin, J. Gordon, *Proc. Natl. Acad. Sci. U.S.A.* **76**, 4350 (1979); J. Symington, M. Green, K. Brackmann, *ibid.* **78**, 177 (1981).
39. L. A. B. M. Otten and R. A. Schilperoort, *Biochim. Biophys. Acta* **527**, 497 (1978).

40. Supported by a grant from the Monsanto Company. P.P.A. was supported by fellowships from the Division of Biology and Biomedical Sciences at Washington University and by program training grants from the National Institutes of Health (GM07067 and GM08036) and the Monsanto Company.

4 February 1986; accepted 27 March 1986

The 54 kDa Protein Encoded by Tobacco Mosaic Virus and Replicase-Mediated Resistance

Milton Zaitlin
Department of Plant Pathology, Cornell University, Ithaca, NY

The observation that plants transformed with a portion of the TMV replicase gene were resistant to TMV was a serendipitous finding in my laboratory. The work was an outcome of our attempts to find the significance of a TMV subgenomic mRNA (I_1-RNA) encoding a 54-kDa protein, discovered in 1977 (Beachy and Zaitlin). Later, Sulzinski *et al.* (1985) discovered that this subgenomic mRNA was found on polyribosomes in TMV-infected leaf tissue, suggesting it was being translated in the TMV-infected plant. However, we could not detect such a protein in infected tissues nor in protoplasts, and moreover, we still cannot find it to this day. To try to resolve the dilemma at that time, we set out to determine if TMV mutants defective in the 54-kDa sequence could be complemented by 54-kDa sequences transformed into the plant genome. But I am getting ahead of myself.

In August 1987, George Lomonossoff, a staff member of the John Innes Institute in Norwich, England came to my laboratory for a study leave. George was a very experienced molecular biologist, and was one of the authors on the paper in which the complete sequence of TMV was first determined (Goelet *et al.*, 1982). My letter to him inviting him to come to Cornell said that I hoped he would work on a project "...in which we seek to isolate plant genes conveying resistance to viruses." I was thinking of the TM-2 gene which conveyed resistance to TMV in tomato. Truly a lofty goal, but those experiments were never undertaken because of subsequent events. George and I had later discussed the possibility of cloning the 54-kDa sequence for the complementation experiment described above; George devised the cloning strategy before he arrived, and thus by 1988 he was able to get well-characterized cloned sequences for my postdoctoral fellow, Dan Golemboski, who used them in *Agrobacterium*-mediated transformation of tobacco. Much to our surprise, all of the 54-kDa-containing transformants expressing the 54-kDa sequence were highly resistant to the tobacco mosaic disease, precluding our anticipated complementation experiments. But we were not displeased with that outcome! Later we discovered that these plants would support some replication in the initially-infected leaf cells (reviewed by Carr and Zaitlin, 1993), but the virus did not spread to produce a systemic infection in the plant; in effect, the plants were resistant to the disease!

Dan created some future problems for us with his choice of a transformation vector. When he was ready to make the construct for the transformation, he went for help to Judy Owen, then a postdoctoral fellow with my colleague Peter Palukaitis. Judy gave him the plant vector expression cassette pMON316; neither she nor Dan realized that Peter had signed a Materials Agreement with the Monsanto Company, the developer of that vector, which: 1) forbade giving it to anyone else without their permission, and 2) gave Monsanto the rights of first refusal for utilization of discoveries that derived from its use. The first restriction was easily resolved, but the second was a problem. Our work was supported in part by the Cornell Biotechnology Program, whose corporate sponsors also had those same rights. Eventually this issue was resolved, but it took some ardent negotiation by Cornell's Research Foundation. Cornell was issued two patents for these discoveries in 1997.

The work with TMV involved using the 54-kDa replicase read-through portion of the 183-kDa replicase gene. Later we undertook a number of studies to characterize the resistance, many of which are described in Carr and Zaitlin (1993). Additionally, we felt that if the concept of replicase-mediated resistance was to be broadly useful, it should be extendible to viruses which do not use a read-through strategy for their replicase. Thus, I teamed up with my departmental colleague, Peter Palukaitis, to examine cucumber mosaic virus as a test case. Peter already had clones of all of the RNAs of that virus, which facilitated the study, and we successfully demonstrated replicase-mediated resistance to CMV through the work of postdoctoral fellow, Joe Anderson (Anderson *et al.*, 1992).

The concept of replicase-mediated resistance has gained considerable acceptance as a method to create virus-resistant plants. It is a complement to the well-known coat protein-mediated resistance considered earlier by Beachy in this volume. A listing of the viruses and host plants with replicase-mediated resistance is given in Palukaitis and Zaitlin (1997); 14 viruses have been shown to exhibit replicase-mediated resistance when their replicase genes or portions of those genes were transformed into a variety of plant species. Replicase-mediated resistance is currently in commercial use, generating potatoes resistant to potato leaf roll virus. Seed containing potato leaf roll virus replicase gene sequences is being sold by

NatureMark, a Monsanto company. In this situation, coat protein protection was not as effective as the replicase strategy in that the coat protein gene did not protect against net necrosis, the tuber storage component of the leaf roll disease, which is particularly important for the widely grown variety, Russett Burbank.

References

Anderson, J. M., Palukaitis, P., and M. Zaitlin. (1992). A defective replicase gene induces resistance to cucumber mosaic virus in transgenic tobacco plants. *Proc. Natl. Acad. Sci. USA* **89**:8759-8763.

Beachy, R. N. and M. Zaitlin. (1977). Characterization and *in vitro* translation of the RNAs from less-than-full-length, virus-related, nucleoprotein rods present in tobacco mosaic virus preparations. *Virology* **81**:160-169.

Carr, J. P. and M. Zaitlin. (1993). Replicase-mediated resistance. *Sem. Virol.* **4**:339-347.

Goelet, P., Lomonossoff, G. P., Butler, P. J. G., Akam, M. E., Gait, M. J. and J. Karn. (1982). Nucleotide sequence of tobacco mosaic virus RNA. *Proc. Natl. Acad. Sci. USA* **79**:5818-5822.

Palukaitis, P. and M. Zaitlin. (1987). Replicase-mediated resistance to plant virus disease. *Adv. Virus Res.* **48**:349-377.

Sulzinski, M. A., Gabard, K. A., Palukaitis, P., and M. Zaitlin. (1985). Replication of tobacco mosaic virus VIII. Characterization of a third subgenomic TMV RNA *Virology* **145**:132-140.

Biographical Sketch

Milton Zaitlin graduated from the University of California, Berkeley with a degree in plant pathology in 1949. He worked as a technician at the California Institute of Technology for one year where he met Sam Wildman, who was moving to a position in the Botany Department at UCLA. He entered graduate school in 1950, and received his Ph.D. under Wildman in 1954 in botany, with a minor in biochemistry. Irving Rappaport and Albert Siegel, postdoctoral fellows in the Wildman laboratory, were also significant in his training. He spent about 4 years working at the Commonwealth Scientific and Industrial Research Organisation in Canberra, Australia, with Keith Boardman, Don Spencer and Paul Whitfeld. He returned to the United States to a position in the Department of Horticulture at the University of Missouri with the late Dan Millikan. In 1960 he worked as a postdoctoral fellow with Albert Siegel at the University of Arizona, Tucson, and rose in the ranks to professor. In 1973 he accepted a position as Professor in the Department of Plant Pathology at Cornell University. He also served as both Associate Director and Director of the Cornell Biotechnology Program. He received Emeritus status in 1997.

Proc. Natl. Acad. Sci. USA
Vol. 87, pp. 6311–6315, August 1990
Agricultural Sciences

Plants transformed with a tobacco mosaic virus nonstructural gene sequence are resistant to the virus

(plant virus/replicase)

DANIEL B. GOLEMBOSKI, GEORGE P. LOMONOSSOFF*, AND MILTON ZAITLIN†

Department of Plant Pathology, Cornell University, Ithaca, NY 14853

Communicated by Myron K. Brakke, May 21, 1990

ABSTRACT ***Nicotiana tabacum*** **cv. Xanthi nn plants were transformed with nucleotides 3472–4916 of tobacco mosaic virus (TMV) strain U1. This sequence contains all but the three 3′ terminal nucleotides of the TMV 54-kDa gene, which encodes a putative component of the replicase complex. These plants were resistant to infection when challenged with either TMV U1 virions or TMV U1 RNA at concentrations of up to 500 μg/ml or 300 μg/ml, respectively, the highest concentrations tested. Resistance was also exhibited when plants were inoculated at 100 μg/ml with the closely related TMV mutant YSI/1 but was not shown in plants challenged at the same concentrations with the more distantly related TMV strains U2 or L or cucumber mosaic virus. Although the copy number of the 54-kDa gene sequence varied in individual transformants from 1 to ≈5, the level of resistance in plants was not dependent on the number of copies of the 54-kDa gene sequence integrated. The transformed plants accumulated a 54-kDa gene sequence-specific RNA transcript of the expected size, but no protein product was detected.**

The organization of the tobacco mosaic virus (TMV) genome is fairly well understood. However, one aspect of the genome strategy that has not been fully elucidated is the exact nature of the replicase enzyme responsible for the synthesis of the genomic and subgenomic RNAs. It is generally accepted that the virus codes for four proteins, two of which are translated from the genomic RNA and two from individual subgenomic RNAs (see review in ref. 1). The 5′ proximal region of the genomic RNA encodes two coinitiated proteins, the 126-kDa and 183-kDa proteins, considered to be components of the replicase (2). The 183-kDa protein is generated by a read-through of the UAG stop codon of the 126-kDa protein. The other two proteins, the 30-kDa cell-to-cell movement protein and the coat protein, are each synthesized from separate subgenomic mRNAs, on which each gene is 5′ proximal. What is not generally accepted, however, is our contention that there is a separate protein (the 54-kDa protein) for which there is an open reading frame in the read-through portion of the gene encoding the 183-kDa protein (1). Our principal evidence for the existence of such a protein comes from the finding that there is a third subgenomic RNA in TMV-infected plants, termed I_1 RNA (3), whose 5′ terminus has been mapped to nucleotide residue 3405 in the TMV genome. This subgenomic RNA contains the open reading frame for a 54-kDa protein, which starts with an AUG codon at position 3495 (4). Support for its function as a mRNA and as a subgenomic RNA is derived from the observation that it is found on polyribosomes (4) and that there is a double-stranded RNA of a size corresponding to the double-stranded version of the I_1 subgenomic RNA (5, 6).

The 54-kDa protein has not been found in infected tissues, however. Saito *et al.* (7) prepared antibodies to a β-galactosidase fusion protein expressed in *Escherichia coli*, which contained 432 amino acids specific to the 183-kDa protein, and could not detect the 54-kDa protein in protoplast extracts by immunoprecipitation or Western blotting under conditions where the antibody would detect the 183-kDa protein. Likewise, we have not been able to detect 54-kDa protein in Western blots using antiserum made to the whole 183-kDa protein (8), although on occasion we have seen inconclusive faint bands in the region of the gel where such a protein would be expected. This is in spite of the fact that our antiserum is capable of precipitating the 54-kDa protein generated from *in vitro* translation products of either TMV RNA or T7 transcripts of the 54-kDa gene.

In an effort to attribute a function to the 54-kDa protein, we have transformed tobacco with the coding sequence for this nonstructural viral protein. Unexpectedly, these plants showed a complete resistance to replication of the U1 strain of TMV from which the 54-kDa gene sequence was derived. These findings and their implications for the production of virus-resistant plants are discussed.

MATERIALS AND METHODS

Plants and Virus Strains. TMV strain U1 was purified from infected *Nicotiana tabacum* cv. Turkish Samsun as described by Asselin and Zaitlin (9). Unless otherwise indicated, subsequent use of the term TMV implies strain U1. Before use, preparations of TMV strain U2 (10) and L (11), which had been stored as laboratory stocks at 4°C, were pelleted at 75,000 × *g* for 25 min and resuspended in water to enrich for full-length virions. A preparation of freshly prepared cucumber mosaic virus (CMV) strain Fny was a gift of M. J. Roossinck (Cornell University). Viral RNA was isolated by phenol extraction and ethanol precipitation. *N. tabacum* cv. Xanthi nn was used as a TMV-susceptible, systemic host and for transformation, and *N. tabacum* cv. Xanthi nc was used as a local lesion host. Virus was sometimes propagated in *N. tabacum* cv. Turkish Samsun. Protoplasts used in the studies to detect the 54-kDa protein were prepared from *N. tabacum* cv. Xanthi NN. The description and origin of these cultivars is discussed by Harrison (12). Plants were maintained in a greenhouse or in a growth chamber with a 14-hr light/10-hr dark cycle at 24°C.

Cloning the 54-kDa Gene. A clone of the TMV 54-kDa gene was obtained by using an oligonucleotide primer consisting of a *Bam*HI site linked to the 5′ end of a sequence complementary to nucleotides 4906–4923 of the TMV RNA sequence. First-strand cDNA was synthesized by Moloney murine leukemia

The publication costs of this article were defrayed in part by page charge payment. This article must therefore be hereby marked "*advertisement*" in accordance with 18 U.S.C. §1734 solely to indicate this fact.

Abbreviations: AlMV, alfalfa mosaic virus; CaMV, cauliflower mosaic virus; CMV, cucumber mosaic virus; TMV, tobacco mosaic virus.

*Permanent address: John Innes Institute, John Innes Centre, Norwich, U.K.
†To whom reprint requests should be addressed.

virus reverse transcriptase and was rendered double stranded by sequential treatment with reverse transcriptase and the Klenow fragment of DNA polymerase I, relying on loop-back synthesis (13). The double-stranded cDNA was digested with *Bam*HI, which cleaves once within the TMV sequence at position 3321, and ligated into the *Bam*HI site of M13mp18. Sequencing showed that all the clones examined contained the sequence from position 3332 but lacked the *Bam*HI site provided by the primer. This resulted in the deletion of the 54-kDa termination codon and the extension of the 54-kDa protein at its C terminus by five amino acids derived from the vector M13mp18. The region of the insert encoding the 54-kDa protein was removed by digestion with *Hae* II, which cleaves at position 3467 in the TMV sequence, treated with Klenow fragment to blunt end the 3′ overhang, and digested with *Pst* I. The *Hae* II/*Pst* I fragment was ligated into *Pst* I/*Sma* I-digested pBS(−) (Stratagene), resulting in a plasmid termed pRTT-1. Sequencing showed that it contained the TMV sequence from nucleotides 3472 to 4916. The orientation of the insert was such that transcription from the T7 promoter gives plus-sense RNA transcripts (Fig. 1*B*).

The TMV 54-kDa gene sequence insert of pRTT-1 was excised by digestion with *Hin*dIII and *Sac* I, made blunt-ended by treatment with the Klenow fragment, and ligated into the *Sma* I or *Xho* I site of pMON316 (14). pMON316 contains a unique *Xho* I site in a polylinker region located between the cauliflower mosaic virus (CaMV) 35S promoter and the nopaline synthase 3′ untranslated region. A *Sma* I site is found in the polylinker region as well as within the Ti plasmid homologous region of pMON316. Plasmid pTS541A (Fig. 1*C*) was generated by insertion of the TMV sequence into the *Sma* I site, which resulted in the deletion of the nopaline synthase 3′ untranslated region and a portion of the Ti plasmid homology region. Insertion of the TMV sequence into the *Xho* I site resulted in the formation of pTS541. Clones containing the 54-kDa gene sequence in the sense or antisense orientation were characterized and isolated. Each construct was transferred to *Agrobacterium tumefaciens* strain GV3111 carrying pTiB6S3-SE by means of a triparental mating system (15), and transconjugates were selected by resistance to kanamycin and streptomycin.

Plant Transformation. Cut pieces of sterile *N. tabacum* cv. Xanthi nn leaves were transformed by the modified *A. tumefaciens* GV3111, containing the TMV 54-kDa gene coding sequence, as described by Horsch *et al.* (16). Transformed calli were selected on regeneration medium supplemented with kanamycin at a concentration of 300 μg/ml. Resistant calli were induced to regenerate shoots and roots, transferred to soil, and maintained in a greenhouse (16).

Nucleic Acid Analysis. DNA was isolated from leaves of plants by a modified procedure of Murray and Thompson (17). The DNA was digested with restriction enzymes, separated in 1.0% agarose gels, transferred to GeneScreen*Plus* (DuPont) nylon membranes, and hybridized to a ^{32}P-labeled probe (18) specific for the TMV 54-kDa gene sequence. RNA was isolated from leaf tissue by a modified procedure of Chirgwin *et al.* (19). Total RNA was separated in a 1.2% agarose/formaldehyde gel and transferred to nitrocellulose (13). The blot was hybridized to a ^{32}P-labeled DNA probe complementary to the 54-kDa gene coding sequence (18).

Immunological Analyses. An antiserum to the 54-kDa protein was made by injecting rabbits with tuberculin coupled to a synthetic polypeptide representing amino acid residues 243–257 of the 54-kDa protein (corresponding to residues 1407–1421 of the 183-kDa protein). An *in vitro* translation product of a T7 transcript from the 54-kDa gene was immunoprecipitable with this antiserum (data not shown). Antiserum to the read-through region of the 183-kDa TMV protein, and thus specific to the 54-kDa protein, was provided by T. Meshi (University of Tokyo). For Western blotting, total

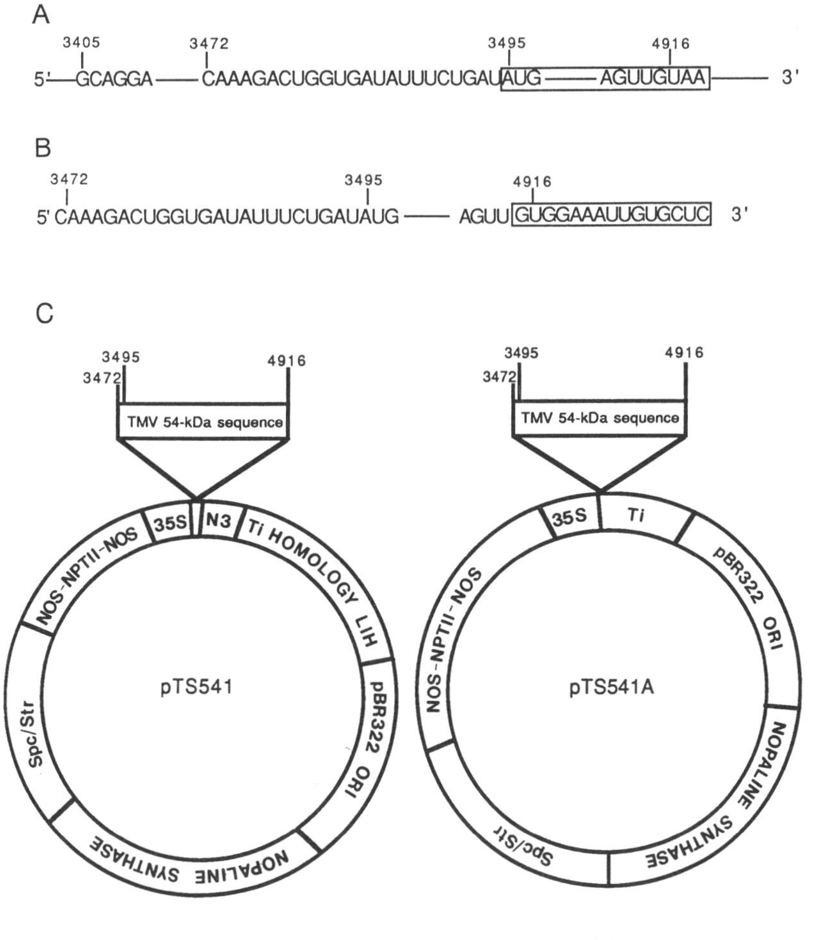

FIG. 1. (*A*) Important features of the sequence of a region of the TMV genome that contains the read-through portion of the 183-kDa protein gene. The I_1 subgenomic RNA begins at nucleotide 3405; the 54-kDa open reading frame extends from nucleotide 3495 to 4919. (*B*) TMV sequence used for plant transformation. The boxed region contains the sequence coding for the five additional amino acids added to the 54-kDa gene product. (*C*) Plant expression vectors containing the TMV 54-kDa gene sequence inserted between the CaMV 35S promoter and the nopaline synthase (NOS) polyadenylylation site. These plasmids were derived by insertion of the TMV cDNA into either the *Xho* I site (pTS541) or the *Sma* I site (pTS541A) in the polylinker region of pMON316. The numbers refer to nucleotides in the TMV genome. The neomycin phosphotransferase II (NPTII) gene confers a selectable kanamycin-resistance marker to transformed plants. LIH, Ti homologus DNA; ORI, origin of replication.

extracts of transformed and untransformed plants were prepared by homogenizing leaf samples in 50 mM Tris·HCl, pH 7.5/0.4 M sucrose/20% (vol/vol) glycerol/5 mM $MgCl_2$/10 mM KCl/10 mM 2-mercaptoethanol. The extracts were centrifuged at 30,000 × *g*. Both the supernatants and the resuspended pellet were subjected to electrophoresis in an SDS/12.5% polyacrylamide gel and transferred to nitrocellulose. The nitrocellulose filter was incubated first with specific antibodies followed by gold-conjugated anti-rabbit antibodies, and the detection was enhanced with silver (8).

In an additional approach to detect the 54-kDa protein, leaf strips (1–2 mm × 50 mm) of TMV-infected Turkish Samsun tobacco were vacuum infiltrated with [^{35}S]methionine at a concentration of 10 μCi/ml (1 Ci = 37 GBq) in 10 mM KH_2PO_4 (pH 7.0) containing chloramphenicol at 1 mg/ml. They were incubated in dim light for 20 hr at 25°C. Protoplasts were also labeled with [^{35}S]methionine; they were prepared from either transgenic *N. tabacum* cv. Xanthi nn or cv. Xanthi NN leaves as described by Hills *et al.* (8). The latter protoplasts were electroporated with TMV RNA at 10 μg/ml. Protoplasts (about 150,000 per ml containing 5–10 μCi of [^{35}S]methionine per ml) were incubated at 25°C in dim light for 40 hr. They were collected by low-speed centrifugation and lysed in 20 mM Tris·HCl, pH 7.5/2 mM EDTA/0.5% SDS/0.2% 2-mercapthethanol containing phenylmethylsulfonyl fluoride as a protease inhibitor. Leaf strips were extracted by grinding them in a mortar and pestle with a similar solution that did not contain the inhibitor. The extracts were centrifuged at 30,000 × *g*, and both the pellet and supernatant fractions were examined for the 54-kDa protein. The extracts of either the labeled leaves or protoplasts were incubated with the antisera described above and immunoprecipitated as described by Berry *et al.* (20). Further analysis was by polyacrylamide gels and autoradiography.

Inoculation of Transformed Plants. R1 seedlings from self-fertilized transgenic plants were routinely inoculated with either 100 μg of TMV (U1 strain) per ml of 50 mM phosphate buffer (pH 7.2), with Celite added as an abrasive, or TMV (U1 strain) RNA at a concentration of 300 μg/ml in 50 mM Tris phosphate buffer (pH 8.6) with Celite. Two leaves of each plant were inoculated. Excess inoculum was washed from the leaves. The volume of the inoculum was not standardized since inoculum concentration is the critical determinant as long as there is sufficient volume for adequate spread. In our more recent experiments, we used a closely related TMV mutant YSI/1 (mutant b6 in ref. 21), which is easier to score as a consequence of a bright yellow symptoms it elicits. Plants were scored daily by visual observation of symptom development. In some cases, presence of viral RNA in inoculated plants was determined by probing leaf extracts with ^{32}P-labeled cDNA to TMV (22).

RESULTS

TMV 54-kDa cDNA Gene Clone. A cDNA clone of part of the TMV genomic RNA was synthesized using a 22-base primer complementary to nucleotides 4906–4923 at the 3′ end of the 183-kDa gene sequence (23). The cloning procedure used resulted in the production of a cDNA clone, pRTT-1, which contains the entire 54-kDa open reading frame downstream from the T7 promoter in the vector pBS(−). The clone also contains an additional 12 nucleotides upstream of the AUG codon at position 3495, which is believed to be the initiator for the synthesis of the 54-kDa protein. The presence of an intact 54-kDa open reading frame in pRTT-1 was verified by synthesizing T7 transcripts *in vitro* and translating them in a rabbit reticulocyte lysate system (Promega). *In vitro* translation yielded a 54-kDa product, which confirmed the presence of an intact open reading frame and suggested that the AUG at position 3495 can function as an initiation codon. Immunoprecipitation with 54-kDa antiserum verified that the translation product is indeed the 54-kDa protein (data not shown).

Four transformed plants were generated with pTS541, and four plants were generated with pTS541A, which lacks the 3′ nopaline synthase untranslated region and a portion of the Ti homology region that is located immediately downstream from the 54-kDa open reading frame. This deletion did not interfere with integration of the chimeric TMV 54-kDa gene sequence into the plant genome. Progeny seed was collected from each self-fertilized plant. Additionally, plants were transformed with the chimeric TMV gene such that 54-kDa gene antisense RNAs were produced. Two independent antisense transformants were regenerated.

TMV 54-kDa Gene DNA Analysis. Six of the independently transformed plants (specified in Fig. 2) were analyzed for expression of the chimeric gene. Genomic DNA was isolated from transformed and untransformed *N. tabacum* cv. Xanthi nn. *Bam*HI digests of the genomic DNA were hybridized to a ^{32}P-labeled TMV 54-kDa gene sequence-specific probe. Hybridization to a 3.0-kilobase (kb) fragment verified the presence of a full-length 54-kDa gene coding sequence (data not shown). The 54-kDa gene sequence insert is 1.44 kb, and another 1.59 kb is contributed by flanking vector DNA. The copy number of the 54-kDa gene sequence in the transgenic plants, as determined by slot blot hybridization analysis, varied from one to five copies per diploid genome between different transgenic plants (Fig. 2). No copies of the 54-kDa gene sequence were detected in nontransformed plants or in plants transformed with pMON316 lacking the 54-kDa gene sequence insert.

TMV 54-kDa Gene RNA Analysis. RNA extracted from transformed plants was examined by Northern blot hybridization analysis (Fig. 3). RNA of the expected size for the chimeric mRNA of 1.6 kb was identified in total RNA from all eight transgenic plants containing the TMV 54-kDa gene sequence in the sense orientation. Plants containing the integrate plasmid that lacks the 3′ nopaline synthase untranslated region and the Ti homologous region (541A11, 541A12, 541A20, and 541A21) synthesized a 1.6-kb transcript in lower amounts (Fig. 3); in addition, a larger transcript was synthesized, which might result from the lack of the transcription termination sequence usually contributed by the nopaline synthase 3′ sequence. In all plants a number of smaller unidentified transcripts were also detected. Plants transformed with the vector alone did not produce any transcripts that hybridized with the probe for the 54-kDa gene sequence (lane 316 in Fig. 3).

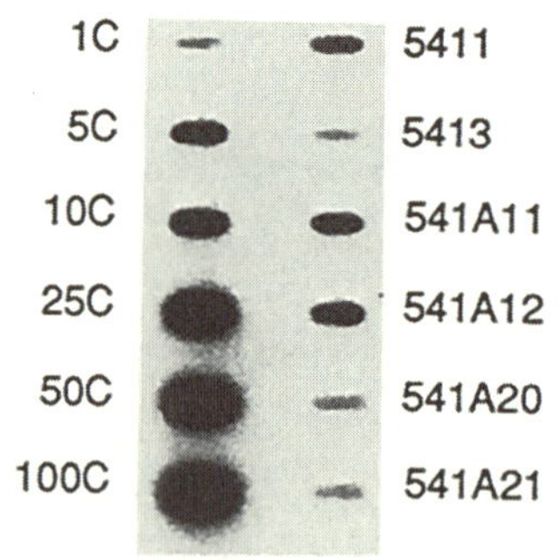

FIG. 2. Determination of the copy number of TMV 54-kDa gene sequences in transgenic tobacco plants. Ten micrograms of total DNA from each transgenic plant was spotted onto a nylon filter and hybridized to a ^{32}P-labeled probe specific for the 54-kDa gene sequence. Copy number was determined by comparison to reconstructions of 10 μg of tobacco DNA mixed with genome equivalents of the 54-kDa sequence insert for the different copy numbers. 1C, 5C, 10C, 25C, 50C, and 100C refer to 1, 5, 10, 25, 50, and 100 copies of the 54-kDa gene sequence per diploid genome, respectively.

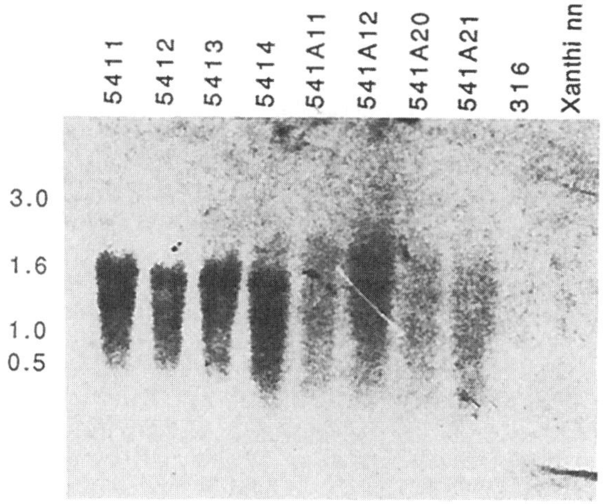

FIG. 3. Analysis of RNA in transformed tobacco plants. Total RNA was isolated from leaves of transgenic plants and separated on a 1.2% agarose/formaldehyde gel, transferred to GeneScreen*Plus*, and hybridized with a 54-kDa gene sequence-specific probe. RNA was isolated from pTS541-transformed tobacco (5411, 5412, 5413, and 5414), pTS541A-transformed tobacco (541A11, 541A12, 541A20, and 541A21), controls of tobacco transformed with pMON316 (316), and untransformed tobacco, Xanthi nn. The sizes of the RNAs (in kb) are given at left.

Analysis of Plants for the 54-kDa Protein. Leaf extracts of transformed plants and of infected untransformed controls were incubated with antiserum raised against a synthetic oligopeptide prepared to an internal region of the 54-kDa protein sequence and with antiserum specific for the 54-kDa read-through portion of the TMV 183-kDa protein. A 54-kDa protein could not be detected either by Western blotting or by immunoprecipitation of ^{35}S-labeled extracts from 54-kDa transgenic plants, protoplasts prepared from 54-kDa transgenic plants, or the infected untransformed control plants.

Resistance of Transformed Plants. In the first experiments to assay for resistance to infection by TMV, plants were inoculated with 50 μg of TMV (U1 strain) per ml. Four rooted cuttings from each of the eight independently transformed plants containing the 54-kDa coding sequence in the sense orientation, controls transformed with the vector alone, and several untransformed Xanthi nn were inoculated. At 5 days after inoculation, the transgenic controls and the untransformed controls had clearly developed characteristic mosaic symptoms, while the transformed plants containing the 54-kDa gene sequence showed no sign of symptom development. No symptoms had developed on these transgenic plants by 48 days after inoculation, at which point the experiment was terminated. A homogenate of the inoculated and the upper leaves of those plants was used to inoculate the local lesion host *N. tabacum* cv. Xanthi nc to determine if the transgenic plants had a symptomless infection. No local lesions developed, indicating the absence of detectable virus in the transgenic plants. All regenerated plants were resistant to TMV regardless of whether they were transformed with pTS541, which has the TMV sequence inserted into the complete pMON316, or pTS541A, which lacks the nopaline synthase 3′ untranslated region and the Ti homologous region. Typical plants showing resistance and susceptibility are shown in Fig. 4. Plants transformed with the chimeric gene in the orientation that resulted in synthesis of the 54-kDa gene antisense RNA were also challenged with TMV; these plants were not resistant, but in a single experiment, there was a delay in symptom development as compared to the vector-transformed control (data not shown). These plants were not examined further; we have concentrated our efforts on the plus-sense plants exhibiting complete resistance.

FIG. 4. Resistance of transformed tobacco plants to TMV infection. Shown are transgenic tobacco plants 15 days after inoculation with 100 μg of TMV, strain YSI/1, per ml. The plant on the left was transformed with the vector alone, and the plant on the right was transformed with the TMV 54-kDa gene sequence. Typical mosaic symptoms have developed only on the control plant.

Progeny seedlings from self-fertilized transgenic plants were also analyzed for inheritance of the resistance phenomenon. R1 seeds were germinated on tissue culture medium containing 300 μg of kanamycin per ml. Kanamycin-sensitive seedlings were considered to be those that were chlorotic and did not grow beyond the cotyledon stage. The segregation ratio of the seedlings expressing kanamycin resistance to those susceptible to kanamycin indicates that in each of the original transformants the neomycin phosphotransferase II (NPTII) gene (which confers resistance to kanamycin) was integrated at multiple loci. On the other hand, the resistance to TMV segregated at ≈3:1, even though some of the transformed plants contained multiple copies of the sequence (data not shown). Thus, the multiple inserts must all be on one chromosome. The large number of kanamycin resistant "escapes" makes this an unreliable means of screening progeny seedlings for expressors of the integrated chimeric TMV gene (24). All subsequent infection experiments described in this report were done with the segregating population of line 541A11-derived R1 seedlings.

Resistance was observed at inoculum concentrations up to 500 μg of TMV per ml or 300 μg of TMV RNA per ml (the highest concentrations tested). The resistant plants were maintained for 30 days after inoculation without any subsequent development of symptoms. To assay for virus replication and spread of the virus, extracts of the leaf samples were probed with cDNA prepared from purified TMV RNA. Viral RNA could not be detected in either the inoculated leaves or in the systemic leaves of the plants that demonstrated resistance; thus, the resistance was not merely a suppression of symptom development.

In another experiment, some plants were transferred immediately after inoculation to a growth chamber maintained at 31°C, to determine if the 54-kDa-induced resistance to TMV is temperature sensitive. Five of the seven inoculated plants in the segregating population, which carry the 54-kDa gene sequence, did not develop symptoms at 31°C, whereas all control plants developed symptoms typical of those kept at 24°C, indicating that the resistance was not temperature sensitive.

Resistance to Other TMV Strains and to CMV. Tobacco plants transformed with the 54-kDa gene sequence and the vector-transformed controls were inoculated with the U2 or L strains of TMV or with CMV. Each virus or strain was tested at concentrations of 2.5–100 μg/ml. No resistance was observed in any of the tests (data not shown).

DISCUSSION

Since the seminal paper of Powell-Abel *et al.* (25) showing that plants transformed with and expressing the coat protein gene of TMV are resistant to TMV, there have been other examples of this concept, and it will undoubtedly have important implications for the protection of many crop species from many viruses. To date, coat protein-mediated resistance has been shown with alfalfa mosaic virus (AlMV) (26, 27), tobacco rattle virus (28, 29), potato virus X (30, 31), CMV (32), potyviruses (33, 34), and plants transformed with both potato virus X and Y coat protein (34). The findings reported here represent an entirely different approach to virus-induced resistance. In this study we have demonstrated that transgenic plants containing the TMV 54-kDa gene coding sequence are resistant to infection with TMV. Presence of the 54-kDa gene sequence prevents the development of local chlorosis and any systemic development of symptoms. There was no detectable virus in the inoculated or systemic leaves of resistant transgenic plants, whereas virus was easily detectable in all of the controls. The 54-kDa-induced resistance is absolute and is not simply a delay in symptom development. This resistance is also not as "fragile" as coat protein-induced resistance, which may break down when high concentrations of inoculum are used (25). In contrast, with the 54-kDa gene sequence, complete resistance is observed in plants challenged with high concentrations of virus or viral RNA. Protection mediated by the coat proteins of TMV and AlMV can be overcome by inoculating with viral RNA (26, 34, 35), whereas 54-kDa gene sequence-induced resistance remains uncompromised when challenged with RNA.

The level of resistance in plants transformed with the 54-kDa gene sequence does not appear to depend upon the number of copies of the inserted sequence. Plants with only one copy of the gene sequence did not show a decrease in resistance to intact virions or viral RNA. On the other hand, a single copy of the TMV coat protein gene was also sufficient to protect (25), whereas one copy of the AlMV coat protein did not provide protection (26). The absence of a polyadenylylation site immediately downstream from the 54-kDa gene sequence was also shown not to have any effect on expression of resistance.

The question remains as to whether the 54-kDa gene sequence-mediated resistance is due to the 54-kDa protein or its nucleic acid. Attempts to detect the 54-kDa protein in TMV-infected leaf tissue or in protoplasts have not been successful. Surprisingly, we have also not been able to detect a 54-kDa gene product in the 54-kDa transgenic plants. This is unexpected because the 54-kDa open reading frame is preceded by the CaMV 35S promoter, which is known to function quite well in the plant genome (14). It is possible that the 54-kDa protein is very labile and rapidly turned over in the cell or that it is made at such low levels that it is below our limit of detection. Based on the suggested role of the 183-kDa protein in replication of the TMV genome (2), one might hypothesize that the 54-kDa protein is also a component of a membrane-bound replicase. In addition, evidence suggests that the TMV 126-kDa protein is also a component of the TMV replicase complex (2). We have produced transgenic plants containing the complete 126-kDa gene sequence and have found these plants to be as susceptible to infection as untransformed plants (data not shown). The predicted 54-kDa protein contains a Gly-Asp-Asp motif that is conserved among many positive-strand viruses (36–38) and has been shown to be necessary for replicase activity in Q_β bacteriophage (39). Constitutive synthesis of either the 54-kDa RNA transcript or the 54-kDa protein, which includes the Gly-Asp-Asp sequence, probably interferes with TMV replication. If the 54-kDa protein itself is not a constituent of the replication complex, it could induce resistance by competing with the 183-kDa protein, which is most probably a constituent of that complex.

We thank Drs. José Osmar Gaspar and John Carr, who participated in studies to identify the 54-kDa protein in plant tissues, and Mary Woo for help with the nucleic acid analysis. We also thank Pat Barker and Neville Huskisson of the Institute of Animal Physiology and Genetics Research (Babraham, U.K.) for producing the anti-54-kDa protein antiserum. This work was supported by the Cornell Biotechnology Program, which is sponsored by the New York State Science and Technology Foundation, a consortium of industries, the U.S. Army Research Office, and the National Science Foundation. G.P.L. was a recipient of a Fulbright Scholarship.

1. Palukaitis, P. & Zaitlin, M. (1986) in *The Plant Viruses*, eds. Van Regenmortel, M. H. V. & Frankel-Conrat, H. (Plenum, New York), Vol. 2, pp. 105–131.
2. Young, N. D., Forney, J. A. & Zaitlin, M. (1987) *J. Cell Sci. Suppl.* **7**, 277–285.
3. Beachy, R. N. & Zaitlin, M. (1977) *Virology* **81**, 160–169.
4. Sulzinski, M. A., Gabard, K. A., Palukaitis, P. & Zaitlin, M. (1985) *Virology* **145**, 132–140.
5. Zelcer, A., Weaber, K. F., Balaczs, E. & Zaitlin, M. (1981) *Virology* **113**, 417–427.
6. Palukaitis, P., Garcia-Arenal, F., Sulzinski, M. A. & Zaitlin, M. (1983) *Virology* **131**, 533–545.
7. Saito, T., Watanabe, Y., Meishi, T. & Okada, Y. (1986) *Mol. Gen. Genet.* **205**, 82–89.
8. Hills, G. J., Plaskitt, K. A., Young, N. D., Watts, J. W., Wilson, T. M. A. & Zaitlin, M. (1987) *Virology* **161**, 488–496.
9. Asselin, A. & Zaitlin, M. (1978) *Virology* **91**, 173–181.
10. Siegel, A. & Wildman, S. G. (1954) *Phytopathology* **44**, 277–282.
11. Ohno, T., Aoyagi, M., Yamanashi, Y., Saito, H., Ikawa, S., Meishi, T. & Okada, Y. (1984) *J. Biochem. (Tokyo)* **96**, 1915–1923.
12. Harrison, B. D. (1987) *Ciba Found. Symp.* **133**, 130–131.
13. Maniatis, T., Fritsch, E. F. & Sambrook, J. (1982) *Molecular Cloning: A Laboratory Manual* (Cold Spring Harbor Lab., Cold Spring Harbor, NY).
14. Sanders, P. R., Winter, S. A., Barnason, A. R., Rogers, S. G. & Fraley, R. T. (1987) *Nucleic Acids Res.* **15**, 1543–1558.
15. Rogers, S. G., Horsch, R. B. & Fraley, R. T. (1986) *Methods Enzymol.* **118**, 627–640.
16. Horsch, R. B., Fry, J. E., Hoffman, N. L., Eichholtz, D., Rogers, S. D. & Fraley, R. T. (1985) *Science* **227**, 1229–1231.
17. Murray, M. G. & Thompson, W. F. (1980) *Nucleic Acids Res.* **8**, 4321–4325.
18. Feinberg, A. P. & Vogelstein, B. (1983) *Anal. Biochem.* **132**, 6–13.
19. Chirgwin, J. M., Przybyla, A. E., MacDonald, R. J. & Rutter, W. J. (1979) *Biochemistry* **18**, 5294–5299.
20. Berry, J. O., Nikolau, B. J., Carr, J. P. & Klessig, D. F. (1985) *Mol. Cell. Biol.* **5**, 2238–2246.
21. Garcia-Arenal, F., Palukaitis, P. & Zaitlin, M. (1984) *Virology* **132**, 131–137.
22. Palukaitis, P. (1988) in *Methods for Plant Molecular Biology*, eds. Weissbach, A. & Weissbach, H. (Academic, New York), pp. 487–506.
23. Goelet, P., Lomonossoff, G. P., Butler, P. J. G., Akam, M. E., Gait, M. J. & Karn, J. (1982) *Proc. Natl. Acad. Sci. USA* **79**, 5818–5822.
24. Lee, B., Murdoch, K., Kreis, M. & Jones, M. G. K. (1989) *Plant Mol. Biol. Rep.* **7**, 129–134.
25. Powell-Abel, P., Nelson, R. S., De, B., Hoffman, N., Rogers, S. G., Fraley, R. T. & Beachy, R. N. (1986) *Science* **223**, 738–743.
26. Loesch-Fries, L. S., Merlo, D., Zinnen, T., Burhop, L., Hill, K., Krahn, K., Jarvis, N., Nelson, S. & Halk, E. (1987) *EMBO J.* **6**, 1845–1851.
27. Tumer, N. E., O'Connell, K. M., Nelson, R. S., Sanders, P. R., Beachy, R. N., Fraley, R. T. & Shah, D. M. (1987) *EMBO J.* **6**, 1181–1188.
28. Van Dun, C. M. P. & Bol. J. F. (1988) *Virology* **167**, 649–652.
29. Angenent, G. C., van den Ouweld, J. M. W. & Bol, J. F. (1990) *Virology* **175**, 191–198.
30. Hemenway, C., Fang, R.-X., Kaniewski, W., Chua, N.-H. & Tumer, N. E. (1988) *EMBO J.* **7**, 1273–1280.
31. Hoekema, A., Huismann, M. J., Molendijk, L., van den Elzen, P. J. M. & Cornelissen, B. J. C. (1989) *Bio/Technology* **7**, 273–278.
32. Cuozzo, M., O'Connell, K. M., Kaniewski, W., Fang, R.-X., Chua, N.-H. & Tumer, N. (1988) *Bio/Technology* **6**, 549–557.
33. Stark, D. M. & Beachy, R. N. (1989) *Bio/Technology* **7**, 1257–1262.
34. Lawson, C., Kaniewski, W., Haley, L., Rozman, R., Newell, C., Sanders, P. & Tumer, N. E. (1990) *Bio/Technology* **8**, 127–134.
35. Van Dun, C. M. P., Bol, J. F. & Van Vloten-Doting, L. (1987) *Virology* **159**, 299–305.
36. Kramer, G. & Agros, P. (1984) *Nucleic Acids Res.* **12**, 7269–7282.
37. Goldbach, R. (1987) *Microbiol. Sci.* **4**, 197–202.
38. Habili, N. & Symons, R. H. (1989) *Nucleic Acids Res.* **17**, 9543–9555.
39. Inokuchi, Y. & Hirashima, A. (1987) *J. Virol.* **61**, 3946–3949.

Acknowledgments

Permission for reproduction of the copyrighted research articles presented in this Anthology were granted as follows:

Abel, P. P., Nelson, R. S., De, B., Hoffmann, N., Rogers, S. G., Fraley, R. T.,and Beachy. R. N. 1986. Delay of disease development in transgenic plants that express the tobacco mosaic virus coat protein gene. *Science* 232:738-743. Copyright 1986, American Association for the Advancement of Science. Reproduced with permission.

Anon. 1898. Prof. Beijerinck, on a contagium vivum fluidum, causing the spot-disease of tobacco leaves. *Nature* 59:216. Copyright 1898, Macmillan Magazines Ltd. Reproduced by copyright permission from *Nature*.

Bawden, F. C., Pirie, N. W., Bernal, J. D., and Fankuchen, I. 1936. Liquid crystalline substances from virus-infected plants. *Nature* 138:1051-1052. Copyright 1936, Macmillan Magazines Ltd. Reproduced by copyright permission from *Nature*.

Beijerinck, M. W. 1898. Concerning a contagium vivum fluidum as cause of the spot disease of tobacco leaves. *In*: Phytopathological Classics, No. 7. (J. Johnson, ed.). American Phytopathological Society, pp. 33-52. Reproduced by copyright permission from the American Phytopathological Society Press.

Bernal, J. D. and Fankuchen, I. 1937. Structure types of protein 'crystals' from virus-infected plants. *Nature* 139:923-924. Copyright 1937, Macmillan Magazines Ltd. Reproduced by copyright permission from *Nature*.

Bernal, J. D. and Fankuchen, I. 1941. X-ray and crystallographic studies of plant virus preparations. III. *J. Gen. Physiol.* 25:147-165. Reproduced by copyright permission from the Rockefeller University Press.

Butler, P. J. G., Finch, J. T., and Zimmern, D. 1977. Configuration of tobacco mosaic virus RNA during virus assembly. *Nature* 265:217-219. Copyright 1977, Macmillan Magazines Ltd. Reproduced by copyright permission from *Nature*.

Caspar, D. L. D. 1956. Radial density distribution in the tobacco mosaic virus particle. *Nature* 177:928. Copyright 1956, Macmillan Magazines Ltd. Reproduced by copyright permission from *Nature*.

Dawson, W. O., Beck, D. L., Knorr, D. A., and Grantham, G. L. 1986. cDNA cloning of the complete genome of tobacco mosaic virus and production of infectious transcripts. *Proc. Natl. Acad. Sci. USA*. 83:1832-1836. Reproduced with copyright permission from the author.

Fraenkel-Conrat, H. and Williams, R. C. 1955. Reconstitution of active tobacco mosaic virus from its inactive protein and nucleic acid components. *Proc. Natl. Acad. Sci. USA*. 41:690-698. Reproduced by permission from the author.

Franklin, R. E. 1956. Location of the ribonucleic acid in the tobacco mosaic virus particle. *Nature* 177:928-930. Copyright 1956, Macmillan Magazines Ltd. Reproduced by copyright permission from *Nature*.

Gierer, A. and Schramm, G. 1956. Infectivity of ribonucleic acid from tobacco mosaic virus. *Nature* 177:702-703. Copyright 1956, Macmillan Magazines Ltd. Reproduced by copyright permission from *Nature*.

Gierer, A. and Mundry, K.-W. 1958. Production of mutants of tobacco mosaic virus by chemical alteration of its ribonucleic acid *in vitro*. *Nature* 182:1457-1458. Copyright 1958, Macmillan Magazines Ltd. Reproduced by copyright permission from *Nature*.

Goelet, P., Lomonossoff, G. P., Butler, P. J. G., Akam, M. E., Gait, M. J., and Karn, J. 1982. Nucleotide sequence of tobacco mosaic virus RNA. *Proc. Natl. Acad. Sci. USA.* 79:5818-5822. Reproduced with copyright permission from the author.

Golemboski, D. B., Lomonossoff, G. P., and Zaitlin, M. 1990. Plants transformed with a tobacco mosaic virus nonstructural gene sequence are resistant to the virus. *Proc. Natl. Acad. Sci. USA.* 87:6311-6315. Reproduced with copyright permission from the author.

Holmes, F. O. 1929. Local lesions in tobacco mosaic. *Bot. Gaz.* 87:39-55. Reproduced by copyright permission from the University of Chicago Press.

Hunter, T. R., Hunt, T., Knowland, J., and Zimmern, D. 1976. Messenger RNA for the coat protein of tobacco mosaic virus. *Nature* 260:759-764. Copyright 1976, Macmillan Magazines Ltd. Reproduced by copyright permission from *Nature.*

Kausche, G. A., Pfankuch, E., and Ruska, H. 1939. Die Sichtbarmachung von pflanzlichem Virus im Übermikroskop. *Naturwissenschaften* 27:292-299. Reproduced by copyright (1939) permission from Springer-Verlag GmbH & Co.

Lebeurier, G., Nicolaieff, A., and Richards, K. E. 1977. Inside-out model for self-assembly of tobacco mosaic virus. *Proc. Natl. Acad. Sci. USA.* 74:149-153. Reproduced by copyright permission from the author.

Nishiguchi, M., Motoyoshi, F., and Oshima, N. 1978. Behaviour of a temperature sensitive strain of tobacco mosaic virus in tomato leaves and protoplasts. *J. Gen. Virol.* 39:53-61. Reproduced by copyright permission from the Society for General Microbiology.

Siegel, A. and Wildman, S. G. 1956. The inactivation of the infectious centers of tobacco mosaic virus by ultraviolet light. *Virology* 2:69-82. Reproduced by permission from the authors and Academic Press, Inc.

Stanley, W. M. 1935. Isolation of a crystalline protein possessing the properties of tobacco-mosaic virus. *Science* 81:644-645. Copyright 1935, American Association for the Advancement of Science. Reproduced with permission.

Takebe, I. and Otsuki, Y. 1969. Infection of tobacco mesophyll protoplasts by tobacco mosaic virus. *Proc. Natl. Acad. Sci. USA.* 64:843-848. Reproduced by copyright permission from the author.

Wilson, T. M. A. 1984. Cotranslational disassembly of tobacco mosaic virus *in vitro*. *Virology* 137:255-265. Reproduced with copyright permission from the author and Academic Press, Inc.

Wittmann, H. G. 1960. Comparison of the tryptic peptides of chemically induced and spontaneous mutants of tobacco mosaic virus. *Virology* 12:609-612. Reproduced by permission from Academic Press, Inc.

The photograph of M. W. Beijerinck's notebook page from 4 December 1898 was provided by Dr. Lesley A. Robertson, the curator of Beijerinck's papers at Delft University of Technology, and with permission from the Kluyver Institute, the Netherlands.

We appreciate the helpful editorial comments and suggestions provided by Dr. Cleora J. D'Arcy during the preparation of this manuscript.